Genetics – Research and Issues Series

THE HUMAN GENOME: FEATURES, VARIATIONS AND GENETIC DISORDERS

Genetics – Research and Issues Series

Sex Chromosomes: Genetics, Abnormalities, and Disorders
Cynthia N. Weingarten and Sally E. Jefferson (Editors)
2009. ISBN: 978-1-60741-304-2

Genetic Diversity
Conner L. Mahoney and Douglas A. Springer (Editors)
2009. ISBN: 978-1-60741-176-5

Bacterial DNA, DNA Polymerase and DNA Helicases
Walter D. Knudsen and Sam S. Bruns (Editors)
2009. ISBN: 978-1-60741-094-2

The Human Genome: Features, Variations and Genetic Disorders
Akio Matsumoto and Mai Nakano (Editors)
2009. 978-1-60741-695-1

Genetics – Research and Issues Series

The Human Genome: Features, Variations and Genetic Disorders

Akio Matsumoto
and
Mai Nakano
Editors

Nova Biomedical Books
New York

For permission to use material from this book please contact us:
Telephone 631-231-7269; Fax 631-231-8175
Web Site: http://www.novapublishers.com

Library of Congress Cataloging-in-Publication Data

The human genome : features, variations, and genetic disorders / [edited by] Akio Matsumoto and Mai Nakano.
p. ; cm.
Includes bibliographical references and index.
ISBN 978-1-60741-695-1 (hardcover)
1. Medical genetics. 2. Human genome. I. Matsumoto, Akio, 1962- II. Nakano, Mai.
[DNLM: 1. Genome, Human. 2. Genetic Predisposition to Disease. 3. Genetics, Medical. QU 470 H918 2009]
RB155.H8475 2009
616'.042--dc22

2009024611

Published by Nova Science Publishers, Inc. ✝ New York

Contents

Preface

The sequencing of the human genome reveals our complete complement of genetic material. The sequenced human genome is one of the most international biomedical research projects ever, which is important in our current often all-too-fractured world. This book defines the function of all "unknown" human genes, delineates the functional and phenotypic significance of human genetic variants in humans, and explores the functions of the vast non-genic regions of the human genome. Human genome sequencing has revealed a great opportunity to deeply investigate the biology and evolution of the sex chromosome pair at a more global level, allowing new frontiers in the genetics research, such as a detailed knowledge of the sequence and the gene content of these chromosomes. This book provides the most current research done in this area.

Chapter 1 - CpG islands (CGIs), the CpG-rich regions in the human genome, are frequently found in the promoter regions and considered gene markers. The promoter-associated CGIs usually remain unmethylated in cells, an important feature in gene regulation. The past two decades have witnessed the development of several computational algorithms on identification of CGIs in genomic sequences and their extensive applications to biological studies. We summarized and compared the major algorithms in this review and suggested that Takai and Jones' algorithm (2002) is the most appropriate for finding CGIs associated with promoter regions and excluding CGIs from repetitive sequences. We further reviewed recently published large-scale methylation profiling studies that focused on CGI regions which have been subsequently used for developing computational algorithms for prediction of methylation status based on the sequence attributes around CGIs.

A large number of studies have been published on investigation of the features, mutation patterns and molecular evolution of CGIs in the human and other genomes. Specifically to the distribution and features of CGIs in the human genome, the density of CGIs per megabase pairs (Mb) was found to be highly positively correlated with many genomic factors such as gene density, GC content, ratio of the observed over the expected CpG dinucleotides and recombination rate. Moreover, housekeeping genes are more likely to be associated with CGIs in the promoter regions than tissue-specific genes, although this difference is not as strong as previously thought. Next, we reviewed mutation patterns in human CGIs. Recent studies of point mutation data revealed that the mutation rate of G/C→A/T in CGIs is lower than that in intergenic regions having similar GC content, and, importantly, methylation-dependent transition rates in CGIs and other genomic regions are dependent on local sequence length, GC content, and genomic regions. Finally, comparative genomics studies found that promoter-associated CGIs have been under loss in the course of genome evolution in a decay pattern starting from both edges of CGIs. The "loss of CGI" scenario has been observed in the rodent, dog and human lineages, suggesting that it is a universal evolutionary mechanism in mammalian or vertebrate genomes.

The maintenance and change of methylation status in CGIs plays important roles in gene regulation and functions. We reviewed the mechanisms of methylation (e.g., hypomethylation and hypermethylation) on CGIs affecting gene expression and functions. We further reviewed recent reports on the abnormal methylation of CGIs that cause diseases, especially cancers. The disease genes affected by aberrant methylation involved in many basic cellular functions. Finally, we reviewed recent advances in genome-wide mapping methylation profiling technologies. We expect an era of epigenomics is coming.

Chapter 2 - Human sex chromosomes display significant differences from autosomes in both structure and function. Particularly, as the human X and Y chromosomes show a unique biology, they have long attracted special attention among geneticists. The human genome sequencing has revealed a great opportunity to deeply investigate the biology and evolution of the sex chromosome pair at a more global level, allowing new frontiers in the genetics research, such as a detailed knowledge of the sequence and the gene content of these chromosomes. Comparison of the human X and Y chromosome sequences, has made possible a reconstruction of their evolutionary history. Their sequence comparison has revealed they have become isolated from each other in a stepwise fashion over hundreds of millions of years, due to the lack of recombination events. The sequencing of the human Y chromosome has revealed that the Male Specific Region (MSR) contains, in addition to the Y-chromosomal male-determining gene SRY, a number of genes that have become specialized for spermatogenesis.

The sex chromosomes hold a unique place in the history of medical genetics. It has been widely demonstrated that a significant fraction of genetic diseases in humans results from point mutations and/or structural anomalies involving the sex chromosomes. This is a consequence of the haploid presence (hemizygosity) of the X and Y chromosomes in males. This phenomenon has prompted decades of intensive study, mainly focused on X-linked inherited disorders. Many of the X-linked diseases currently actively investigated, are discussed in depth in this review.

Furthermore, although the past decades have witnessed many advances in the understanding of molecular processes underlying dosage compensation between sexes in mammals, the mechanism of X chromosome inactivation still continues to puzzle investigators. There is clear evidence that the expression of X-linked mutations in females is fine-tuned, and highly influenced, by these processes. Indeed, X-linked dominant male-lethal disorders represent a paradigmatic example of such influences. The observations reviewed here emphasize the importance of studying in depth the sex chromosomes, in order to better understand the evolution of human chromosomes and the pathological mechanisms related to the sex chromosomes.

Chapter 3 - Gene amplification is a cytogenetic abnormality frequently observed in cancer cells. It often occurs as a double minute chromosomes (dmin) consisting of a repeat of unit (amplicon) whose origin stems from a different chromosome. The number of dmins in each cell is unstable because they do not contain the centromeres. We found a large extrachromosomal element (LEE) that is a novel form of gene amplification in human myeloid leukemia HL-60 cells. In fact, HL-60 cells contained dmins, on which the *MYC* oncogene was amplified at early passages of cell culture; however, cells harboring LEEs emerged during long continuous culture, and finally replaced dmin-containing cells in late passages. Although the LEE resembled dmin in that it lacked alphoid sequence, it persisted in cells stably, unlike what is observed for dmin. The stabilization of the extrachromosomal element is believed to be involved in the progress of malignancy in many cancers. Thus, we think that LEEs provide a good experimental model to study the potential relationship between the stabilization of extrachromosomal elements and the progression of cancers. Here we summarize the molecular nature of LEE and its possible therapeutic use of LEEs for neoplastic cells.

Chapter 4 - Recent advances in technology have greatly improved our understanding of the molecular basis of inherited disorders. Prior to the development of recombinant DNA technology, only inherited disorders where there was an abundance of gene product were conducive to study. With the advent of recombinant DNA technology, specific hybridization probes became available that allowed the detection and diagnosis of rarer disorders using Southern blot analysis and manual DNA sequencing of cloned DNA fragments. The development of the polymerase chain reaction and automated DNA sequence analysis has since extended predictive DNA testing to most inherited disorders where the non-functional gene has been identified. Further technological advances led to the completion of the Human Genome Project (HGP) in 2003. Since that time, next-generation DNA sequencers hold the potential of the fast and cheap determination of an individual's genomic sequence leading to the promise of personalized medicine. In this review, we discuss the application of these advancing technologies on the molecular basis of human coagulopathies including (1) the relatively frequent X-linked hemophilias including the gene rearrangement that is the cause of up to 50% of the cases of hemophilia A, (2) the rarer autosomal bleeding disorders, and (3) the complex coagulopathies caused by indirect effects of mutations in non-clotting factor genes. These studies have revealed a multitude of genetic mutations that give rise to coagulopathies. When individual genomic sequencing becomes routine, the coagulopathies will be amenable to diagnostic testing with subsequent genetic counseling on the consequences of the findings.

Chapter 5 - An estimate of the information content of human genome is presented. The evolutionary processes are pictured as an information channel and the loss/gain of the information content is discussed. The evolutionary mechanisms such as nucleotide replacements, insertions, deletions and replication slippages that are shaping the human genome are discussed. The nucleotide replacement processes are modeled in terms of minimal number of parameters such as the transition/transversion ratio and some context dependent parameters. The variation of DNA sequences is scrutinized for the period since mammalian origin and for the time since the last common ancestor to human and chimpanzee. A special emphasis is given to the distinction between four sequence categories that emerge when the genomic sequences are partitioned into i) CpG rich islands, ii) Alu - type short interspersed repeats, iii) the cross section of these two categories and iv) the class of all the remaining genomic sequences. It is shown that although the Alu sequences seem to be a typical representative of "junk DNA" they participate in CpG islands, which are known to embed the regulatory regions and are thus the sequences of supreme functional importance.

Chapter 6 - Alternative splicing is an important mechanism mediating the function and complexity of genes in multicellular organisms. Recently, a new splice-junction wobbling mechanism is discovered that generates subtle alterations in mRNA by indiscriminately selecting tandem donor sites (GTNGT) or acceptor sites (NAGNAG). It results in tri-nucleotides insertion/deletion in the transcripts that can escape from the nonsense-mediated decay surveillance due to the in-frame InDel event occurred in these mRNA without generating new and premature stop codon. The reading frame is not altered by the insertion or deletion of tri-nucleotides in the transcripts, therefore the resulting protein isoforms would be highly similar in sequences. Nonetheless, most of subtly changes in protein generated by wobble splicing could increase functional diversity of protein and some of theses wobble splicing isoforms might have functional impacts and disease implications in terms of cellular functions and regulations. Therefore, the wobble splicing phenomenon occurs mostly in a tissue and developmental stage–independent manner. Only a few wobble splicing genes are proven to be differentially spliced in tissues or developmental stages. Remarkably, most of this wobble-splicing process is likely due to stochastic splice site selection at tandem motif sequence. Here, we review recent progress in understanding functional aspects as well as the mechanism of wobble splicing at tandem motifs.

Chapter 7 - Subtle alternative splicing at short distance is defined as the selection between two or more adjacent splicing signals during mRNA processing. The co-existence of expressed sequences with such minute distance differences is typified as Wobble Splicing, which implies the ambiguous recognition of splicing complexes in selection between the nearby splicing sites. Successive tandem splice acceptors arrangement, such as NAGNAG, is the most common wobble splicing type in eukaryotic genomes. In addition to the short distance tandem splicing sites previously reported, DNA segmental duplication or multiplication also has the potential to duplicate the splicing signals *in situ*, once it located at the boundaries between intron and exon. Herein, junctional tandem repetitive sequences in human are mapped and investigated in this study. The splicing site selection among the homologous duplicated splicing sequences was also detected and characterized. Junctions with duplicated 5' splicing signatures are more prevalent than 3' ones by summation of total numbers of splicing sites observed. The distance between the duplicated repetitive sequences seems to play a critical role in the alternative splicing site selection as observed by ESTs. This reveals the additional control mechanism of splicing acceptor site selection, which contain branch site sequence and polypyrimidine tract aside from the minimal dinucleotide - AG signal.

Chapter 8 - Susceptibility to complex traits, by definition, involves aetiological polymorphisms at multiple genetic loci combined with variable contributions by environmental factors. However, the approaches taken to identifying genetic loci implicated in susceptibility to complex traits frequently overlooks the compounding contribution of multiple loci in favour of highlighting a single gene solely responsible for predisposition. It is only in a small minority of cases that this has resulted in clear disease heritability associated with polymorphisms in a single gene. More often, this approach has led to an accumulation of single-gene associations with minor contributions to disease susceptibility. As the genomic era advances and genome-wide screens become higher in resolution and throughput, the need for simultaneous consideration of multiple loci is becoming more important. With special reference to non-Hodgkin's lymphoma (NHL), this chapter will overview the current progress made in elucidating genetic polymorphisms associated with disease susceptibility. We also present novel data from a high-resolution single nucleotide polymorphism (SNP) microarray screen for susceptibility loci that are involved in NHL. Using an 'informed approach', the findings are highlighted within the context of cellular pathways, and provide insight and new ideas for methods of analysis for genome-wide screens for susceptibility.

Chapter 9 - The twenty-first century is the era of the personal genome, enabled by a trilogy of scientific achievements: the Human Genome Project, the international HapMap project, and large-scale genome-wide association studies. Based on these achievements, service providers emerge to offer ordinary people an unprecedented opportunity to view their genetic heritage. These services usually include personal ancestry analysis and lifetime risk estimations for various common complex diseases. The era of the personal genome will have a great impact on many aspects of life. It promises a better health-care system featuring preventive and personalized medicine. In this chapter, we introduce multiple aspects of the personal genome, including the science, technology, applications and concerns.

Chapter 10 - Microsatellites are abundant repetitive sequences accounting for 3% of the human genome. The list of diseases that are triggered by the unstable expansion of some of these sequences continues to increase, and includes disorders such as Huntington's Disease or Fragile X Syndrome. Diseases of unstable repeat expansion share peculiar genetic features. The size of the repetitive array correlates with the severity and the age of onset of the disease. Moreover, the microsatellite has a strong tendency to expand promoting earlier and more severe expression of the disease in successive generations. The most important factors determining this repeat instability seem to be related to structural properties. After DNA slippage, the more stable non-B DNA conformations serve as substrates for DNA repair and might therefore be excised. In contrast, some non-B DNA conformations could avoid the DNA repair systems. Furthermore, a recent study suggests the recruitment of microsatellites by genes that encode transcription factors and other regulatory genes, particularly in the nervous system. Therefore, some of these repeat polymorphisms may be associated with phenotypic traits that have the potential to increase fitness, but also susceptibility to unstable expansion diseases. The general pathogenic mechanisms involve altered protein function or aberrant RNA–protein interactions. Therapeutic strategies target the protein or counteract cellular defects reversing metabolic abnormalities. Additionally, gene therapy holds great promise hampering allele expansion or reducing the expression of the expanded allele using small interfering RNAs or viral-mediated approaches. Although much effort has been devoted to understanding the full disease process and the development of an effective therapy, many aspects of these disorders still remain to be fully understood and are addressed in this chapter.

Chapter 11 - The genetic make up of an individual, at least in part, determines disease susceptibility and response to drug treatment. It is because of this reason a tremendous progress has been made in cataloging human sequence variations. It is thought that a high-density map of variations will provide necessary tools to develop genetic based diagnostic and therapeutic options. The most common type of variation is called single nucleotide polymorphism (SNP). These are highly abundant, stable and distributed throughout the genome. They are also associated with diversity in the population, individuality, susceptibility to diseases and response to medicine. It has been suggested that SNPs can be used for heterogeneity testing, pharmacogenetic studies and to identify and map complex, common diseases such as high blood pressure, diabetes and heart diseases. Consistent with this proposal is the identification of the patterns of SNPs in conditions such as schizophrenia, blood pressure homeostasis and diabetes. Recently, a new form of genetic variation known as copy number variation (CNV) has also been identified. By using different types of genome wide scanning procedures, CNVs have been shown to be associated with several complex and common disorders including nervous system disorders. One of the common features of the regions associated with the complex and common disorders identified so far is the presence of CNVs and segmental duplications. Segmental duplications lead to genome instability. Because of their location and nature (several of them contain genes) many CNVs have functional consequences such as gene dosage alteration, disruption of genes and modulation of activities of other genes. Therefore, these genetic variations will have influence on phenotypes, susceptibility of an individual to disease, drug response and human genome evolution. These types of variants (gain and loss of DNA) are not restricted to humans but they have also been identified in other organisms. Because most common, complex disorders are caused by the combined effects of multiple genes and non-genetic environmental factors, it is likely that sequence variation alone is not sufficient to predict the risk of disease susceptibility, particularly in homeostatic organisms like humans. Nevertheless, these variations (SNPs) may provide a starting point for future inquiry. Our current knowledge on CNVs and their heritability is still rudimentary because of their location in regions of complex genomic structure. Future advances in the technology will help in constructing a new CNV map that can be used to (a) find genes underlying common diseases (b) understand the familial genetic conditions (c) uncover the severe developmental defects in humans and other organisms and (d) genome evolution.

Chapter 12 - Uterine neoplasms are common tumors, formed by endometrial and cervical cancers; endometrial cancer is the fourth most frequently diagnosed cancer in developed countries and the eighth leading cause of cancer death in women, and cervical cancer is the second most common cancer in women worldwide and is a leading cause of cancer-related death in women in underdeveloped countries.

Cervical cancer arises by HPV DNA damaging; in fact cervical cancer starts in the cells on the surface of the cervix, exposed to viral infective agents, as HPV, founded in 80% of patients affected by cervical cancer. Thus, more than 99% of cervical uterine cancer cases show HPV presence.

Nevertheless, Endometrial cancer involves cancerous growth of the endometrium, and increasing evidence indicates that different biological and genetic factors play relevant roles its onset so as carcinogenesis generally develops by hormonal modifications.

Both tumors can be safely and feasibly managed from minimally invasive surgical techniques till to endoscopic radical operations, such as hysterectomy, bilateral salpingo-oophorectomy, pelvic and para-aortic lymphadenectomy for surgical treatment.

The authors reviewed several excellent reviews and studies in the area of hormonal, viral and genetical risk factors associated with endometrial and cervical cancer risk and development, analyzing the area of biologic markers, all papers dealing with serum and plasma markers involved in uterine cancer detection, development, progression and minimally invasive treatment.

Chapter 13 - It is a common assumption that we currently live in an *information society*. Control and access to information are seen by the public as crucial means for social knowledge and power. Similarly, in biology, the recent completion of the Human Genome Project has led to the consideration as fundamental scientific knowledge the sequence of information in our genes. The emergence of information as a key concept in both biology and society has a history of more than sixty years, which is not that generally acknowledged. This chapter will explore such history by investigating the development of the first computerized biological databases and their connection with the understanding of information as a valuable social resource. By studying two European database initiatives, one of them developed in the 1960s and the other in the early 1980s, I will argue that the emergence of the personal computer and the increasing perception of data gathering as an essential social and scientific activity marked the different fate of each project. Whereas the 1960s database faced financial difficulties, the 1980s effort—devoted to the storage of DNA sequences—was perceived as priority and cutting-edge science, associated with the new discipline of genomics and given unprecedentedly large funding.

Chapter 14 - For the past 11 years, fresh frozen and paraffin embedded human normal and tumor tissues have been banked in a collaborative effort between various clinical departments including surgery, pathology and clinical oncology in the Indiana University Simon Cancer Center in order to study the translational relationship between genes and proteins which are altered in various types of neoplasms and compared to translational expression in normal human tissues.This review is a compilation of the work between several departments within the School of Medicine, from IRB protocol approval, patient informed consent document, and HIPAA consent sign off process, frozen tissue collection, both frozen and fixed tissue sample processing, storage of the tissue, and database tracking of the specimens as they are sent to the researchers. This review highlights the banking processes as well as the quality control of the tissues and the lessons learned in the tissue bank process over the past 11 years as it relates to using high quality biospecimens for translational genomic research. High quality biospecimens is the key for best practices in all translational genomic and proteomic research.

The success of genomic research and its application in both the clinical and basic research for translational medicine and drug discovery is strongly dependent on the best practices for collection, handling, and storage of human tissue samples for research'

(Farkas, Kaul et al. 1996), (Naber 1996)(Holland et al)

Chapter 15 - Justifiably great fanfare accompanied the two original publications announcing the sequencing of the human genome some eight years ago and much progress has been made since in exploiting this new resource! However, biochemists and molecular biologists have not fully embraced the need for exploration of this new field. Thus, this may be a good time to take stock and to remind ourselves of underexplored areas that require the attention of all biomedical scientists. Therefore, we here wish to draw particular attention to three unanswered questions: i) the large fraction of "unknown" genes, ii) the need for understanding human and other genetic variation mechanistically and iii) the function of the large portion of non-genic DNA in humans. Accordingly, we propose that there is an urgent need for detailed large-scale functional characterization, e.g. by biochemical, bioinformatic and molecular analyses, of both hitherto "unknown" genes the plethora of human and other genetic variants, such as SNPs (single nucleotide polymorphisms), haplotypes or larger structural variants and the function of non-genic areas.

In: The Human Genome: Features, Variations… ISBN: 978-1-60741-695-1
Editor: Akio Matsumoto and Mai Nakano

Chapter 1

CpG Islands in the Human Genome: Identification, Features, Mutations and Diseases

***Zhongming Zhao*[1,2,3*] *and Leng Han*[1]**

[1]Department of Biomedical Informatics, Vanderbilt University School of Medicine, Nashville, TN 37203, USA
[2]Department of Psychiatry, Vanderbilt University School of Medicine, Nashville, TN 37203, USA
[3]Department of Cancer Biology, Vanderbilt-Ingram Cancer Center, Nashville, TN 37211, USA

Abstract

CpG islands (CGIs), the CpG-rich regions in the human genome, are frequently found in the promoter regions and considered gene markers. The promoter-associated CGIs usually remain unmethylated in cells, an important feature in gene regulation. The past two decades have witnessed the development of several computational algorithms on identification of CGIs in genomic sequences and their extensive applications to biological studies. We summarized and compared the major algorithms in this review and suggested that Takai and Jones' algorithm (2002) is the most appropriate for finding CGIs associated with promoter regions and excluding CGIs from repetitive sequences. We further reviewed recently published large-scale methylation profiling studies that focused on CGI regions which have been subsequently used for developing computational algorithms for prediction of methylation status based on the sequence attributes around CGIs.

* Corresponding author: Zhongming Zhao, PhD, Department of Biomedical Informatics, Vanderbilt University School of Medicine, 2525 West End Avenue, Suite 600, Nashville, TN 37203, USA, Phone: (615) 343-9158, FAX: (615) 936-8545, Email: zhongming.zhao@vanderbilt.edu

A large number of studies have been published on investigation of the features, mutation patterns and molecular evolution of CGIs in the human and other genomes. Specifically to the distribution and features of CGIs in the human genome, the density of CGIs per megabase pairs (Mb) was found to be highly positively correlated with many genomic factors such as gene density, GC content, ratio of the observed over the expected CpG dinucleotides and recombination rate. Moreover, housekeeping genes are more likely to be associated with CGIs in the promoter regions than tissue-specific genes, although this difference is not as strong as previously thought. Next, we reviewed mutation patterns in human CGIs. Recent studies of point mutation data revealed that the mutation rate of G/C→A/T in CGIs is lower than that in intergenic regions having similar GC content, and, importantly, methylation-dependent transition rates in CGIs and other genomic regions are dependent on local sequence length, GC content, and genomic regions. Finally, comparative genomics studies found that promoter-associated CGIs have been under loss in the course of genome evolution in a decay pattern starting from both edges of CGIs. The "loss of CGI" scenario has been observed in the rodent, dog and human lineages, suggesting that it is a universal evolutionary mechanism in mammalian or vertebrate genomes.

The maintenance and change of methylation status in CGIs plays important roles in gene regulation and functions. We reviewed the mechanisms of methylation (e.g., hypomethylation and hypermethylation) on CGIs affecting gene expression and functions. We further reviewed recent reports on the abnormal methylation of CGIs that cause diseases, especially cancers. The disease genes affected by aberrant methylation involved in many basic cellular functions. Finally, we reviewed recent advances in genome-wide mapping methylation profiling technologies. We expect an era of epigenomics is coming.

Introduction

CpG islands (CGIs), the CpG-rich regions in the human genome, are frequently observed in the promoter regions and considered gene markers [1-3]. Because of their functional importance, several algorithms have been developed for identification and classification of CGIs. These algorithms are mainly based on two approaches. One is based on the traditional three sequence parameters (length, GC content and Obs_{CpG}/Exp_{CpG} ratio), first proposed by Gardiner-Garden and Frommer in 1987 [4], and then refined in several other studies [5-7]. Here, Obs_{CpG}/Exp_{CpG} ratio is the ratio of the observed over the expected CpG dinucleotides in a sequence or genome, as originally defined in Gardiner-Garden and Frommer [4]. The other is to detect clusters of CpGs (CpGclusters) by statistical significance based on the physical distance between neighboring CpGs on a chromosome or sequence [8, 9]. It is worthy noting that identification of CGIs in a genome is usually an initial step to examine large-scale methylation status of genomic regions [10-14] or to predict methylation status combined with other genomic features [15-17].

Having been considered as gene markers, the number of CGIs in a chromosome is positively correlated with its number of genes in the human genome [5]. Moreover, the density of CGIs is correlated with several other genomic factors [18]. During the past decade, millions of single nucleotide polymorphisms (SNPs) have been discovered in the human genome. These data allowed investigators to examine mutation patterns in CGIs in the human genome [19-22]. Comparative genomics approach has also been a powerful tool for studying evolution of CGIs among vertebrates, especially mammals [18, 23-25].

CpG dinucleotides in CGIs tend to be unmethylated, in contrast to the prevalent methylation at the CpG sites in non-CGI regions. Experiments indicated that methylation of CGIs plays critical roles in gene silencing, genomic imprinting, X-chromosome inactivation, carcinogenesis, and silencing of intragenomic parasites [26]. More recent experimental work found that abnormal methylation (hypomethylation or hypermethylation) is associated with a large number of human diseases or disorders such as cancers including breast cancer, cervix cancer, colon cancer, kidney cancer, and other disorders including autoimmune disorders, imprinting disorders, cardiovascular diseases, and mental disorders including schizophrenia [27-31]. A large number of genes involved in these diseases have been identified by recently developed methylation profiling technologies. These genes involved in many basic cellular functions and pathways [28, 32].

Identification of CGIs and Methylation

Computational Algorithms for Identification of CGIs

CpG dinucleotides are frequently methylated in mammalian genomes. It has been commonly estimated that ~80% of CpGs are methylated in mammalian genomes and, because of the hypermutability of methylated CpGs to TpGs/CpAs, CpGs have been observed only ~20-25% as what expected in most sequenced mammalian genomes [5, 18, 33, 34]. In contrast, CpG dinucleotides in GC-rich regions (e.g., CGIs) are usually unmethylated, which is an important epigenetic feature in the promoter regions [26, 35]. Because of their functional importance, multiple algorithms have been developed for identifying CGIs in a genome or a sequence. In 1987, Gardiner-Garden and Frommer proposed the first algorithm based on three sequence parameters (length, GC content, and Obs_{CpG}/Exp_{CpG} ratio) [4]. The specific criteria are: length > 200 bp; GC content > 50%, and $Obs_{CpG}/Exp_{CpG} > 0.60$. Based on these criteria, a total of 221,538 CGIs were found in the human genome, many of which were actually in repetitive sequences. This number of CGIs is remarkably greater than the number of genes (e.g., as estimated to be 20,000-25,000 [36]). The large difference is because Gardiner-Garden and Frommer's algorithm could not exclude short interspersed repeats (e.g., *Alu*), which typically have a sequence length of 80-400 bp and are GC-rich. Because of this limitation, researchers often applied this relaxed algorithm to non-repeat portion of the genomes or sequences. This modified strategy could identify similar number of CGIs to that of genes [5, 37]. However, it also completely excluded the repeats, some of which are functional.

In 2002, Takai and Jones proposed stringent criteria after their evaluation of CGI characteristics and distribution features with genes and Alu repeats. Their specific criteria are: length ≥ 500 bp, GC content ≥ 55%, and $Obs_{CpG}/Exp_{CpG} \geq 0.65$. This refined algorithm is highly effective on excluding short interspersed repeats [7]. Based on this algorithm, a total of 37,729 CGIs were identified in the human genome, which is close to the number of human genes. One advantage of this algorithm is that it scans the CGIs on all continuous sequences, not only the non-repeat part. In the same year, Ponger and Mouchiroud (2002) proposed intermediately stringent criteria and named CpGPRoD: length $\geq$ 500 bp, GC content $\geq$ 50%, and $Obs_{CpG}/Exp_{CpG} \geq 0.60$. The CpGPRoD identified 79,690 CGIs in the human genome, more than three times the number of human genes [6]. Among these traditional algorithms, Takai and Jones' stringent algorithm outperforms the others because it can effectively exclude short interspersed elements and it can identify CGIs that are more likely associated with the 5' regions of genes (Table 1) [7, 38, 39]. Interestingly, Gardiner-Garden and Frommer set the three parameters above the threshold values (e.g., GC content > 50%, excluding the threshold value 50%), but the refined algorithms have the parameters greater or equal to the threshold values (e.g., GC content $\geq$ 50%, including the threshold value 50%). This slight difference actually has a major effect on identification of CGIs. For example, there were 351,604 CGIs in the human genome using the inclusive threshold criteria (length $\geq$ 200 bp, GC content $\geq$ 50%, and $Obs_{CpG}/Exp_{CpG} \geq 0.60$), but only 221,538 CGIs using the exclusive threshold criteria (length > 200 bp, GC content > 50%, and $Obs_{CpG}/Exp_{CpG} > 0.60$). This large difference indicates that many CGIs in the human genome carry marginal parameter: length = 200 bp, GC content = 50% or Obs_{CpG}/Exp_{CpG} =0.6. This feature has been ignored by investigators.

Alternatively, Hackenberg *et al.* (2006) developed a new algorithm, CpGcluster, to detect clusters of CpGs (i.e., CpG clusters) by statistical significance based on the physical distance between neighboring CpGs on a chromosome [8]. Instead of employing the three parameters typically used in traditional algorithms, CpGcluster is based on the assumption that the distance distributions between neighboring CpGs differ in CGIs from bulk DNA sequences. It is highly sensitive to detect short CGIs, even as short as 8 bp. Surprisingly, this algorithm identified as many as 198,702 CpG clusters in the human genome, 7-10 times the number of human genes [18]. Therefore, it introduces a large number of false positives that are actually not located in the promoter regions (table 1) [39]. Similarly, Glass *et al.* (2007) defined CG-dense fragments as CG clusters without imposing a base compositional *a priori* assumption [9]. They obtained species-specific benchmarks by sequence analysis. In the human genome, the benchmark is a minimum of 27 CpG dinucleotides in a DNA sequence fragment of no more than 531 bp in length. This resulted in annotated 41,487 CGIs [9]. Compared to the simple and efficient computing in other algorithms including CpGcluster and traditional algorithms, CG clusters in Glass *et al.* is not straightforward, though recent evaluation indicated similar performance on identifying CGIs to Takai and Jones' algorithm (table 1) [39].

Mapping and Prediction of Methylation Status Using CGIs

It is important to map DNA methylation to large-scale genomic regions because the epigenetic effect on gene regulation and function is critical in biological studies [40]. DNA methylation studies in genomic regions, chromosomes and genomes have greatly accelerated during the past three years thanks to the rapid advancement of high throughput technologies such as microarray and next-generation sequencing [41, 42]. Bisulphite conversion, methylation-sensitive restriction enzymes and affinity purifications have been used in the sample pretreatment of high-throughput DNA methylation analysis, and then, DNA microarrays or direct sequencing have been used to generate and read the data [41]. These large-scale DNA methylation mappings always target on GC-rich regions (e.g., CpG islands), which can be summarized as the length of regions ranging 200 - 500 bp, GC content 50 - 65%, and Obs_{CpG}/Exp_{CpG} ratio 0.5 - 0.8 (table 1) [10-14]. Bock *et al.* (2008) compared inter-individual variation of DNA methylation. They suggested that it is necessary to map methylation at single-CpG resolution in the CpG-poor regions of the human genome, while it is sufficient to measure average methylation levels in CpG-rich regions [43]. Therefore, in our view, an appropriate identification of CGIs in a DNA sequence is always an initial step in mapping methylation profiling and saves much cost and labor in such investigations.

Table 1. Benchmarks for CpG island (CGI) identification

Identification of CGI		Number of CGIs		
		Total	Promoter (%)	Non-repeat (%)
Gardiner-Garden and Frommer (1987) [4]	Length > 200 bp, GC > 50%, $Obs_{CpG}/Exp_{CpG} > 0.60$	221,538	19,040 (8.6)	77,111 (34.8)
Takai and Jones (2002) [7]	Length ≥ 500 bp, GC ≥ 55%, $Obs_{CpG}/Exp_{CpG} \geq 0.65$	37,729	13,196 (35.0)	28,380 (75.2)
CpGPRoD (2002) [6]	Length > 500 bp, GC > 50%, $Obs_{CpG}/Exp_{CpG} > 0.60$	79,690	14,695 (18.4)	38,737 (48.6)
CpGcluster (2006) [8]	Clusters of CpGs separated by median distance, significance P-value $<10^{-5}$	198,702	29,156 (14.7)	
CG clusters (2007) [9]	Human ≥ 27 CpGs in DNA fragment ≤ 531 bp	41,487	13,414 (32.3)	
Recent large-scale methylation studies focusing on CGIs				
Eckhardt et al. (2006) [10]	Length > 400 bp, GC > 50%, $Obs_{CpG}/Exp_{CpG} > 0.60$			
Keshet et al. (2006) [11]	Length ≥ 200 bp, GC > 65%, $Obs_{CpG}/Exp_{CpG} > 0.80$			
Khulan et al. (2006) [12]	Length > 200 bp, GC ≥ 50%, $Obs_{CpG}/Exp_{CpG} > 0.60$			
Rollines et al. (2006) [13]	Length > 300 bp, GC > 55%, $Obs_{CpG}/Exp_{CpG} > 0.50$			
Weber et al. (2007) [14]	Length ≥ 500 bp, GC > 55%, $Obs_{CpG}/Exp_{CpG} > 0.75$			
Methylation status prediction based on CGIs		Sequence attributes		
Bock et al. (2006) [15]	Length > 400 bp, GC > 50%, $Obs_{CpG}/Exp_{CpG} > 0.60$	DNA sequence properties and patterns, repeat frequency and distribution, predicted DNA structure		
Das et al. (2006) [16]	Length ≥ 500 bp, GC ≥ 55%, $Obs_{CpG}/Exp_{CpG} \geq 0.65$	GC content, di- and tri-nucleotide count, Alu coverage, hexamers		
Fang et al. (2006) [17]	Length > 400 bp, GC > 50%, $Obs_{CpG}/Exp_{CpG} > 0.60$	GC content, CpG ratio, TpG content, distribution of Alu Y, transcription factor binding sites		

Using experimentally verified DNA methylation data, Bock *et al.* (2006) found that CGI methylation in human lymphocytes is highly correlated with DNA sequence, repeats, and predicted DNA structure [15]. Two recent studies were published for predicting methylation status around CGIs. First, Das *et al.* (2006) predicted the methylation status based on GC content, di- and tri-nucleotide count, *Alu* coverage, and hexamers distribution [16]. Second, Fang *et al.* (2006) applied GC content, CpG ratio, TpG content, distribution of *Alu* Y, transcription factor binding sites (TFBSs) in their methylation prediction [17] (table 1). These studies greatly improved our understanding of the inherent relationship between CGIs, DNA composition (sequence, repeats and structure) and methylation. More prediction algorithms or methods are expected to be published in the near future.

Features, Mutation Patterns and Evolution of CGIs

Distribution and Features of CGIs in the Human Genome

Using Takai and Jones' (2002) algorithm [7], Han *et al.* (2008) identified 37,729 CGIs in the human genome [18]. These CGIs had an average length of 1090 bp, average GC content of 62.0%, and average Obs_{CpG}/Exp_{CpG} ratio of 0.743. The average GC content and Obs_{CpG}/Exp_{CpG} ratio of human CGIs were much higher than the corresponding average values of the whole genome. That is, the average Obs_{CpG}/Exp_{CpG} ratio was only 0.24, and the average GC content was only 40.9% in the human genome. These human CGIs covered a total length of 41.1 Mb, which accounted for 1.4% of the human genome sequence. On average, the human genome has a density of 13.2 CGIs per Mb with a high standard deviation (±16.8 CGIs/Mb). Among these CGIs, 28,380 (75.2%) located in the non-repeat portion of the human genome; they accounted for 2.0% of the non-repeat genome. This larger portion suggests that Takai and Jones's algorithm can effectively exclude the short repeats, especially GC-rich *Alu* repeats [7]. Correspondingly, CGI density in the non-repeat portion of the human genome (18.7 /Mb) is much higher than that (13.2 /Mb) of the whole genome.

We further examined the distribution and features of CGIs on each human chromosome (NCBI human genome assembly build 36) using Takai and Jones' (2002) algorithm. The details are summarized in table 2. The number of CGIs and CGI density varied greatly. Chromosome 1, the largest autosome, had the largest number of CGIs (3223 CGIs), while chromosome 21, the smallest autosome, had only 447 CGIs. The chromosome 19, the most gene-dense chromosome [5], had the highest CGI density (53.1 CGIs/Mb). This density is 6.3 times that of chromosome 4, which had the lowest CGI density (8.3 CGIs/Mb). The initial analysis of the non-repeat portion of the human genome using Gardiner-Garden and Frommer (1987) algorithm revealed four chromosomes (16, 17, 19, and 22) had notably higher CGI density than other chromosomes [5]; this feature was confirmed in our reanalysis of the whole human genome using Takai and Jones' (2002) algorithm (table 2). As expected, we observed that CGI density was highly correlated to gene density in the human genome (linear regression, $r = 0.96$, $P = 5.9 \times 10^{-13}$), supporting the notion that CGIs function as gene markers. Moreover, CGI density in a human chromosome was highly correlated with other genomic factors such as GC content ($r = 0.88$, $P = 1.5 \times 10^{-8}$) and Obs_{CpG}/Exp_{CpG} ratio ($r = 0.92$, $P =$

1.4×10^{-10}), indicating that CGIs depend on local genomic features. These correlations were summarized in table 3.

Table 2. Distribution and features of CpG islands on human chromosomes

	Chromosome				CpG island (CGI)				
Chr	Gene #	Length (Mb)[a]	GC content	Obs_{CpG}/Exp_{CpG}	CGI #	Length (bp)	GC content	Obs_{CpG}/Exp_{CpG}	CGI density/ Mb
1	2393	222.8	41.7	0.233	3223	1103.7	62.2	0.743	14.5
2	1652	237.5	40.2	0.224	2390	1129.1	61.8	0.756	10.1
3	1282	194.6	39.7	0.212	1717	1078.7	61.4	0.747	8.8
4	1023	187.2	38.2	0.214	1546	1096.8	61.3	0.750	8.3
5	1047	177.7	39.5	0.217	1797	1084.4	61.2	0.751	10.1
6	1322	167.3	39.6	0.225	1859	1063.3	60.8	0.753	11.1
7	1152	154.8	40.7	0.243	2082	1073.9	61.8	0.742	13.5
8	856	142.6	40.2	0.227	1452	1127.5	62.1	0.747	10.2
9	1006	117.8	41.4	0.239	1671	1081.8	62.3	0.745	14.2
10	980	131.6	41.6	0.238	1604	1136.6	62.0	0.749	12.2
11	1643	131.1	41.6	0.228	1687	1119.6	62.9	0.735	12.9
12	1208	130.3	40.8	0.235	1729	1042.7	61.4	0.738	13.3
13	472	95.6	38.5	0.226	909	1045.2	61.0	0.756	9.5
14	967	88.3	40.9	0.233	1114	1103.8	62.2	0.751	12.6
15	802	81.3	42.2	0.240	1112	1151.4	61.9	0.756	13.7
16	967	78.9	44.8	0.277	1772	1094.4	63.0	0.740	22.5
17	1263	77.8	45.5	0.286	2071	1124.3	62.8	0.736	26.6
18	379	74.7	39.8	0.229	754	1135.3	61.4	0.761	10.1
19	1505	55.8	48.4	0.324	2960	1037.9	62.8	0.716	53.1
20	643	59.5	44.1	0.248	1048	1089.0	63.0	0.741	17.6
21	292	34.2	40.9	0.261	447	1075.1	62.3	0.758	13.1
22	551	34.8	48.0	0.288	911	1107.6	63.8	0.745	26.2
X	1150	150.4	39.5	0.211	1403	1020.8	61.2	0.729	9.3

[a]Chromosome length after nucleotides "N" in the assembly sequence were excluded.

Table 3. Linear correlation between CpG islands and genomic factors in the human genome

Genomic factor	r	P
Gene density (/Mb)	0.96	5.9×10^{-13}
GC content (%)	0.88	1.5×10^{-8}
Obs_{CpG}/Exp_{CpG}	0.92	1.4×10^{-10}
Recombination rate		
1-Mb window	0.18	1.1×10^{-22}
5-Mb window	0.33	5.9×10^{-16}
10-Mb window	0.40	1.7×10^{-12}
Combined bins	0.71	1.1×10^{-16}

Recombination Rate and CGIs

Pardo-Manuel de Villena and Sapienza (2001) reported that recombination rate correlates with the number of chromosome arms [44]. Meunier and Duret (2004) reported that recombination elevates GC content [45]. Fine-scale recombination rates vary extensively among populations [46, 47], genomic regions [48], or the homologous regions between two closely related organisms (human and chimpanzee) [49, 50], suggesting a rapid evolution of local pattern of recombination rates. Based on high-resolution recombination map, Kong *et al.* (2002) reported that recombination rate is positively correlated with the fraction of CpG dinucleotides (r= 0.40) and GC content (r= 0.39) [51]. Jensen-Seaman *et al.* (2004) compared the genetic maps in the human, mouse and rat genomes and estimated the local recombination rates across these genomes. They found that recombination rate was significantly correlated with several genomic factors, including GC content, CpG density, repetitive elements, and the neutral mutation rate [52]. Using the large-scale recombination rate data in 1-Mb windows from UCSC Genome Browser (http://genome.ucsc.edu/) and 5-Mb and 10-Mb windows from the dataset prepared by Jensen-Seaman *et al.* [52], Han *et al.* (2008) further found a significant positive correlation between CGI density and recombination rate (table 3), no matter which window size was used [18]. However, when window size increased, the correlation coefficient increased, for example, from 0.18 by 1-Mb window to 0.40 by 10-Mb window (table 3). In an alternative approach, when we separated the recombination data into 100 bins according to the recombination rates and then calculated the averaged CGI density in each bin, we found even a stronger correlation between recombination rate and CGI density (r= 0.71, P= 1.1×10^{-16}, table 3). The correlation supports a rapid evolution of local pattern of recombination rates. It also revealed that the correlation between CGIs and recombination rate became stronger in a larger scale.

Interestingly, as recombination rates were found to increase from the centromeric towards telomeric regions [5], Han *et al.* (2008) also observed a trend of higher CGI density in the telomeric regions in many human chromosomes [18]. This feature further supports a positive correlation between CGI density and recombination rate. It is worthy mentioning that this finding is opposite to a previous observation of no correlation between CGI features and chromosomal telomere positions using a small gene dataset [24]. More investigations are warranted.

Gene Expression and CGIs

In 1992, Larsen *et al.* analyzed 375 human genes and found that all housekeeping and widely expressed genes, but only 25% of tissue-specific genes, had a CGI covering the transcription start site (TSS) [3]. This distribution of CGIs in genes has been widely accepted until Ponger *et al.* (2001) revised that 90% of housekeeping genes had promoter-associated CGIs by using expressed sequence tag (EST) data [53]. More recently, Jiang *et al.* (2007) reexamined the distribution of CGIs in the promoter regions of genes with different gene expression level. In that work, the promoters were defined as a 2 kb sequence upstream of the most 5' start codon (ATG) and found only 67% of the human housekeeping genes having at least one CGI in the promoter region. When the broadly expressed genes were considered, i.e., expressed in more than 80% of the tissues, the frequency was 61%. The results significantly revised the previous view of that all housekeeping genes have promoter associated CGIs [25]. However, these results actually depend on the CGI searching algorithm, the definition of promoter-associated CGIs, the complexity of promoters (e.g., alternative promoters), and the gene expression data used. Due to these concerns, we reexamined this issue using an alternative definition of promoters: a 2 kb interval around the furthest 5' transcriptional start site (TSS) of a gene, i.e., from -1500 to + 500 bp relative to that TSS [39]. We used the second version of Gene Expression Atlas, which surveyed genome-wide gene expression in 79 human tissues [54]. This gene expression dataset was widely accepted for gene expression studies. According to the definitions that housekeeping genes were those expressed in all tissues, widely expressed genes expressed in more than 80% of tissues and tissue-specific genes expressed in less than 20% of tissues [25], we found a higher proportion of genes associated with promoters in all categories of human genes based on annotation of TSS than annotation of ATG (figure 1). For the human housekeeping genes, 86.1% genes contain promoter-associated CGIs according to TSS annotation, while only 66.7% of genes according to ATG annotation. This may largely be due to the variance of the length of 5'-UTR. No matter how we defined the promoter regions, we found much higher proportion of tissue specific genes associated with CGIs than previously thought (> 45%). These results indicated less difference between the association of CGIs with the promoter regions of housekeeping genes and that of tissue-specific genes (e.g., 25% based on TSS annotation, and 22% based on ATG annotation, versus 75% previously estimated). Overall, housekeeping genes are more likely to have CGIs in their promoter regions, supporting a strong correlation between expression and CGIs.

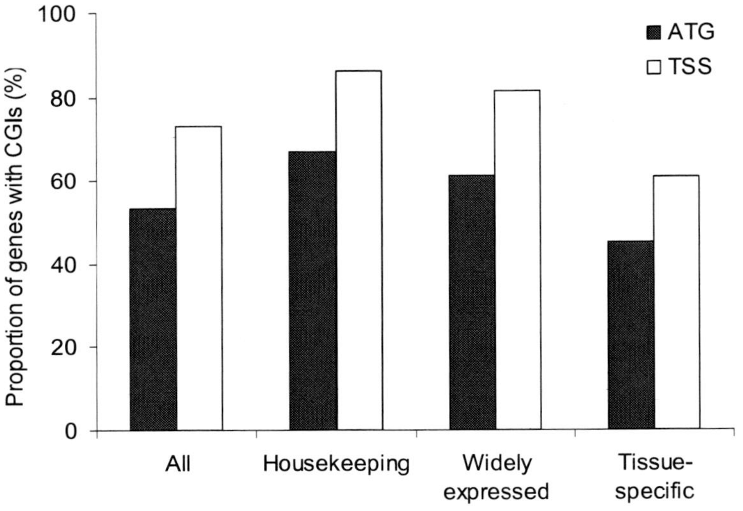

Figure 1. Promoter associated CpG islands (CGIs) in human genes with different expression level. ATG represents the promoter regions roughly defined based on translation start site (start codon ATG). TSS represents the promoters roughly defined based on transcriptional start site (TSS). Promoter associated CGIs were identified according to the promoter regions using ATG or TSS information.

Isochores and Mutation Patterns in CGIs

CGIs share some features with isochores. Isochores are defined as large genomic DNA fragments (> 100 kb) with GC content varying from 35 to 65% [5, 55]. This sequence pattern was first discovered by Bernardi *et al.* (1985) [56]. Subsequent studies found the variation of long-range base composition affects all regions of the genomes, including intergenic regions, introns, exons, and especially the third position of codons [57]. These isochores are correlated with genomic features, such as density of LINE and *Alu* repeats, methylation level, recombination, and gene density [5, 58-63]. Therefore, Bernardi *et al.* (2000) proposed a model of mammalian genomes for these isochores [64]. Later, the complete sequence of the human genome revealed that the GC content varies continuously without clear boundaries between GC-poor and GC-rich regions, although most of the structure detectable at a large scale (e.g., ~300 kb) [5]. The GC-rich isochores originated in the amniotes lineage, after the split with amphibians, but before the divergence of mammals, birds and reptiles [65-67]. Several models have been proposed to explain the origination and maintenance of the GC isochores: 1) selection for a high GC content in some regions of isochores [56, 64], 2) variable mutational bias (VMB) along genomes [68, 69] and, 3) biased gene conversion (BGC) which links to the process of recombination [70-72]. To understand the evolution of GC-rich isochores, Duret *et al.* (2002) analyzed synonymous substitution pattern in coding sequences from closely related mammals. They found GC-rich isochores have been slowly vanishing from the human genome [73]. Subsequently, Gu and Li (2006) found that GC-rich isochores vanished in those genomes only if the recombination rate became low (e.g., marsupials) [74].

CGIs, observed in a shorter scale (e.g., ~1 kb), contain higher GC content (> 50%) and higher Obs_{CpG}/Exp_{CpG} ratio (> 0.60) than isochores [2, 4]. Jiang and Zhao (2006) used human SNPs to systematically examine the mutational spectrum in the whole genome and categorized regions, including the non-coding and CGI regions [21]. Among all mutations in CGI regions, the frequency of G/C→A/T was 57.0%, substantially higher than that (23.3%) of A/T→G/C. The difference was slightly smaller in the promoter-associated CGIs: 56.4% were G/C→A/T mutations, compared to 22.9% A/T→G/C mutations. In the intergenic regions with similar GC content (e.g., ⩾60%), the frequency difference became larger. They observed 60.7% of point mutations being G/C→A/T but only 23.7% being A/T→G/C. Considering that the frequency of G/C→A/T increased while A/T→G/C decreased with the increasing local GC content in the sequence, CGIs (average GC content: 62.07%) and promoter-associated CGIs (average GC content: 63.38%) are expected to have a higher frequency of G/C→A/T and lower frequency of A/T→G/C. Their results indicated that, in the recent human genome, there is a trend to have more G or C changes to A or T, suggesting that the point mutations may tend to decrease the GC content, which results in the loss of CGIs [21]. Although the human genome has been shifting to being AT-rich, the shifting rate in CGIs is much slow.

Deamination of unmethylated cytosine produces uracil (U), which can be removed by uracil glycosylase [75, 76]. On the other hand, deamination of 5mC, which is frequently found at the methylated CpG sites in the human genome, produces thymine (T), can not be removed by this enzyme. This results in that the transition rate of 5mCpG to TpG/CpA is approximately 10- to 50-fold higher than other transitional changes [77-80]. CpG dinucleotides are more abundant at the polymorphic sites than expected in the human genome, but less prevalent than expected in the CGI sequences [19]. Similar patterns were observed in short n-mers (n=3-8) [33]. These studies revealed that 5mC deamination plays an important role for genetic variability in the different genomic regions in the human genome, and more specifically, the variation at CpGs is suppressed in CGIs. A recent survey found that CpG mutation rates in the human genome are highly dependent on local GC content [20] and follow-up study found that CpG mutation rates also depend on both the local sequence length and the genomic region [22]. CpG mutation rate in CGI regions is much different from that in other regions including intergenic regions [22].

Molecular Evolution of CGIs

CpG dinucleotides in CGIs are often protected from methylation. This mechanism helps maintain CGIs in genome evolution. An early study by Antequera and Bird (1993) suggested that CGIs have been likely under loss during evolution [23]. Their study was preliminary, as only based on the analysis of three genes. At the same time, Matsuo *et al.* (1993) analyzed 23 orthologous genes and found evidence for erosion of CGIs [24]. Compared to the loss of CGIs in the human genome, both studies suggested that rodent lineages underwent faster rate of erosion, therefore, the number of CGIs in the mouse genome is much smaller than that in the human genome [23, 24]. Based on these studies, Jiang *et al.* (2007) performed a genome-wide systematic survey of CGI characteristics in the human and mouse genomes [25]. They examined 1,257 pairs of promoter-associated CGIs in human-mouse homologous genes and found much stronger CGI features in human CGIs, including CGI length, GC content (%), Obs_{CpG}/Exp_{CpG} ratio, and CpG density. They found many losses of CGIs in both genomes, but the loss rate in the human genome is smaller than that in the mouse genome. For the human genes that were more likely to lose their promoter-associated CGIs than their mouse homologues, they likely involved in 'enzyme regulator activity', 'chromosome', 'chromatin binding', 'transport', 'DNA binding', 'transport activity', and 'cytoplasmic membrane-bound vesicle' [25].

Several issues remain further investigation. One is how CGIs have been lost in the course of genome evolution. Specifically to the human CGIs, we dissected CGIs into 11 segments and examined CGI characteristics in these segments. Figure 2 shows that erosion of a CGI likely started from both edges and gradually moved towards the center. Similar pattern was found in mouse CGIs or CGIs grouped by different size [25]. Although the strongest CGI characteristics were found in the central segment(s) in most cases, they could be in any segment [25]. Among the three parameters (CpG density, Obs_{CpG}/Exp_{CpG}, and GC content) examined, CpG density decayed at the strongest extent. CpG density in both edge segments was ~30% of that in the center segment. This compared to the GC content and Obs_{CpG}/Exp_{CpG} ratio in both edges being ~60% and ~70% of the center segment (figure 2). This comparison indicates a direct contribution of CpG dinucleotide decay during the erosion of CGIs, which is consistent with the previous studies [24, 80]. Another issue is how fast CGIs have been lost in the human or other genome. Using the dog lineage as the outgroup [81], Jiang *et al.* (2007) made the first attempt to estimate the CGI loss rate. The human lineage had ~1.5 losses per CGI per billion years and the mouse lineage had ~2.8 losses per CGI per billion years. The loss rate in the mouse lineage was ~1.9 times that in the human lineage [25].

Third, it is not clear whether loss of CGIs is universal in mammalian genomes or even in vertebrate genomes. Our recent survey on the dog genome suggests that this is likely true [82]. We found that the dog genome had much higher CGI density than the human genome in non-functional regions, e.g., intergenic regions and intronic regions. However, difference in the functional regions such as promoter regions or 3'-UTR was much smaller. When comparing CGIs in the 10,196 human-dog homologous genes, we found that, surprisingly, there were fewer dog genes (6048, 59.3%) associated with CGIs in the promoter regions than human genes (7466 genes, 73.2%). Similarly, the frequency differences between dog and human genes were 17.4% (housekeeping genes), 16.9% (widely expressed genes) and 9.2% (narrowly expressed genes), respectively. Selective pressure on functional regions could protect CGIs from methylation because abnormal methylation in promoter-associated CGIs might result in severe diseases [83]. Recent studies revealed that domestication process might accumulate the deleterious mutations and relax the selective constraints on both mitochondrial DNA (mtDNA) and nuclear genes [84-86]. The relaxation of selection may weaken the protection of CGIs from erosion in the functional regions. In summary, in our comparison of human-dog homologous genes, we consistently found a weaker extent of presence of promoter-associated CGIs in dog genes than in human genes, regardless of housekeeping or tissue-specific genes. Importantly, the difference in the presence of promoter-associated CGIs between dog and human genes tended to be larger when gene expression level increased.

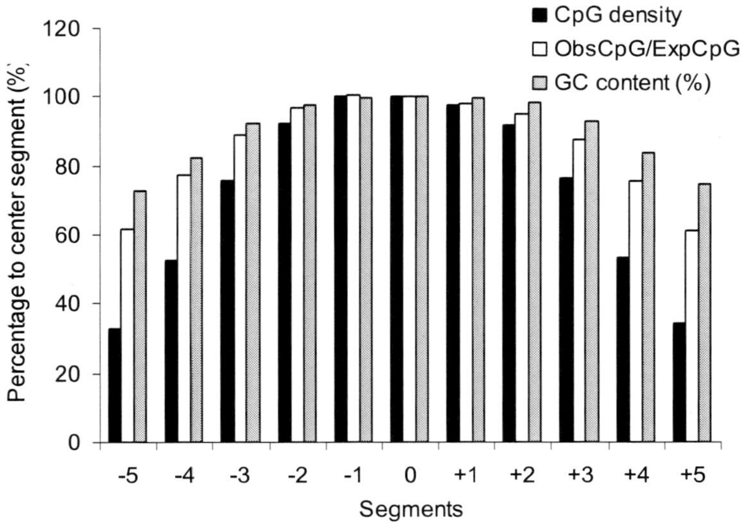

Figure 2. Edge decay of CpG islands (CGIs) in the human genome. On the X-axis, a CGI was dissected into 11 segments (-5 to +5) and 0 represents the center segment. On the Y-axis, values of CGI parameters were measured relative to the center segment. Black, white and grey represent CpG density (per 100 bp), Obs_{CpG}/Exp_{CpG} ratio and GC content (%), respectively.

Methylation Status of CGIs and Diseases

The methylation status in CGIs plays important roles in gene regulation and functions. Hypomethylation, which was the first observed epigenetic abnormality, was discovered in cancer cells as early as in 1983 [87, 88]. It is particularly prominent in pericentromeric regions [89-91]. The global hypomethylation in cancer cells may either aberrantly activate expression of normally silent genes [92-95] or increase genomic stability [96-98], thus, it leads to cancer. Besides, hypomethylation may contribute to the development of cancer by reactivating transposable elements and disrupting imprinting [99]. Hypomethylation, which has been overlooked for many years because of researchers' preference of hypermethylation, was actually prevalent in different cancers, including breast [94, 100], cervix [101, 102], colon [103], kidney [104], lung [105], pancreas [106, 107], and stomach [108, 109] cancers (figure 3).

Besides a massive overall hypomethylation, human tumors also acquire specific hypermethylation in certain promoters [110, 111]. The first evidence for loss of gene function caused by promoter hypermethylation was found in calcitonin gene [112], and then followed by tumor-suppressor genes [113-117], or even inactivation of microRNA (miRNA) genes [118, 119]. Now, we know that cancer genes are involved in cell cycle, DNA repair, apoptosis and all other cellular processes and could be hypermethylated and aberrantly silenced [28, 32]. So far, hypermethylation has linked to various types of cancers, including breast [120], cervix [121], colon [122], lung [123, 124], and prostate [125] cancers (figure 3). A large number of genes have been identified in the involvement of hypermethylation from various human cancers. These genes are involved in nearly all cellular functions or pathways such as DNA repair, cell cycle, transcription factors and apoptosis [28, 32] (figure 3). It is worthy noting that DNA methylation patterns varied among different tumor stages and cell types [126, 127] and different genic sites could be affected by both hypomethylation and hypermethylation in the same type of cancer [125, 128].

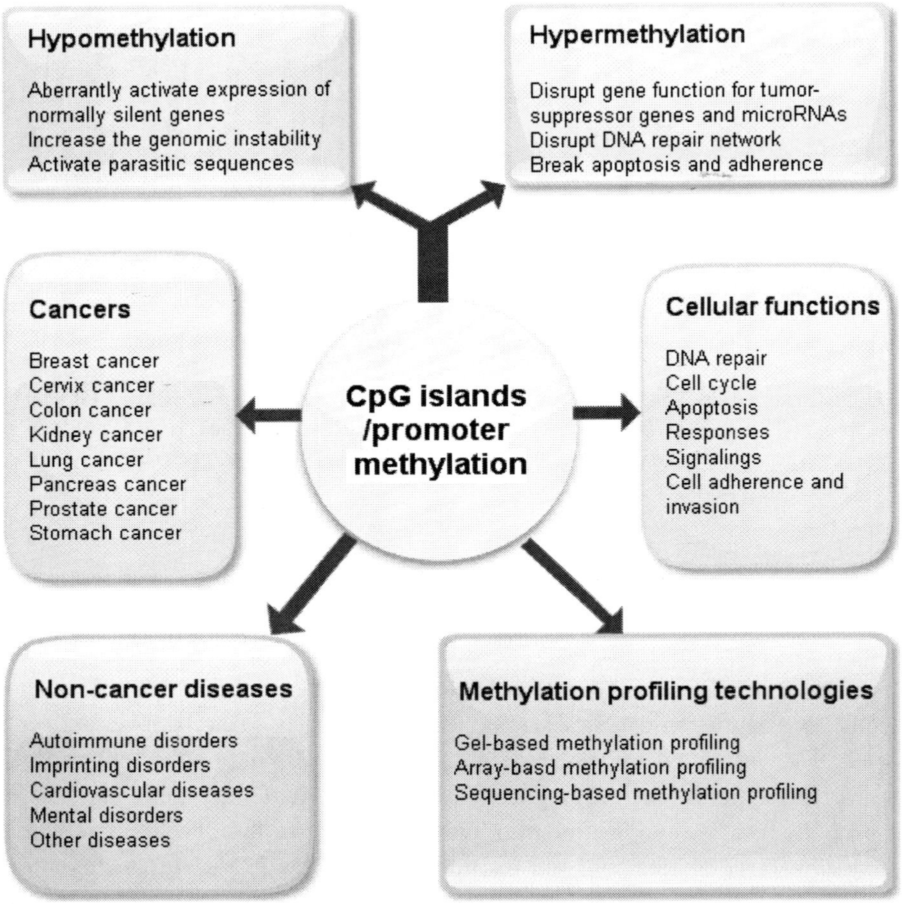

Figure 3. CpG islands (CGIs), methylation and diseases. This figure summarizes the methylation of CGIs/promoters and its mechanisms, relationship to cancers and other diseases, involvement in cellular functions, and methylation profiling mapping technologies.

Besides cancers, abnormal DNA methylation has been found to be involved in many other human diseases. Accumulated evidence indicated that DNA methylation plays important roles in autoimmune disorders (e.g., systemic lupus erythematosus (SLE) [129, 130], immunodeficiency, centromeric instability and facial anomalies syndrome (ICF) [131]), imprinting disorders (e.g., Prader-Willi syndrome (PWS) [132] and Beekwith-Wiedemann syndrome (BWS) [133]), cardiovascular diseases (e.g., coronary artery disease [134] and atherosclerosis [135]), mental disorders (e.g., Alzheimer's disease [136], schizophrenia, bipolar disorder [137-142] and psychiatric morbidity [143]), and other diseases (e.g., αthalassaemia/mental retardation syndrome X-linked (ATR-X) [144], facioscapulohumeral muscular dystrophy (FSHD) [145-147] and even osteoarthritis (OA) [148]) (figure 3). Although being involved in a great variety of diseases, methylation is a common feature in dysregulation of gene expression, either by hypomethylation that aberrantly activates expression of normally silent genes, or by hypermethylation that inactivates the genes that are necessary in normal conditions.

As described above, DNA methylation plays many fundamental roles in gene function, especially in cancers. Therefore, it is necessary to construct large-scale (e.g., genome-wide) and high-resolution DNA methylation profiles. Recently, some cancer methylomes have been available [120, 122, 149]. Rapid development of high throughput technologies shall further accelerate the progress of the comprehensive human and other mammalian methylomes. Gel-based methylation profiling, based on the restriction landmark genomic scanning (RLGS), was the first genome-wide profiling method achieved by either methylation-sensitive endonucleases or computational prediction of the respective restriction fragments [150, 151]. Array-based methylation profiling stands on the development of microarray technology, one of the major products in the revolution of functional genomics. It includes three major techniques – bisulphate conversion, which examines the cytosine to thymine signal ratio [152], methylation-sensitive restriction enzymes, which cut and fractionate genomic DNA into methylated and unmethylated fragments [153] and immunoprecipitation, which shears genomic DNA with an antibody against methylated cytosine [154, 155]. Sequencing-based methylation profiling, the finest methylation resolution to date, achieved in three different approaches – bisulphate PCR sequencing [10], reduced representation bisulphate sequencing (RRBS), which involves restriction and size selection of the genomic DNA followed by linker addition, bisulphate conversion, PCR amplification and cloning [156], and paired-end sequencing of methylated and unmethylated domains [13] (figure 3). The availability of next-generation sequencing platforms, including the Roche (454) GS FL-X sequencer, the Illumina (Solexa) 1G genome analyzer and the Applied Biosystems SOLiD sequencer [157] overcome the limitation in large-scale and high-resolution methylation analysis. Thus, we expect a new era of epigenomics is coming.

Conclusions

In summary, we reviewed several major computational algorithms for identification of CGIs, including those criteria using in mapping genome-wide methylation profiling and predicting methylation status. On the one hand, recent progresses on large-scale methylomes will greatly improve our understanding of the relationship between CGIs and methylation status or epigenetic status, thus help us develop more efficient computational algorithms and tools for identification functional CGIs, with both high sensitivity and specificity. These new algorithms and tools can be applied to new sequences and other genomes. For example, Bock *et al.* (2007) combined traditional algorithms and epigenetic information to select "*bona fide*" CGIs [158]. On the other hand, identifying CGIs in a specific sequence by an appropriate algorithm will ease many potential problems currently available in large-scale mapping methylation status, and finally save much effort and energy of the researchers.

In the human genome there is significant correlation between CGI density and gene density, supporting CGIs being gene markers. Many other genomic factors have been found to be highly correlated with CGI density. Comparative genomics and large-scale SNP data was recently used to investigate the mutation patterns and evolution of CGIs in the human genome. It is now clear that the transition rate of methylated CpG (5mCpG) to TpG is many-fold (e.g., 10- to 50-fold) higher than the rate of other transitional changes. Specifically, the variation has been suppressed in CGIs. Recent studies concluded that CGIs have been under the process of erosion and their erosion started from both edges towards the center.

Although many interesting features and important biological functions have been discovered, methylome is still in its early stage. Therefore, it is necessary to carry out comprehensive methylation profiling in a greater variety of normal and aberrant cell types and from different genomes. Furthermore, recently, cyclical changes in the methylation status of promoter CpGs were found to be important in transcriptional process, beyond the spatial changes [159, 160]. This calls for dynamic methylation profiling. In 2005, a blueprint of human epigenome was launched in the American Association of Cancer Research (AACR) workshop [161]. It was subsequently translated into a comprehensive epigenomics program in 2007 as part of NIH Roadmap for Medical Research (http://nihroadmap.nih.gov/epigenomics/index.asp), which targets on developing comprehensive reference epigenome maps and new technologies for comprehensive epigenomic analyses. The next-generation sequencing will establish the comprehensive reference maps of major epigenetic markers. Bioinformatics and biostatistics are expected to have key roles in the interpretation of the upcoming massive epigenetic information and in its integration with other genetic or genomic information.

Acknowledgements

This work was supported by a NIH grant (LM009598) from the National Library of Medicine, the Thomas F. and Kate Miller Jeffress Memorial Trust Fund, and Institutional Research Grant IRG-73-001-31 from the American Cancer Society.

References

[1] Bird, A. P. (1986). CpG-rich islands and the function of DNA methylation. *Nature*, *321*: 209-213.

[2] Bird, A. P. (1987). CpG islands as gene markers in the vertebrate nucleos. *Trends Genet.*, *3*: 342-347.

[3] Larsen, F., Gundersen, G., Lopez, R. & Prydz, H. (1992). CpG islands as gene markers in the human genome. *Genomics*, *13*: 1095-1107.

[4] Gardiner-Garden, M. & Frommer, M. (1987). CpG islands in vertebrate genomes. *J. Mol. Biol.*, *196*: 261-282.

[5] Lander, E. S., Linton, L. M., Birren, B., Nusbaum, C., Zody, M. C., Baldwin, J., Devon, K., Dewar, K., Doyle, M., FitzHugh, W., et al. (2001). Initial sequencing and analysis of the human genome. *Nature*, *409*: 860-921.

[6] Ponger, L. & Mouchiroud, D. (2002). CpGProD: identifying CpG islands associated with transcription start sites in large genomic mammalian sequences. *Bioinformatics*, *18*: 631-633.

[7] Takai, D. & Jones, P. A. (2002). Comprehensive analysis of CpG islands in human chromosomes 21 and 22. *Proc. Natl. Acad. Sci. USA*, *99*: 3740-3745.

[8] Hackenberg, M., Previti, C., Luque-Escamilla, P. L., Carpena, P., Martinez-Aroza, J. & Oliver, J. L. (2006). CpGcluster: a distance-based algorithm for CpG-island detection. *BMC Bioinformatics*, *7*: 446.

[9] Glass, J. L., Thompson, R. F., Khulan, B., Figueroa, M. E., Olivier, E. N., Oakley, E. J., Van Zant, G., Bouhassira, E. E., Melnick, A., Golden, A., et al. (2007). CG dinucleotide clustering is a species-specific property of the genome. *Nucleic Acids Res.*, *35*: 6798-6807.

[10] Eckhardt, F., Lewin, J., Cortese, R., Rakyan, V. K., Attwood, J., Burger, M., Burton, J., Cox, T. V., Davies, R., Down, T. A., et al. (2006). DNA methylation profiling of human chromosomes 6, 20 and 22. *Nat. Genet.*, *38*: 1378-1385.

[11] Keshet, I., Schlesinger, Y., Farkash, S., Rand, E., Hecht, M., Segal, E., Pikarski, E., Young, R. A., Niveleau, A., Cedar, H., et al. (2006). Evidence for an instructive mechanism of de novo methylation in cancer cells. *Nat. Genet.*, *38*: 149-153.

[12] Khulan, B., Thompson, R. F., Ye, K., Fazzari, M. J., Suzuki, M., Stasiek, E., Figueroa, M. E., Glass, J. L., Chen, Q., Montagna, C., et al. (2006). Comparative isoschizomer profiling of cytosine methylation: the HELP assay. *Genome Res.*, *16*: 1046-1055.

[13] Rollins, R. A., Haghighi, F., Edwards, J. R., Das, R., Zhang, M. Q., Ju, J. & Bestor, T. H. (2006). Large-scale structure of genomic methylation patterns. *Genome Res.*, *16*: 157-163.

[14] Weber, M., Hellmann, I., Stadler, M. B., Ramos, L., Paabo, S., Rebhan, M. & Schubeler, D. (2007). Distribution, silencing potential and evolutionary impact of promoter DNA methylation in the human genome. *Nat. Genet.*, *39*: 457-466.

[15] Bock, C., Paulsen, M., Tierling, S., Mikeska, T., Lengauer, T. & Walter, J. (2006). CpG island methylation in human lymphocytes is highly correlated with DNA sequence, repeats, and predicted DNA structure. *PLoS Genet.*, *2*: e26.

[16] Das, R., Dimitrova, N., Xuan, Z., Rollins, R. A., Haghighi, F., Edwards, J. R., Ju, J., Bestor, T. H. & Zhang, M. Q. (2006). Computational prediction of methylation status in human genomic sequences. *Proc. Natl. Acad. Sci. USA*, *103*: 10713-10716.

[17] Fang, F., Fan, S., Zhang, X. & Zhang, M. Q. (2006). Predicting methylation status of CpG islands in the human brain. *Bioinformatics*, *22*: 2204-2209.

[18] Han, L., Su, B., Li, W. H. & Zhao, Z. (2008). CpG island density and its correlations with genomic features in mammalian genomes. *Genome Biol.*, *9*: R79.

[19] Tomso, D. J. & Bell, D. A. (2003). Sequence context at human single nucleotide polymorphisms: overrepresentation of CpG dinucleotide at polymorphic sites and suppression of variation in CpG islands. *J. Mol. Biol.*, *327*: 303-308.

[20] Fryxell, K. J. & Moon, W. J. (2005). CpG mutation rates in the human genome are highly dependent on local GC content. *Mol. Biol. Evol.*, *22*: 650-658.

[21] Jiang, C. & Zhao, Z. (2006). Mutational spectrum in the recent human genome inferred by single nucleotide polymorphisms. *Genomics*, *88*: 527-534.

[22] Zhao, Z. & Jiang, C. (2007). Methylation-dependent transition rates are dependent on local sequence lengths and genomic regions. *Mol. Biol. Evol.*, *24*: 23-25.

[23] Antequera, F. & Bird, A. (1993). Number of CpG islands and genes in human and mouse. *Proc. Natl. Acad. Sci. USA*, *90*: 11995-11999.

[24] Matsuo, K., Clay, O., Takahashi, T., Silke, J. & Schaffner, W. (1993). Evidence for erosion of mouse CpG islands during mammalian evolution. *Somat. Cell. Mol. Genet.*, *19*: 543-555.

[25] Jiang, C., Han, L., Su, B., Li, W. H. & Zhao, Z. (2007). Features and trend of loss of promoter-associated CpG islands in the human and mouse genomes. *Mol. Biol. Evol.*, *24*: 1991-2000.

[26] Antequera, F. (2003). Structure, function and evolution of CpG island promoters. *Cell. Mol. Life Sci.*, *60*: 1647-1658.

[27] Feinberg, A. P. & Tycko, B. (2004). The history of cancer epigenetics. *Nat. Rev. Cancer*, *4*: 143-153.

[28] Esteller, M. (2007). Epigenetic gene silencing in cancer: the DNA hypermethylome. *Hum. Mol. Genet.*, *16 Spec No 1*: R50-59.

[29] Esteller, M. (2007). Cancer epigenomics: DNA methylomes and histone-modification maps. *Nat. Rev. Genet.*, *8*: 286-298.

[30] Szyf, M. (2009). Epigenetics, DNA methylation, and chromatin modifying drugs. *Annu. Rev. Pharmacol. Toxicol.*, *49*: 243-263.

[31] Iacobuzio-Donahue, C. A. (2009). Epigenetic changes in cancer. *Annu. Rev. Pathol.*, *4*: 229-249.

[32] Herman, J. G. & Baylin, S. B. (2003). Gene silencing in cancer in association with promoter hypermethylation. *N. Engl. J. Med.*, *349*: 2042-2054.

[33] Zhao, Z. & Zhang, F. (2006). Sequence context analysis of 8.2 million single nucleotide polymorphisms in the human genome. *Gene*, *366*: 316-324.

[34] Zhao, Z. & Zhang, F. (2006). Sequence context analysis in the mouse genome: single nucleotide polymorphisms and CpG island sequences. *Genomics*, *87*: 68-74.

[35] Saxonov, S., Berg, P. & Brutlag, D. L. (2006). A genome-wide analysis of CpG dinucleotides in the human genome distinguishes two distinct classes of promoters. *Proc. Natl. Acad. Sci. USA*, *103*: 1412-1417.

[36] International Human Genome Sequencing Consortium. (2004). Finishing the euchromatic sequence of the human genome. *Nature*, *431*: 931-945.

[37] Waterston, R. H., Lindblad-Toh, K., Birney, E., Rogers, J., Abril, J. F., Agarwal, P., Agarwala, R., Ainscough, R., Alexandersson, M., An, P., et al. (2002). Initial sequencing and comparative analysis of the mouse genome. *Nature*, *420*: 520-562.

[38] Wang, Y. & Leung, F. C. (2004). An evaluation of new criteria for CpG islands in the human genome as gene markers. *Bioinformatics*, *20*: 1170-1177.

[39] Han, L. and Zhao, Z. (2009). CpG islands or CpG clusters: how to identify functional GC-rich regions in a genome? *BMC Bioinformatics, 10*: 65.

[40] Esteller, M. (2006). The necessity of a human epigenome project. *Carcinogenesis*, *27*: 1121-1125.
[41] Suzuki, M. M. & Bird, A. (2008). DNA methylation landscapes: provocative insights from epigenomics. *Nat. Rev. Genet.*, *9*: 465-476.
[42] Beck, S. & Rakyan, V. K. (2008). The methylome: approaches for global DNA methylation profiling. *Trends Genet.*, *24*: 231-237.
[43] Bock, C., Walter, J., Paulsen, M. & Lengauer, T. (2008). Inter-individual variation of DNA methylation and its implications for large-scale epigenome mapping. *Nucleic Acids Res.*, *36*: e55.
[44] Pardo-Manuel de Villena, F. & Sapienza, C. (2001). Recombination is proportional to the number of chromosome arms in mammals. *Mamm. Genome*, *12*: 318-322.
[45] Meunier, J. & Duret, L. (2004). Recombination drives the evolution of GC-content in the human genome. *Mol. Biol. Evol.*, *21*: 984-990.
[46] Evans, D. M. & Cardon, L. R. (2005). A comparison of linkage disequilibrium patterns and estimated population recombination rates across multiple populations. *Am. J. Hum. Genet.*, *76*: 681-687.
[47] McVean, G. A., Myers, S. R., Hunt, S., Deloukas, P., Bentley, D. R. & Donnelly, P. (2004). The fine-scale structure of recombination rate variation in the human genome. *Science*, *304*: 581-584.
[48] Myers, S., Bottolo, L., Freeman, C., McVean, G. & Donnelly, P. (2005). A fine-scale map of recombination rates and hotspots across the human genome. *Science*, *310*: 321-324.
[49] Ptak, S. E., Hinds, D. A., Koehler, K., Nickel, B., Patil, N., Ballinger, D. G., Przeworski, M., Frazer, K. A. & Paabo, S. (2005). Fine-scale recombination patterns differ between chimpanzees and humans. *Nat. Genet.*, *37*: 429-434.
[50] Winckler, W., Myers, S. R., Richter, D. J., Onofrio, R. C., McDonald, G. J., Bontrop, R. E., McVean, G. A., Gabriel, S. B., Reich, D., Donnelly, P., et al. (2005). Comparison of fine-scale recombination rates in humans and chimpanzees. *Science*, *308*: 107-111.
[51] Kong, A., Gudbjartsson, D. F., Sainz, J., Jonsdottir, G. M., Gudjonsson, S. A., Richardsson, B., Sigurdardottir, S., Barnard, J., Hallbeck, B., Masson, G., et al. (2002). A high-resolution recombination map of the human genome. *Nat. Genet.*, *31*: 241-247.
[52] Jensen-Seaman, M. I., Furey, T. S., Payseur, B. A., Lu, Y., Roskin, K. M., Chen, C. F., Thomas, M. A., Haussler, D. & Jacob, H. J. (2004). Comparative recombination rates in the rat, mouse, and human genomes. *Genome Res.*, *14*: 528-538.
[53] Ponger, L., Duret, L. & Mouchiroud, D. (2001). Determinants of CpG islands: expression in early embryo and isochore structure. *Genome Res.*, *11*: 1854-1860.
[54] Su, A. I., Wiltshire, T., Batalov, S., Lapp, H., Ching, K. A., Block, D., Zhang, J., Soden, R., Hayakawa, M., Kreiman, G., et al. (2004). A gene atlas of the mouse and human protein-encoding transcriptomes. *Proc. Natl. Acad. Sci. USA*, *101*: 6062-6067.
[55] Nekrutenko, A. & Li, W. H. (2000). Assessment of compositional heterogeneity within and between eukaryotic genomes. *Genome Res.*, *10*: 1986-1995.
[56] Bernardi, G., Olofsson, B., Filipski, J., Zerial, M., Salinas, J., Cuny, G., Meunier-Rotival, M. & Rodier, F. (1985). The mosaic genome of warm-blooded vertebrates. *Science*, *228*: 953-958.

[57] D'Onofrio, G., Mouchiroud, D., Aissani, B., Gautier, C. & Bernardi, G. (1991). Correlations between the compositional properties of human genes, codon usage, and amino acid composition of proteins. *J. Mol. Evol.*, *32*: 504-510.

[58] Mouchiroud, D., D'Onofrio, G., Aissani, B., Macaya, G., Gautier, C. & Bernardi, G. (1991). The distribution of genes in the human genome. *Gene*, *100*: 181-187.

[59] Duret, L., Mouchiroud, D. & Gautier, C. (1995). Statistical analysis of vertebrate sequences reveals that long genes are scarce in GC-rich isochores. *J. Mol. Evol.*, *40*: 308-317.

[60] Jabbari, K. & Bernardi, G. (1998). CpG doublets, CpG islands and Alu repeats in long human DNA sequences from different isochore families. *Gene*, *224*: 123-127.

[61] Smit, A. F. (1999). Interspersed repeats and other mementos of transposable elements in mammalian genomes. *Curr. Opin. Genet. Dev.*, *9*: 657-663.

[62] Fullerton, S. M., Bernardo Carvalho, A. & Clark, A. G. (2001). Local rates of recombination are positively correlated with GC content in the human genome. *Mol. Biol. Evol.*, *18*: 1139-1142.

[63] Montoya-Burgos, J. I., Boursot, P. & Galtier, N. (2003). Recombination explains isochores in mammalian genomes. *Trends Genet.*, *19*: 128-130.

[64] Bernardi, G. (2000). Isochores and the evolutionary genomics of vertebrates. *Gene*, *241*: 3-17.

[65] Bernardi, G. & Bernardi, G. (1990). Compositional transitions in the nuclear genomes of cold-blooded vertebrates. *J. Mol. Evol.*, *31*: 282-293.

[66] Bernardi, G., Hughes, S. & Mouchiroud, D. (1997). The major compositional transitions in the vertebrate genome. *J. Mol. Evol.*, *44 Suppl 1*: S44-51.

[67] Hughes, S., Zelus, D. & Mouchiroud, D. (1999). Warm-blooded isochore structure in Nile crocodile and turtle. *Mol. Biol. Evol.*, *16*: 1521-1527.

[68] Wolfe, K. H., Sharp, P. M. & Li, W. H. (1989). Mutation rates differ among regions of the mammalian genome. *Nature*, *337*: 283-285.

[69] Francino, M. P. & Ochman, H. (1999). Isochores result from mutation not selection. *Nature*, *400*: 30-31.

[70] Holmquist, G. P. (1992). Chromosome bands, their chromatin flavors, and their functional features. *Am. J. Hum. Genet.*, *51*: 17-37.

[71] Eyre-Walker, A. & Hurst, L. D. (2001). The evolution of isochores. *Nat. Rev. Genet.*, *2*: 549-555.

[72] Galtier, N., Piganeau, G., Mouchiroud, D. & Duret, L. (2001). GC-content evolution in mammalian genomes: the biased gene conversion hypothesis. *Genetics*, *159*: 907-911.

[73] Duret, L., Semon, M., Piganeau, G., Mouchiroud, D. & Galtier, N. (2002). Vanishing GC-rich isochores in mammalian genomes. *Genetics*, *162*: 1837-1847.

[74] Gu, J. & Li, W. H. (2006). Are GC-rich isochores vanishing in mammals? *Gene*, *385*: 50-56.

[75] Lindahl, T. (1974). An N-glycosidase from Escherichia coli that releases free uracil from DNA containing deaminated cytosine residues. *Proc. Natl. Acad. Sci. USA*, *71*: 3649-3653.

[76] Lindahl, T., Karran, P. & Wood, R. D. (1997). DNA excision repair pathways. *Curr. Opin. Genet. Dev.*, *7*: 158-169.

[77] Duncan, B. K. & Miller, J. H. (1980). Mutagenic deamination of cytosine residues in DNA. *Nature*, *287*: 560-561.

[78] Bulmer, M. (1986). Neighboring base effects on substitution rates in pseudogenes. *Mol. Biol. Evol.*, *3*: 322-329.

[79] Britten, R. J., Baron, W. F., Stout, D. B. & Davidson, E. H. (1988). Sources and evolution of human Alu repeated sequences. *Proc. Natl. Acad. Sci. USA*, *85*: 4770-4774.

[80] Sved, J. & Bird, A. (1990). The expected equilibrium of the CpG dinucleotide in vertebrate genomes under a mutation model. *Proc. Natl. Acad. Sci. USA*, *87*: 4692-4696.

[81] Springer, M. S., Murphy, W. J., Eizirik, E. & O'Brien, S. J. (2003). Placental mammal diversification and the Cretaceous-Tertiary boundary. *Proc. Natl. Acad. Sci. USA*, *100*: 1056-1061.

[82] Han, L. & Zhao, Z. (2009). Contrast features of CpG islands in the promoter and other regions in the dog genome. *Genomics, 94*: 117-124.

[83] Esteller, M. (2006). Epigenetics provides a new generation of oncogenes and tumour-suppressor genes. *Br. J. Cancer*, *94*: 179-183.

[84] Bjornerfeldt, S., Webster, M. T. & Vila, C. (2006). Relaxation of selective constraint on dog mitochondrial DNA following domestication. *Genome Res.*, *16*: 990-994.

[85] Lu, J., Tang, T., Tang, H., Huang, J., Shi, S. & Wu, C. I. (2006). The accumulation of deleterious mutations in rice genomes: a hypothesis on the cost of domestication. *Trends Genet.*, *22*: 126-131.

[86] Cruz, F., Vila, C. & Webster, M. T. (2008). The legacy of domestication: Accumulation of deleterious mutations in the dog genome. *Mol. Biol. Evol.*, *25*: 2331-2336.

[87] Feinberg, A. P. & Vogelstein, B. (1983). Hypomethylation distinguishes genes of some human cancers from their normal counterparts. *Nature*, *301*: 89-92.

[88] Gama-Sosa, M. A., Slagel, V. A., Trewyn, R. W., Oxenhandler, R., Kuo, K. C., Gehrke, C. W. & Ehrlich, M. (1983). The 5-methylcytosine content of DNA from human tumors. *Nucleic Acids Res.*, *11*: 6883-6894.

[89] Narayan, A., Ji, W., Zhang, X. Y., Marrogi, A., Graff, J. R., Baylin, S. B. & Ehrlich, M. (1998). Hypomethylation of pericentromeric DNA in breast adenocarcinomas. *Int. J. Cancer*, *77*: 833-838.

[90] Qu, G. Z., Grundy, P. E., Narayan, A. & Ehrlich, M. (1999). Frequent hypomethylation in Wilms tumors of pericentromeric DNA in chromosomes 1 and 16. *Cancer Genet. Cytogenet.*, *109*: 34-39.

[91] Tuck-Muller, C. M., Narayan, A., Tsien, F., Smeets, D. F., Sawyer, J., Fiala, E. S., Sohn, O. S. & Ehrlich, M. (2000). DNA hypomethylation and unusual chromosome instability in cell lines from ICF syndrome patients. *Cytogenet. Cell Genet.*, *89*: 121-128.

[92] Cho, B., Lee, H., Jeong, S., Bang, Y. J., Lee, H. J., Hwang, K. S., Kim, H. Y., Lee, Y. S., Kang, G. H. & Jeoung, D. I. (2003). Promoter hypomethylation of a novel cancer/testis antigen gene CAGE is correlated with its aberrant expression and is seen

in premalignant stage of gastric carcinoma. *Biochem. Biophys. Res. Commun.*, *307*: 52-63.

[93] Lee, T. S., Kim, J. W., Kang, G. H., Park, N. H., Song, Y. S., Kang, S. B. & Lee, H. P. (2006). DNA hypomethylation of CAGE promotors in squamous cell carcinoma of uterine cervix. *Ann. N. Y. Acad. Sci.*, *1091*: 218-224.

[94] Kim, S. J., Kang, H. S., Chang, H. L., Jung, Y. C., Sim, H. B., Lee, K. S., Ro, J. & Lee, E. S. (2008). Promoter hypomethylation of the N-acetyltransferase 1 gene in breast cancer. *Oncol. Rep.*, *19*: 663-668.

[95] Watanabe, M., Ogawa, Y., Itoh, K., Koiwa, T., Kadin, M. E., Watanabe, T., Okayasu, I., Higashihara, M. & Horie, R. (2008). Hypomethylation of CD30 CpG islands with aberrant JunB expression drives CD30 induction in Hodgkin lymphoma and anaplastic large cell lymphoma. *Lab. Invest.*, *88*: 48-57.

[96] Chen, R. Z., Pettersson, U., Beard, C., Jackson-Grusby, L. & Jaenisch, R. (1998). DNA hypomethylation leads to elevated mutation rates. *Nature*, *395*: 89-93.

[97] Eden, A., Gaudet, F., Waghmare, A. & Jaenisch, R. (2003). Chromosomal instability and tumors promoted by DNA hypomethylation. *Science*, *300*: 455.

[98] Gaudet, F., Hodgson, J. G., Eden, A., Jackson-Grusby, L., Dausman, J., Gray, J. W., Leonhardt, H. & Jaenisch, R. (2003). Induction of tumors in mice by genomic hypomethylation. *Science*, *300*: 489-492.

[99] Esteller, M. (2008). Epigenetics in cancer. *N. Engl. J. Med.*, *358*: 1148-1159.

[100] Ito, Y., Koessler, T., Ibrahim, A. E., Rai, S., Vowler, S. L., Abu-Amero, S., Silva, A. L., Maia, A. T., Huddleston, J. E., Uribe-Lewis, S., et al. (2008). Somatically acquired hypomethylation of IGF2 in breast and colorectal cancer. *Hum. Mol. Genet.*, *17*: 2633-2643.

[101] Badal, V., Chuang, L. S., Tan, E. H., Badal, S., Villa, L. L., Wheeler, C. M., Li, B. F. & Bernard, H. U. (2003). CpG methylation of human papillomavirus type 16 DNA in cervical cancer cell lines and in clinical specimens: genomic hypomethylation correlates with carcinogenic progression. *J. Virol.*, *77*: 6227-6234.

[102] de Capoa, A., Musolino, A., Della Rosa, S., Caiafa, P., Mariani, L., Del Nonno, F., Vocaturo, A., Donnorso, R. P., Niveleau, A. & Grappelli, C. (2003). DNA demethylation is directly related to tumour progression: evidence in normal, pre-malignant and malignant cells from uterine cervix samples. *Oncol. Rep.*, *10*: 545-549.

[103] Nakamura, N. & Takenaga, K. (1998). Hypomethylation of the metastasis-associated S100A4 gene correlates with gene activation in human colon adenocarcinoma cell lines. *Clin. Exp. Metastasis*, *16*: 471-479.

[104] Cho, M., Uemura, H., Kim, S. C., Kawada, Y., Yoshida, K., Hirao, Y., Konishi, N., Saga, S. & Yoshikawa, K. (2001). Hypomethylation of the MN/CA9 promoter and upregulated MN/CA9 expression in human renal cell carcinoma. *Br. J. Cancer*, *85*: 563-567.

[105] Piyathilake, C. J., Henao, O., Frost, A. R., Macaluso, M., Bell, W. C., Johanning, G. L., Heimburger, D. C., Niveleau, A. & Grizzle, W. E. (2003). Race- and age-dependent

alterations in global methylation of DNA in squamous cell carcinoma of the lung (United States). *Cancer Causes Control*, *14*: 37-42.

[106] Iacobuzio-Donahue, C. A., Maitra, A., Olsen, M., Lowe, A. W., van Heek, N. T., Rosty, C., Walter, K., Sato, N., Parker, A., Ashfaq, R., et al. (2003). Exploration of global gene expression patterns in pancreatic adenocarcinoma using cDNA microarrays. *Am. J. Pathol.*, *162*: 1151-1162.

[107] Sato, N., Maitra, A., Fukushima, N., van Heek, N. T., Matsubayashi, H., Iacobuzio-Donahue, C. A., Rosty, C. & Goggins, M. (2003). Frequent hypomethylation of multiple genes overexpressed in pancreatic ductal adenocarcinoma. *Cancer Res.*, *63*: 4158-4166.

[108] Akiyama, Y., Maesawa, C., Ogasawara, S., Terashima, M. & Masuda, T. (2003). Cell-type-specific repression of the maspin gene is disrupted frequently by demethylation at the promoter region in gastric intestinal metaplasia and cancer cells. *Am. J. Pathol.*, *163*: 1911-1919.

[109] Oshimo, Y., Nakayama, H., Ito, R., Kitadai, Y., Yoshida, K., Chayama, K. & Yasui, W. (2003). Promoter methylation of cyclin D2 gene in gastric carcinoma. *Int. J. Oncol.*, *23*: 1663-1670.

[110] Baylin, S. B. & Herman, J. G. (2000). DNA hypermethylation in tumorigenesis: epigenetics joins genetics. *Trends Genet.*, *16*: 168-174.

[111] Robertson, K. D. & Wolffe, A. P. (2000). DNA methylation in health and disease. *Nat. Rev. Genet.*, *1*: 11-19.

[112] Baylin, S. B., Fearon, E. R., Vogelstein, B., de Bustros, A., Sharkis, S. J., Burke, P. J., Staal, S. P. & Nelkin, B. D. (1987). Hypermethylation of the 5' region of the calcitonin gene is a property of human lymphoid and acute myeloid malignancies. *Blood*, *70*: 412-417.

[113] Sakai, T., Toguchida, J., Ohtani, N., Yandell, D. W., Rapaport, J. M. & Dryja, T. P. (1991). Allele-specific hypermethylation of the retinoblastoma tumor-suppressor gene. *Am. J. Hum. Genet.*, *48*: 880-888.

[114] Herman, J. G., Latif, F., Weng, Y., Lerman, M. I., Zbar, B., Liu, S., Samid, D., Duan, D. S., Gnarra, J. R., Linehan, W. M., et al. (1994). Silencing of the VHL tumor-suppressor gene by DNA methylation in renal carcinoma. *Proc. Natl. Acad. Sci. USA*, *91*: 9700-9704.

[115] Gonzalez-Zulueta, M., Bender, C. M., Yang, A. S., Nguyen, T., Beart, R. W., Van Tornout, J. M. & Jones, P. A. (1995). Methylation of the 5' CpG island of the p16/CDKN2 tumor suppressor gene in normal and transformed human tissues correlates with gene silencing. *Cancer Res.*, *55*: 4531-4535.

[116] Herman, J. G., Merlo, A., Mao, L., Lapidus, R. G., Issa, J. P., Davidson, N. E., Sidransky, D. & Baylin, S. B. (1995). Inactivation of the CDKN2/p16/MTS1 gene is frequently associated with aberrant DNA methylation in all common human cancers. *Cancer Res.*, *55*: 4525-4530.

[117] Merlo, A., Herman, J. G., Mao, L., Lee, D. J., Gabrielson, E., Burger, P. C., Baylin, S. B. & Sidransky, D. (1995). 5' CpG island methylation is associated with transcriptional silencing of the tumour suppressor p16/CDKN2/MTS1 in human cancers. *Nat. Med.*, *1*: 686-692.

[118] Saito, Y., Liang, G., Egger, G., Friedman, J. M., Chuang, J. C., Coetzee, G. A. & Jones, P. A. (2006). Specific activation of microRNA-127 with downregulation of the proto-oncogene BCL6 by chromatin-modifying drugs in human cancer cells. *Cancer Cell*, *9*: 435-443.

[119] Lujambio, A., Ropero, S., Ballestar, E., Fraga, M. F., Cerrato, C., Setien, F., Casado, S., Suarez-Gauthier, A., Sanchez-Cespedes, M., Git, A., et al. (2007). Genetic unmasking of an epigenetically silenced microRNA in human cancer cells. *Cancer Res.*, *67*: 1424-1429.

[120] Shann, Y. J., Cheng, C., Chiao, C. H., Chen, D. T., Li, P. H. & Hsu, M. T. (2008). Genome-wide mapping and characterization of hypomethylated sites in human tissues and breast cancer cell lines. *Genome Res.*, *18*: 791-801.

[121] Feng, Q., Balasubramanian, A., Hawes, S. E., Toure, P., Sow, P. S., Dem, A., Dembele, B., Critchlow, C. W., Xi, L., Lu, H., et al. (2005). Detection of hypermethylated genes in women with and without cervical neoplasia. *J. Natl. Cancer. Inst.*, *97*: 273-282.

[122] Weber, M., Davies, J. J., Wittig, D., Oakeley, E. J., Haase, M., Lam, W. L. & Schubeler, D. (2005). Chromosome-wide and promoter-specific analyses identify sites of differential DNA methylation in normal and transformed human cells. *Nat. Genet.*, *37*: 853-862.

[123] Anglim, P. P., Alonzo, T. A. & Laird-Offringa, I. A. (2008). DNA methylation-based biomarkers for early detection of non-small cell lung cancer: an update. *Mol. Cancer*, *7*: 81.

[124] Tessema, M. & Belinsky, S. A. (2008). Mining the epigenome for methylated genes in lung cancer. *Proc. Am. Thorac. Soc.*, *5*: 806-810.

[125] Yegnasubramanian, S., Haffner, M. C., Zhang, Y., Gurel, B., Cornish, T. C., Wu, Z., Irizarry, R. A., Morgan, J., Hicks, J., DeWeese, T. L., et al. (2008). DNA hypomethylation arises later in prostate cancer progression than CpG island hypermethylation and contributes to metastatic tumor heterogeneity. *Cancer Res.*, *68*: 8954-8967.

[126] Costello, J. F., Fruhwald, M. C., Smiraglia, D. J., Rush, L. J., Robertson, G. P., Gao, X., Wright, F. A., Feramisco, J. D., Peltomaki, P., Lang, J. C., et al. (2000). Aberrant CpG-island methylation has non-random and tumour-type-specific patterns. *Nat. Genet.*, *24*: 132-138.

[127] Hu, M., Yao, J., Cai, L., Bachman, K. E., van den Brule, F., Velculescu, V. & Polyak, K. (2005). Distinct epigenetic changes in the stromal cells of breast cancers. *Nat. Genet.*, *37*: 899-905.

[128] Ehrlich, M., Jiang, G., Fiala, E., Dome, J. S., Yu, M. C., Long, T. I., Youn, B., Sohn, O. S., Widschwendter, M., Tomlinson, G. E., et al. (2002). Hypomethylation and hypermethylation of DNA in Wilms tumors. *Oncogene*, *21*: 6694-6702.

[129] Sekigawa, I., Okada, M., Ogasawara, H., Kaneko, H., Hishikawa, T. & Hashimoto, H. (2003). DNA methylation in systemic lupus erythematosus. *Lupus*, *12*: 79-85.

[130] Richardson, B., Ray, D. & Yung, R. (2004). Murine models of lupus induced by hypomethylated T cells. *Methods Mol. Med.*, *102*: 285-294.

[131] Yehezkel, S., Segev, Y., Viegas-Pequignot, E., Skorecki, K. & Selig, S. (2008). Hypomethylation of subtelomeric regions in ICF syndrome is associated with abnormally short telomeres and enhanced transcription from telomeric regions. *Hum. Mol. Genet.*, *17*: 2776-2789.

[132] Lewis, A. & Reik, W. (2006). How imprinting centres work. *Cytogenet. Genome Res.*, *113*: 81-89.

[133] Thorvaldsen, J. L., Duran, K. L. & Bartolomei, M. S. (1998). Deletion of the H19 differentially methylated domain results in loss of imprinted expression of H19 and Igf2. *Genes Dev.*, *12*: 3693-3702.

[134] Sharma, P., Kumar, J., Garg, G., Kumar, A., Patowary, A., Karthikeyan, G., Ramakrishnan, L., Brahmachari, V. & Sengupta, S. (2008). Detection of altered global DNA methylation in coronary artery disease patients. *DNA Cell Biol.*, *27*: 357-365.

[135] DeRuiter, M. C., Alkemade, F. E., Gittenberger-de Groot, A. C., Poelmann, R. E., Havekes, L. M. & van Dijk, K. W. (2008). Maternal transmission of risk for atherosclerosis. *Curr. Opin. Lipidol.*, *19*: 333-337.

[136] West, R. L., Lee, J. M. & Maroun, L. E. (1995). Hypomethylation of the amyloid precursor protein gene in the brain of an Alzheimer's disease patient. *J. Mol. Neurosci.*, *6*: 141-146.

[137] Polesskaya, O. O. & Sokolov, B. P. (2002). Differential expression of the "C" and "T" alleles of the 5-HT2A receptor gene in the temporal cortex of normal individuals and schizophrenics. *J. Neurosci. Res.*, *67*: 812-822.

[138] Abdolmaleky, H. M., Cheng, K. H., Russo, A., Smith, C. L., Faraone, S. V., Wilcox, M., Shafa, R., Glatt, S. J., Nguyen, G., Ponte, J. F., et al. (2005). Hypermethylation of the reelin (RELN) promoter in the brain of schizophrenic patients: a preliminary report. *Am. J. Med. Genet. B Neuropsychiatr. Genet.*, *134B*: 60-66.

[139] Grayson, D. R., Jia, X., Chen, Y., Sharma, R. P., Mitchell, C. P., Guidotti, A. & Costa, E. (2005). Reelin promoter hypermethylation in schizophrenia. *Proc. Natl. Acad. Sci. USA*, *102*: 9341-9346.

[140] Abdolmaleky, H. M., Cheng, K. H., Faraone, S. V., Wilcox, M., Glatt, S. J., Gao, F., Smith, C. L., Shafa, R., Aeali, B., Carnevale, J., et al. (2006). Hypomethylation of MB-COMT promoter is a major risk factor for schizophrenia and bipolar disorder. *Hum. Mol. Genet.*, *15*: 3132-3145.

[141] Guidotti, A., Ruzicka, W., Grayson, D. R., Veldic, M., Pinna, G., Davis, J. M. & Costa, E. (2007). S-adenosyl methionine and DNA methyltransferase-1 mRNA overexpression in psychosis. *Neuroreport*, *18*: 57-60.

[142] Popendikyte, V., Laurinavicius, A., Paterson, A. D., Macciardi, F., Kennedy, J. L. & Petronis, A. (1999). DNA methylation at the putative promoter region of the human dopamine D2 receptor gene. *Neuroreport*, *10*: 1249-1255.

[143] Tassone, F., Hagerman, R. J., Loesch, D. Z., Lachiewicz, A., Taylor, A. K. & Hagerman, P. J. (2000). Fragile X males with unmethylated, full mutation trinucleotide repeat expansions have elevated levels of FMR1 messenger RNA. *Am. J. Med. Genet.*, *94*: 232-236.

[144] Gibbons, R. J., McDowell, T. L., Raman, S., O'Rourke, D. M., Garrick, D., Ayyub, H. & Higgs, D. R. (2000). Mutations in ATRX, encoding a SWI/SNF-like protein, cause diverse changes in the pattern of DNA methylation. *Nat. Genet.*, *24*: 368-371.

[145] Tsien, F., Sun, B., Hopkins, N. E., Vedanarayanan, V., Figlewicz, D., Winokur, S. & Ehrlich, M. (2001). Methylation of the FSHD syndrome-linked subtelomeric repeat in normal and FSHD cell cultures and tissues. *Mol. Genet. Metab.*, *74*: 322-331.

[146] van Overveld, P. G., Enthoven, L., Ricci, E., Rossi, M., Felicetti, L., Jeanpierre, M., Winokur, S. T., Frants, R. R., Padberg, G. W. & van der Maarel, S. M. (2005). Variable hypomethylation of D4Z4 in facioscapulohumeral muscular dystrophy. *Ann. Neurol.*, *58*: 569-576.

[147] de Greef, J. C., Wohlgemuth, M., Chan, O. A., Hansson, K. B., Smeets, D., Frants, R. R., Weemaes, C. M., Padberg, G. W. & van der Maarel, S. M. (2007). Hypomethylation is restricted to the D4Z4 repeat array in phenotypic FSHD. *Neurology*, *69*: 1018-1026.

[148] Roach, H. I. & Aigner, T. (2007). DNA methylation in osteoarthritic chondrocytes: a new molecular target. *Osteoarthr. Cartil.*, *15*: 128-137.

[149] Schuebel, K. E., Chen, W., Cope, L., Glockner, S. C., Suzuki, H., Yi, J. M., Chan, T. A., Van Neste, L., Van Criekinge, W., van den Bosch, S., et al. (2007). Comparing the DNA hypermethylome with gene mutations in human colorectal cancer. *PLoS Genet.*, *3*: 1709-1723.

[150] Hatada, I., Hayashizaki, Y., Hirotsune, S., Komatsubara, H. & Mukai, T. (1991). A genomic scanning method for higher organisms using restriction sites as landmarks. *Proc. Natl. Acad. Sci. USA*, *88*: 9523-9527.

[151] Rouillard, J. M., Erson, A. E., Kuick, R., Asakawa, J., Wimmer, K., Muleris, M., Petty, E. M. & Hanash, S. (2001). Virtual genome scan: a tool for restriction landmark-based scanning of the human genome. *Genome Res.*, *11*: 1453-1459.

[152] Reinders, J., Delucinge Vivier, C., Theiler, G., Chollet, D., Descombes, P. & Paszkowski, J. (2008). Genome-wide, high-resolution DNA methylation profiling using bisulfite-mediated cytosine conversion. *Genome Res.*, *18*: 469-476.

[153] Murrell, A., Rakyan, V. K. & Beck, S. (2005). From genome to epigenome. *Hum. Mol. Genet.*, *14 Spec No 1*: R3-R10.

[154] Zhang, X., Yazaki, J., Sundaresan, A., Cokus, S., Chan, S. W., Chen, H., Henderson, I. R., Shinn, P., Pellegrini, M., Jacobsen, S. E., et al. (2006). Genome-wide high-resolution mapping and functional analysis of DNA methylation in arabidopsis. *Cell*, *126*: 1189-1201.

[155] Zilberman, D., Gehring, M., Tran, R. K., Ballinger, T. & Henikoff, S. (2007). Genome-wide analysis of Arabidopsis thaliana DNA methylation uncovers an interdependence between methylation and transcription. *Nat. Genet.*, *39*: 61-69.

[156] Meissner, A., Gnirke, A., Bell, G. W., Ramsahoye, B., Lander, E. S. & Jaenisch, R. (2005). Reduced representation bisulfite sequencing for comparative high-resolution DNA methylation analysis. *Nucleic Acids Res.*, *33*: 5868-5877.

[157] Mardis, E. R. (2008). The impact of next-generation sequencing technology on genetics. *Trends Genet.*, *24*: 133-141.

[158] Bock, C., Walter, J., Paulsen, M. & Lengauer, T. (2007). CpG island mapping by epigenome prediction. *PLoS Comput. Biol.*, *3*: e110.

[159] Kangaspeska, S., Stride, B., Metivier, R., Polycarpou-Schwarz, M., Ibberson, D., Carmouche, R. P., Benes, V., Gannon, F. & Reid, G. (2008). Transient cyclical methylation of promoter DNA. *Nature*, *452*: 112-115.

[160] Metivier, R., Gallais, R., Tiffoche, C., Le Peron, C., Jurkowska, R. Z., Carmouche, R. P., Ibberson, D., Barath, P., Demay, F., Reid, G., et al. (2008). Cyclical DNA methylation of a transcriptionally active promoter. *Nature*, *452*: 45-50.

[161] Jones, P. A. & Martienssen, R. (2005). A blueprint for a Human Epigenome Project: the AACR Human Epigenome Workshop. *Cancer Res.*, *65*: 11241-11246.

In: The Human Genome: Features, Variations...
Editor: Akio Matsumoto and Mai Nakano
ISBN: 978-1-60741-695-1

Chapter 2

The Sex Chromosomes: Sequence, Evolution and Human Diseases

***Alfredo Ciccodicola*[1,2*]*, Valerio Costa*[1]*, Teresa Esposito*[1]**
***and Fernando Gianfrancesco*[1]**
[1]Institute of Genetics and Biophysics "Adriano Buzzati-Traverso", CNR, Naples, Italy.
[2]Faculty of Science and Technology, University of Naples "Parthenope", Naples, Italy.

Abstract

Human sex chromosomes display significant differences from autosomes in both structure and function. Particularly, as the human X and Y chromosomes show a unique biology, they have long attracted special attention among geneticists. The human genome sequencing has revealed a great opportunity to deeply investigate the biology and evolution of the sex chromosome pair at a more global level, allowing new frontiers in the genetics research, such as a detailed knowledge of the sequence and the gene content of these chromosomes. Comparison of the human X and Y chromosome sequences, has made possible a reconstruction of their evolutionary history. Their sequence comparison has revealed they have become isolated from each other in a stepwise fashion over hundreds of millions of years, due to the lack of recombination events. The sequencing of the human Y chromosome has revealed that the Male Specific Region (MSR) contains, in addition to the Y-chromosomal male-determining gene SRY, a number of genes that have become specialized for spermatogenesis.

The sex chromosomes hold a unique place in the history of medical genetics. It has been widely demonstrated that a significant fraction of genetic diseases in humans results from point mutations and/or structural anomalies involving the sex chromosomes. This is a consequence of the haploid presence (hemizygosity) of the X and Y chromosomes in males. This phenomenon has prompted decades of intensive study, mainly focused on X-

* Corresponding Author:: Professor of Genetics - University of Naples "Parthenope", Institute of Genetics and Biophysics "Adriano Buzzati-Traverso" (IGB), CNR, Via P. Castellino, 111, 80131 Naples, Italy. Tel: +39-0816132-259, Fax: +39-0816132-617, e-mail: ciccodic@igb.cnr.it

linked inherited disorders. Many of the X-linked diseases currently actively investigated, are discussed in depth in this review.

Furthermore, although the past decades have witnessed many advances in the understanding of molecular processes underlying dosage compensation between sexes in mammals, the mechanism of X chromosome inactivation still continues to puzzle investigators. There is clear evidence that the expression of X-linked mutations in females is fine-tuned, and highly influenced, by these processes. Indeed, X-linked dominant male-lethal disorders represent a paradigmatic example of such influences. The observations reviewed here emphasize the importance of studying in depth the sex chromosomes, in order to better understand the evolution of human chromosomes and the pathological mechanisms related to the sex chromosomes.

Introduction

Mammalian sex chromosomes display significant differences from autosomes in both structure and function. The involvement of the Y in sex determination and fertility; the need for correct pairing and segregation of two grossly different chromosomes during male meiosis; the ability to dosage compensate X-linked genes by epigenetic means; the viability of sex chromosome aneuploidies; and the different consequences of mutations in X-linked genes in males and females as a result of the mosaic nature of mammalian females: these are just some of the unique properties that evolution has donated to our sex chromosomes. This review focuses on recent findings on sex chromosome organization and evolution, on X chromosome inactivation and on diseases caused by sex chromosome abnormalities.

The emerging picture is that each of the above-mentioned features reveals a considerable number of exceptions, suggesting that in our evolutionary history the presence of heteromorphic sex chromosomes might represent a transition phase we are still trying to adjust to.

X and Y chromosomes have long attracted special attention among geneticists. The hemizygosity of the human X chromosome in males exposes recessive disease alleles, and this phenomenon has prompted decades of intensive study of X-linked disorders. By contrast, the small size of the human Y chromosome, and its prominent heterochromatic long-arm suggested the absence of function beyond sex determination. As detailed in the review, both human sex chromosomes have now been sequenced. The X chromosome remains at the forefront of research into human genetic disease - research now aided by detailed knowledge of the sequence and gene content of this chromosome. The sequencing of the human Y chromosome and related studies have shown that, in addition to the male-determining gene SRY, the Y chromosome bears numerous genes essential for normal levels of sperm production.

The review also discusses the fascinating reconstruction of the evolutionary history of the human X and Y chromosomes, which was made possible by comparison of their sequences. Comparison of Y and X chromosome sequence shows that the X and Y chromosomes have become recombinationally isolated from each other in stepwise fashion over hundreds of millions of years. All available evidence indicates that whenever a region of the Y became recombinationally isolated, most of the genes in that region were subsequently inactivated by nonsense mutations and deletions. However, a few genes have persisted for tens or hundreds of millions of years, sometimes for reasons that are not clear. In other cases, it is clear that the Y-chromosomal isoforms have become specialized for spermatogenesis and that often this specialization was accompanied by wholesale duplication of the gene. As a consequence, large interstitial deletions of the human Y chromosome - ranging in size from 0.8 to 7.7 Mb – have no clinical phenotype beyond loss of spermatogenesis.

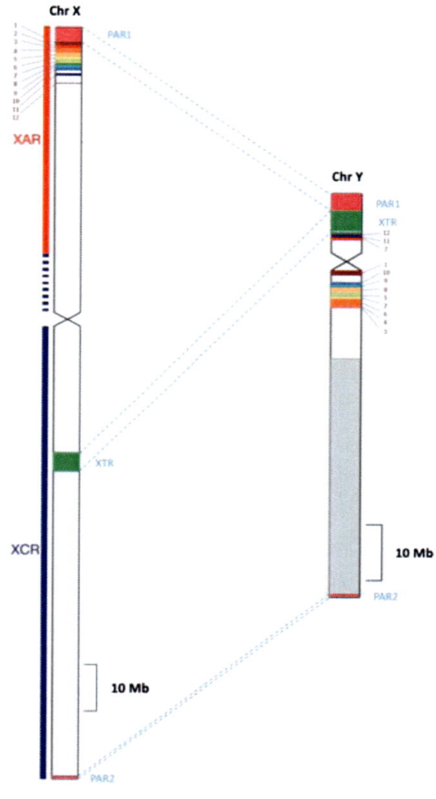

Figure 1. Schematic representation of major homologies between the human sex chromosomes. The major heterochromatic region on Yq is indicated by the pale grey box proximal to PAR2. Homologies coloured in the figure are either part of the XAR (PAR1 and blocks 1–12), or were duplicated from the X chromosome to the Y chromosome since the divergence of human and chimpanzee lineages (XTR and PAR2). Major blocks of homology remaining between the XAR and the YAR. Expansion of the BLASTN plot from 0–12 Mb on the X chromosome and 0–20 Mb on the Y chromosome. On the X chromosome, the major homologies lie in the terminal 8.5 Mb of Xp: PAR1 (magenta line) and numbered blocks 1–10. Lesser homologies 11 and 12 contain the TBL1X/TBL1Y and AMELX/AMELY genes, respectively.

Furthermore, a significant fraction of genetic disease in humans results from mutations and other anomalies involving the sex chromosomes. This is a consequence of the haploid presence of the X and Y chromosomes in males, and of the more frequent survival of individuals with altered numbers of sex chromosomes (aneuploidy). Other sex-linked disorders can exhibit unusual features that result from the unique process of X inactivation in the female and from consequences to tissues that are mosaic for mutant and normal cells.

We describe current understanding of the causes of chromosome segregation errors that result in sex chromosome aneuploidy, and the consequences to meiosis in individuals carrying these anomalies. In particular, we focused on Turner syndrome and premature ovarian failure, for which several X-linked *loci* have been found to be responsible. We describe recent advances that offer a number of explanations for these disorders, ranging from single gene variation to chromosome anomalies.

Moreover, we focused on recent advances in the study of a set of X-linked disease: X-linked mental retardation, incontinentia pigmenti that are instrumental to show the peculiarity of X linked disease.

A large fraction of the male mental retardation among humans is thought to result from lesions in X-linked genes, and the pursuit of the genes responsible has been highly successful in the past decade. Among the X-linked mental retardations, fragile X syndrome stands out as a very common cause. As the first discovered triplet repeat disorder, this fragile site-associated mutation provided the first glimpses of the peculiar features of mutations that confer instability on themselves.

Another gene that generates variable phenotypes when mutated is the *NEMO* gene. Curiously, there exists a particularly common mutation of this gene in individuals with incontinentia pigmenti, despite high variation in the disease severity. This reflects the significant contribution of X inactivation in females with incontinentia pigmenti. Here we updates the state of knowledge about *NEMO* mutations and function, describing very recent work that implicates the gene in nuclear–cytoplasmic signaling of DNA damage; the consequences of abrogating this function remain unexplored.

Finally, a considerable fraction of the male retinal dystrophies in humans is thought to result from mutations in X-linked genes. Indeed, different human X-linked retinal degenerations have been described, and the pursuit of disease-causing genes has been highly successful in the past decade. Although these pathologies comprise a wide group of blinding diseases, the X-linked degenerative forms fall into three main categories: Retinitis pigmentosa (RP), retinoschisis (RS), and color blindness (protanopia and deuteranopia).

The Sequences of the Human Sex Chromosomes

The finished[1] sequences of both human sex chromosomes have been published in 2003 [1] and 2005 [2], Y and X respectively. The euchromatic portions are nearly complete (X >99%, Y ~97%) and highly accurate (>99.99%). The availability of these sequences has contributed to our understanding of chromosomes' biology and of the disorders associated with them (Figure 1). As the rest of the reviews in this issue illustrate, the sequences provide

---

[1] No human chromosome can be completely sequenced with current technologies. 'Finishing' is a phase in the sequencing procedure through which the highest quality sequence is ensured. A 'finished' sequence is one which has been through this phase successfully. Therefore, the 'finished sequence of the X chromosome' is not the same as 'the completed sequence of the human X chromosome'. The euchromatic portion of the human X is covered to a level >99% with finished sequence, whereas the Y chromosome-specific euchromatin is covered to a level >97%.

a common reference point for diverse studies, supplying fundamental information such as the physical locations of markers (e.g. single nucleotide polymorphisms [SNPs]) and near-complete lists of genes. They thus facilitate investigations ranging from searches for genetic factors underlying human disorders to analyses of mammalian evolution. In addition, they reveal features that would not otherwise be apparent, such as large palindromic structures or unusual densities of retroposons. Each reference sequence is, however, a single representative, and now needs to be complemented by a more complete understanding of the associated variation.

The X chromosome sequence

The X chromosome sequence is a mosaic of several individuals [2]. A map of the X chromosome was constructed using predominantly P1-artificial chromosome (PAC) and bacterial artificial chromosome (BAC) clones. These were assembled into contigs [2] using restriction-enzyme fingerprinting and integrated with earlier maps using sequence-tagged site (STS) content analysis [3]. Were selected 1,832 clones [2] from the map for shotgun and direct sequencing using established procedures [4]. Gaps were closed by targeted screening of clone libraries in bacteria or yeast, and by assessing BAC and fosmid end-sequence data for evidence of spanning clones [2].

Finished sequences were estimated to be more than 99.99% accurate by independent assessment [5]. The sequence of the X chromosome extend into the telomeric $(TTAGGG)_n$ repeat arrays at the ends of the chromosome arms, and include both pseudoautosomal regions (PARs) [2,6]. Ross et al [2] have determined a total of 151,005,926 bp of sequence. The remaining 14 gaps were estimated to have a combined size of less than 1Mb, and the sequence therefore covers at least 99.3% of the X chromosome euchromatin. There is also a single heterochromatic gap corresponding to the polymorphic 3.0 (±0.4) Mb array [7] of alpha satellite DNA at the centromere. On this basis, the X chromosome is approximately 155Mb in length [2].

The analysis of the sequence reveals a gene-poor chromosome that is highly enriched in interspersed repeats and has a low (G+C) content (39%) compared with the genome average (41%). Based on a manual assessment of all publicly available human expressed sequences and genes from other organisms, were annotated 1,098 genes (7.1 genes per Mb) across four different categories: 1) known genes (699), 2) novel coding sequences (132), 3) novel transcripts (166), and 4) putative transcripts (101). Ross et al. [2] have also identified 700 pseudogenes in the sequence (4.6 pseudogenes per Mb). The gene density (excluding pseudogenes) on the X chromosome is among the lowest for the chromosomes that have been annotated to date. This might simply reflect a low gene density on the ancestral autosomes. Alternatively, selection may have favoured transposition of particular classes of gene from the X chromosome to the autosomes during mammalian evolution. These could include developmental genes for which the protein products are required in double dose in males (or in females after XCI has occurred), or genes for which mutation in male somatic tissues is lethal.

The X chromosome contains the largest known gene in the human genome, the dystrophin (*DMD*) *locus* in Xp21.1, which spans 2,220,223 bp. Consistent with its low gene density, the frequency of predicted CpG islands on the X chromosome is only 5.25 per Mb, which is exactly half of the estimated genome average [4]. During the analysis have also been identified 4,493 evolutionarily conserved regions (ECRs), by comparing the X chromosome sequence to the genomes of mouse, rat, zebrafish and the pufferfishes Tetraodon nigroviridis and Fugu rubripes. Of these, 4,393 ECRs overlap with 4,373 annotated exons. The remaining 100 are most likely to be unannotated exons, although some could be highly conserved control or structural elements. From these data, Ross et al. [2] conclude that at least 97.8% of the protein-coding exons on the X chromosome have been annotated.

The X chromosome gene set includes 173 predicted non-coding RNA (ncRNA) genes and/or pseudogenes only when there is supporting evidence of expression from complementary DNA or expressed-sequence-tag (EST) sources. Furthermore, were predicted only two transfer RNA genes on the X chromosome, out of the several hundred predicted in the human genome [4]. Thirteen microRNAs from the microRNA registry [8] have also been mapped onto the sequence. The most prominent of the ncRNA genes on the X chromosome is *XIST* (X (inactive)-specific transcript) [9], which is critical for X-chromosome inactivation (XCI). The *XIST locus* spans 32,103 bp in Xq13, and its untranslated transcript coats and transcriptionally silences one X chromosome in cis. There is also evidence for shorter XIST transcripts generated by alternative splicing, particularly in the 3' region of the gene [10]. In the mouse, *Tsix* is antisense to *Xist* [11], and its transcript (or the process of its transcription) is believed to repress the accumulation of *Xist* RNA. Despite the evidence for a transcription antisense to *XIST* in human [12,13], in the public databases, there are no corresponding expressed sequences of the human *TSIX* gene. In the human sequence, two other ncRNA genes are annotated in the 400 kb region distal to *XIST*, which are orthologues of the mouse genes described previously as Jpx and Ftx [14].

Finally, the most prominent finding was the presence of the MAGE domain (IPR002190) in 32 genes, whereas only four other MAGE genes are reported in the rest of the genome (*MAGEF1* on chr 3, and *MAGEL2*, *NDN* and *NDNL2* on chr 15). The MAGE gene products are members of the cancer-testis (CT) antigen group, which are characterized by their expression in a number of cancer types, while their expression in normal tissues is solely or predominantly in testis. This expression profile has led to the suggestion that the CT-antigens are potential targets for tumour immunotherapy. The X chromosome gene set contains 99 CT-antigen genes and includes novel members of the MAGE, GAGE, SSX, LAGE, CSAGE and NXF families. Therefore, Ross et al [2] predict that approximately 10% of the genes on the X chromosome are of the CT-antigen type. The remarkable enrichment for CT-antigen genes on the X chromosome relative to the rest of the genome might be indicative of a male advantage associated with these genes. The CT-antigen genes on the X chromosome are also notable for the expansion of various gene families by duplication. This degree of duplication is perhaps an indication of selection in males for increased copy number.

Interspersed repeats account for 56% of the euchromatic X chromosome sequence, compared with a genome average of 45%. Within this, the Alu family of short interspersed nuclear elements (SINEs) is below average, in keeping with the gene-poor nature of the chromosome. Conversely, long terminal repeat (LTR) retroposon coverage is above average; but the most remarkable enrichment is for long interspersed nuclear elements (LINEs) of the L1 family, which account for 29% of the X chromosome sequence compared to a genome average of only 17%. Applying the criterion of at least 90% sequence identity over at least 5 kb [15], was estimated that intrachromosomal segmental duplications account for 2.59% of the X chromosome. In contrast, interchromosomal segmental duplications indicated by sequence matches to the autosomes account for a very small fraction (0.24%) of the X chromosome.

Among the intrachromosomal duplications are well-described cases that are associated with genomic disorders [16]. In Xp22.32, deletions of the steroid sulphatase (STS) gene, causing X-linked ichthyosis (Online Mendelian Inheritance in Man (OMIM) [17] entry number 308100), result from recombination between flanking duplications that contain copies of the *VCX* gene. Also, some instances of Hunter syndrome (OMIM 309900), red-green colour blindness (OMIM 303800), Emery-Dreifuss muscular dystrophy (OMIM 310300), incontinentia pigmenti (OMIM 308300) and haemophilia A (OMIM 306700) result from rearrangements involving duplicated sequences in Xq28. In haemophilia A, mutations are frequently the result of inversions between a sequence in intron 22 of the F8 gene and one of two more distally located copies.

The X chromosome sequence extends from both arms into centromeric, higher-order repeat sequences, which are known to be functionally associated with the X centromere [18-20]. The most proximal 494 kb and 360 kb of the Xp and Xq sequences, respectively, consist of extensive regions of satellite DNA, adjacent to euchromatin of the chromosome arms that is exceptionally high in L1 content. The satellite region on Xp contains small amounts of other satellite families [18], whereas that on Xq entirely consists of alpha satellite. Similar to all other human chromosome arms that have been examined [20,21], these transition regions consist of monomeric alpha satellite not associated with centromere function. Both the Xp and Xq contigs, though, extend more proximally and reach into highly homogeneous, higher-order repeat alpha satellite (DXZ1). Critically, the Xp and Xq contig copies of the DXZ1 repeat are themselves 98–100% identical in sequence, and are oriented in the same direction along the chromosome. On this basis, the two contigs reach the "end" of each chromosome arm and thus also reach the centromeric locus from either side. This represents a logical endpoint for efforts to complete the sequence of chromosome arms in the human genome, and the first demonstration of this endpoint is provided by the X chromosome sequence.

Finally, a total of 153,146 candidate single-nucleotide polymorphisms (SNPs) have been mapped onto the X chromosome sequence. These include 901 SNPs that result in non-synonymous changes in protein-coding regions, and are therefore candidate functional protein variants. The heterozygosity level on the X chromosome is known to be well below that of the autosomes, and this difference can be explained partly or entirely by population genetic factors [22]. Using comparable sequence data for chromosome 20, it was calculated that the heterozygosity level on the X chromosome is approximately 57% of that observed for the autosome.

The Y chromosome sequence

The male-specific region of the Y chromosome, the MSY, differentiates the sexes and comprises 95% of the chromosome's length. The MSY is a mosaic of heterochromatic sequences and three classes of euchromatic sequences: X-transposed, X-degenerate and ampliconic [1].

The MSY's euchromatic DNA sequences total roughly 23 megabases (Mb), including 8Mb on the short arm (Yp) and 14.5 Mb on the long arm (Yq; see Figure 2) [1]. In the finished nucleotide sequence, was obtained roughly 97% of the MSY euchromatin and contains an estimated error rate of about 1 per 105 nucleotides, with two known exceptions. First, there remain two gaps, each of which is roughly 50 kilobases (kb) long. Second, was obtained a representative but incomplete sequence for a tandem array that spans roughly 0.7Mb on Yp [1].

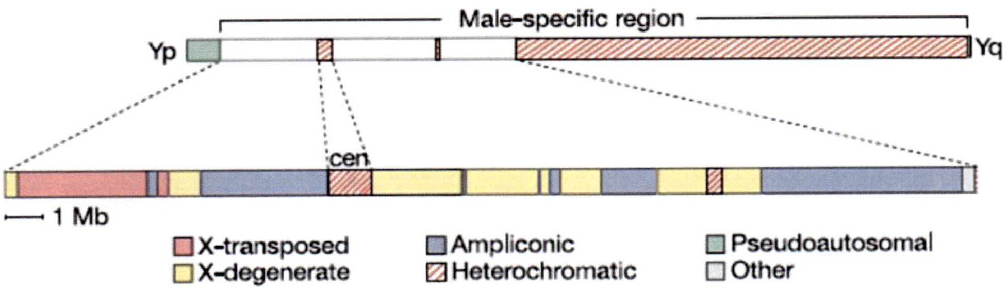

Figure 2. The male-specific region of the Y chromosome. **A**) Schematic representation of the whole chromosome, including the pseudoautosomal and heterochromatic regions. **B**) Enlarged view of a 24-Mb portion of the MSY, extending from the proximal boundary of the Yp pseudoautosomal region to the proximal boundary of the large heterochromatic region of Yq. Shown are three classes of euchromatic sequences, as well as heterochromatic sequences.

So far, efforts to gain sequence-based understanding of human chromosomes have largely by-passed heterochromatic regions [4,23], including a large block of heterochromatic sequences found in the centromeric region of every nuclear chromosome [24]. In addition to its centromeric heterochromatin (approximately 1 Mb), [25], the Y chromosome was previously shown to contain a second, much longer heterochromatic block (roughly 40 Mb) that comprises the bulk of the distal long arm (Figure 1). In the course of the chromosome Y sequencing project, was discovered and characterized a third heterochromatic block [1] - a sharply demarcated island that spans approximately 400 kb, comprises >3,000 tandem repeats of 125 bp, and interrupts the euchromatic sequences of proximal Yq (Figure 2). The other two heterochromatic blocks also consist of massively amplified tandem repeats of low sequence complexity. In total, was found that the heterochromatin of MSY encompasses at least six distinct sequence species, each of which form long, homogeneous tandem arrays [1].

The MSY includes at least 156 transcripts, half of which probably encode proteins, and all the transcripts are located in euchromatic sequences. There is no evidence of transcription of the MSY heterochromatin. Of the approximately 78 protein-coding units, about 60 are members of nine different MSY-specific gene families, each characterized by > 98% nucleotide identity among family members, in both exons and introns. The remaining 18 protein-coding genes are present in one copy each in the MSY. These include two genes, *RPS4Y1* and *RPS4Y2*, exhibiting 93.6% of nucleotide identity in coding exons but are much more diverged in introns [1]. Thus, the MSY seems to encode at least 27 distinct proteins or protein families.

Furthermore, the MSY includes at least 78 transcripts for which strong evidence of protein coding is lacking; many of these transcription units are probably non-coding. Of these 78 transcripts, 13 occur in single copy in the MSY and the remaining 65 are members of 15 MSY-specific families. Considering together both coding and non-coding transcripts, the MSY appears to contain 24 MSY-specific families, which collectively account for 125 of the 156 MSY transcripts identified [1]. On the basis of earlier experiments, most of the MSY genes were thought to fall into two functional classes, with genes in the first group expressed throughout the body, in many organs, and genes in the second group predominantly or exclusively expressed in testes [26].

The MSY euchromatic sequences fall into three classes: X-transposed, X-degenerate and ampliconic. The MSY euchromatin is a patchwork of these three sequence classes [1]. The X-transposed sequences are 99% identical to DNA sequences in Xq21, a band in the midst of the long arm of the human X chromosome. The X-transposed sequences are so named because their presence in the human MSY is the result of a massive X-to-Y transposition that occurred about 3–4 million years ago, after the divergence of the human and chimpanzee lineages [27-29]. Subsequently, an inversion within the MSY short arm cleaved the X-transposed block into two non-contiguous segments, as observed in the modern MSY [27,28]. The X-transposed sequences do not participate in X–Y crossing over during male meiosis, distinguishing them from the pseudoautosomal sequences. Within the X-transposed segments (which have a combined length of 3.4 Mb) were identified only two genes, both of which have homologues in Xq21. Thus the X-transposed exhibit the lowest density of genes among the MSY euchromatin sequences, as well as the highest density of interspersed repeat elements. In particular, long interspersed nuclear element 1 (LINE1) elements account for 36% of all X-transposed sequence, or nearly twice the genome average of 20% [4,23]. As expected, low gene density and high repeat density also characterize the homologous sequence block in Xq21.

In contrast to the X-transposed sequence blocks, the X-degenerate segments of the MSY are dotted with single-copy gene or pseudogene homologues of 27 different X-linked genes [1]. These single-copy MSY genes and pseudogenes display 60% to 96% nucleotide sequence identity to their X-linked homologues, and they seem to be surviving relics of ancient autosomes from which the X and Y chromosomes co-evolved. In 13 cases, the MSY homologue is a pseudogene with sequence similarity to exons and introns of the functional X homologue [1]. In the remaining 14 cases, the MSY homologue seems to be a transcribed, functional gene, and the X- and Y-linked genes encode very similar but non-identical protein isoforms [1]. Together, the X-degenerate sequences encode 16 of the MSY's 27 distinct proteins or protein families. Notably, all 12 ubiquitously expressed MSY genes reside in the X-degenerate regions; no such genes have been identified elsewhere in the MSY. Conversely, among the 11 MSY genes were found to be expressed predominantly in testes, only one gene, the sex-determining SRY, is X-degenerate [1].

The third class of euchromatic sequences, the ampliconic segments, are largely composed of sequences that exhibit marked similarity - as much as 99.9% identity over tens or hundreds of kilobases - to other sequences in the MSY. The amplicons are located in seven segments scattered across the euchromatic long arm and proximal short arm, and whose combined length is 10.2Mb [1]. Notably, 60% (6.1 Mb) of the ampliconic sequences exhibit intrachromosomal identities of 99.9% or greater. After heterochromatic and LINE1 repeats have been accounted for, the MSY is seen to contain many long stretches of sequence similar to those elsewhere in the MSY. The ampliconic sequences exhibit by far the highest density of genes among the three sequence classes in the MSY euchromatin [1]. Considering together both coding and non-coding elements, the ampliconic sequences contain 135 of the 156 MSY transcripts identified [1]. About 60 transcripts represent nine distinct MSY-specific protein-coding gene families. Furthermore, the ampliconic sequences include at least 75 other transcripts for which strong evidence of protein coding is lacking. Of these 75 putative non-coding transcripts, 65 are members of 15 MSY-specific families, and the remaining 10 occur in single copy [1].

In contrast to the ubiquitous expression of most X-degenerate genes, the ampliconic genes and transcripts show highly restricted expression. All nine protein-coding families in the ampliconic regions are predominantly or exclusively expressed in testes, as are most of the regions' non-coding transcripts [1].

The most pronounced structural features of the ampliconic regions of Yq are eight massive palindromes. In all eight palindromes, the arms are highly symmetrical, with arm-to-arm nucleotide identities of 99.94–99.997% [1]. (By convention, these percentage identities refer only to nucleotide substitutions and do not take account of insertions and deletions by which palindrome arms differ.) The palindromes are long, their arms ranging from 9 kb to 1.45 Mb in length. They are imperfect in that each contains a unique, non-duplicated spacer, 2–170 kb in length, at its centre [1].

Six of the eight palindromes carry recognized protein-coding genes, all of which seem to be specifically expressed in testes. In all known cases of genes on MSY palindromes, identical or nearly identical gene copies exist on opposite arms of the palindrome [1]. Of the nine multi-copy, protein-coding gene families, eight have members on palindromes. Indeed, six families are located exclusively in palindromes. These include the DAZ genes and the CDY genes, which both occur in four copies. In addition, the palindromes contain at least seven families of apparently non-coding transcription units, all expressed exclusively or predominantly in testes [1]. In addition to the eight palindromes, the ampliconic regions of Yq and Yp contain five sets of more widely spaced inverted repeats (IR) with repeat lengths of 62–298 kb. The IR1, IR2 and IR3 exhibit nucleotide identities of 99.66–99.95%. Inversion of the IR3 repeats, both located on Yp, was probably a direct consequence of the molecular evolutionary event that cleaved the X-transposed sequences into two non-contiguous segments [1]. Subsequent homologous recombination between inverted IR3 repeats was responsible for a 3.6 Mb inversion polymorphism observed on the short arm of the modern Y chromosome [30].

In addition, the ampliconic regions of Yq and Yp contain a variety of long tandem arrays. Prominent among these are the newly identified NORF (no long open reading frame) clusters, which in aggregate account for about 622 kb. The NORF arrays are based on a repeat unit of 2.48 kb [1]. A consensus sequence for the repeat is readily identifiable but the sequence of individual repeat elements typically diverges from that consensus by 14–20%. The NORF arrays are so named because they harbour a great diversity of spliced but apparently non-coding transcription units. The TSPY arrays, which comprise about 700 kb, are based on a 20.4-kb repeat unit [31] that encodes, on one strand, a previously identified protein, TSPY. A newly identified transcript, CYorf16, is found on the opposite strand; its protein coding potential remains to be tested [1]. Approximately 35 copies of this repeat unit - and hence 35 TSPY genes and 35 CYorf16 transcripts - are found in a single, highly regular tandem array in proximal Yp; the sequences of individual repeat units rarely differ from the consensus by more than 1%. Furthermore, a single, isolated TSPY repeat unit, whose sequence diverges 3% from the consensus, is located more distally in Yp, embedded in the distal IR3 inverted repeat. The 35-unit TSPY cluster is the largest and most homogeneous protein-coding tandem array identified so far in the human genome [1].

Origin and Evolution of Human Sex Chromosomes

Sex-determination system

In many animal species, males and females genetically differ by only one chromosome. There are several chromosomal sex-determination systems found in nature, with the XY sex-determination system that is present in most mammals, including humans and Drosophila.

Males are the heterogametic sex (XY), and females are homogametic (XX). In the ZW sex-determination system, which is found in birds, butterflies, and many reptiles, females are the heterogametic sex (ZW), while males are homogametic (ZZ). Comparative mapping of human X-borne genes in chicken confirms that these two regions were separated in a common ancestor of birds and mammals 310 Million Years Ago (MYA; Figure 3).

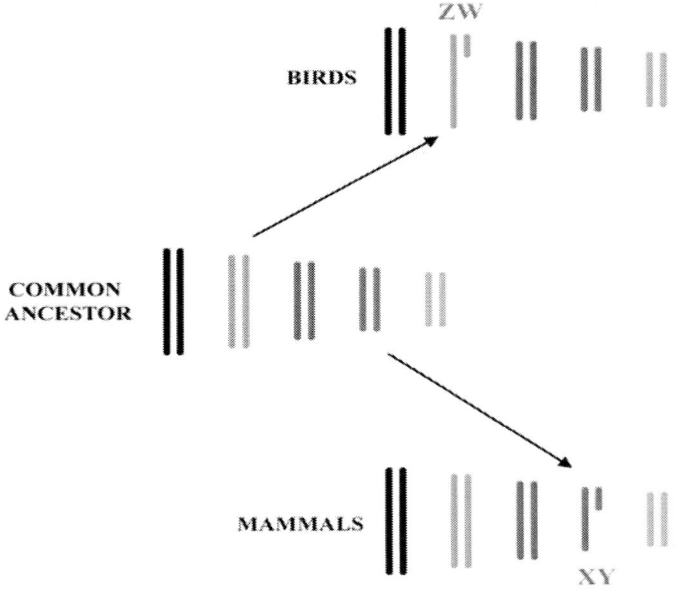

Figure 3. The XY and ZW sex-determination systems are derived from different autosomal pairs of a common ancestor of birds and mammals (310 MYA).

How sex chromosomes evolved has been debated for decades. The X and Y chromosome are thought to have evolved from an ordinary pair of autosomes that stopped recombining with each other after acquiring a sex-determining role [32,33]. The accumulation of genes that are beneficial in one sex but detrimental in the other, favors the suppression of recombination between the nascent sex chromosomes. In the absence of recombination, these originally homologous chromosomes continue to differentiate (Figure 4). It is generally thought that the morphological differentiation between the sex chromosomes is a product of the chromosome degeneration that is present only in the heterogametic sex (Y or W) and is thus completely sheltered from genetic recombination.

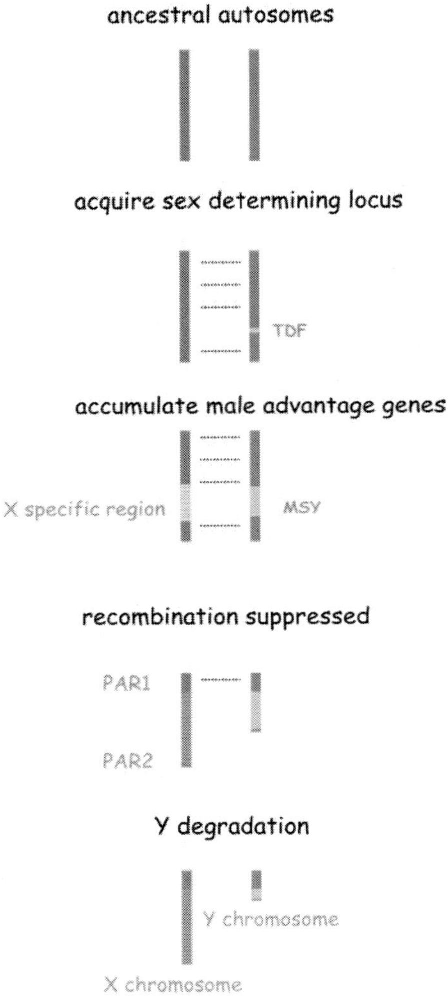

Figure 4. The differentiation of the X and Y Chromosomes is initiated when one partner acquires a sex-determining locus such as the testis-determining factor (TDF). Accumulation of male-specific alleles selects for repression of recombination, creating an X-specific region and a male-specific region on the Y (MSY). Soppression of recombination leads to the degradation of the MSY leaving only two small pseudoautosomal regions (PAR1 and PAR2).

The difference in gene content evolution between the degrading Y/W chromosome and the autosomes is immediately apparent and has extensively been studied [34]. In contrast, X/Z chromosomes have been viewed as similar to the ancient autosomes, with little change occurring, even if recent studies have found that the gene content of X/Z chromosomes can be quite different from that of autosomes [35].

Human Y chromosome has excited particular attention because of its small size, and the paucity of genes it bears. The theory that the Y chromosome degrades rapidly is supported by comparative studies in insects and vertebrates including mammals. A crude calculation of the average rate of loss of active genes from the human Y predicted its extinction in 10 million years [36].

Human sex chromosomes

The human X and Y chromosomes are highly differentiated. The 155Mb X chromosome represents about 5% of the haploid genome and contains 1100 genes [2]. The mammal X is highly conserved between species, the gene order and content is almost identical between species with the exception of the murid rodent X, in which gene order is scrambled [37,38]. Although most genes on the human X chromosome are not involved with sex, there is an increased frequency of sex and reproduction related genes [39]. The human Y chromosome is much smaller, and rich in repetitive sequence. Although it looks completely different from the X chromosome, it shares a small region of homology with the X (Pseudoautosomal Region 1, PAR1), within which there is an obligatory recombination event during male meiosis that mediates X and Y segregation. A second small PAR2 lies at the other end of the human X and Y. The rest of the Y is male specific (MSY) and represents about 2% of the haploid genome (60 Mb). Much of the MSY is composed of simple-sequence repetitive DNA, and contains no genes. Even the euchromatic 24Mb contains few active genes. Of 172 transcriptional units on the MSY, many are untranslatable pseudogenes and others are amplification products. The MSY encodes for only 27 distinct proteins [1]. Of these 27 protein-coding genes, 20 have a partner on the X chromosome. X–Y homology in the PAR, and the preponderance of XY shared genes, support the proposal that the mammalian X and Y originated from an autosomal pair.

Of the 1100 genes on the ancestral Y only a total of 45 survive. Being confined to males, many of these evolved a function in male reproduction, and came under positive selection. The deletion and inactivation of most of the Y led to problems of chromosome segregation and gene dosage. Chromosome pairing at meiosis was compromised, such that special mechanisms that ensure high recombination in the tiny PAR evolved. The Y is also unusual in the "functional coherence" of the genes it bears, many of which have functions in sex or fertility [26]. The human Y is replete with repetitive sequences of diverse origins, and many multicopy gene arrays are embedded in palindromes [1,40].

Evolution of human sex chromosomes

The human X and Y have completely been sequenced over the past two years, enabling detailed molecular comparisons of gene content and copy number as well as repeat content and chromosome structure [1,2]. To understand the organization and function of genes on the human X and Y chromosomes, it is essential to deduce how they evolved.

Evidence that the mammalian X and Y chromosomes differentiated from an ordinary autosomal pair is found in the homology between the X and the Y, both within the PAR and in Y-borne genes with homologues on the X. Differentiation of the X and Y occurred during the 300 million years since mammals diverged from birds and reptiles. After the acquisition of a male-determining gene, the mammalian Y chromosome was progressively degraded (Figure 4). Ancestral chromosome blocks that fused to form the X-Y pair, are revealed by comparing the gene content of the X and Y in humans and distantly related mammals, like marsupials and monotremes (which diverged 180 and 210 million years ago respectively) and the gene contents of orthologous regions in other vertebrates. Only part of the human X (Xqter-p11.2) is shared with the marsupial and monotreme X chromosomes, and the rest is autosomal. This defines an ancient X-conserved region (XCR) and an X-added region (XAR), which was added to the placental X after the marsupial–placental divergence but before the placental radiation [41] (Figure 5). The Y chromosome comprises a corresponding ancient conserved region, YCR, and an added region, YAR [42]. The ancient region on the human Y contains only four genes, whereas the added region, including the PAR, accounts for the rest.

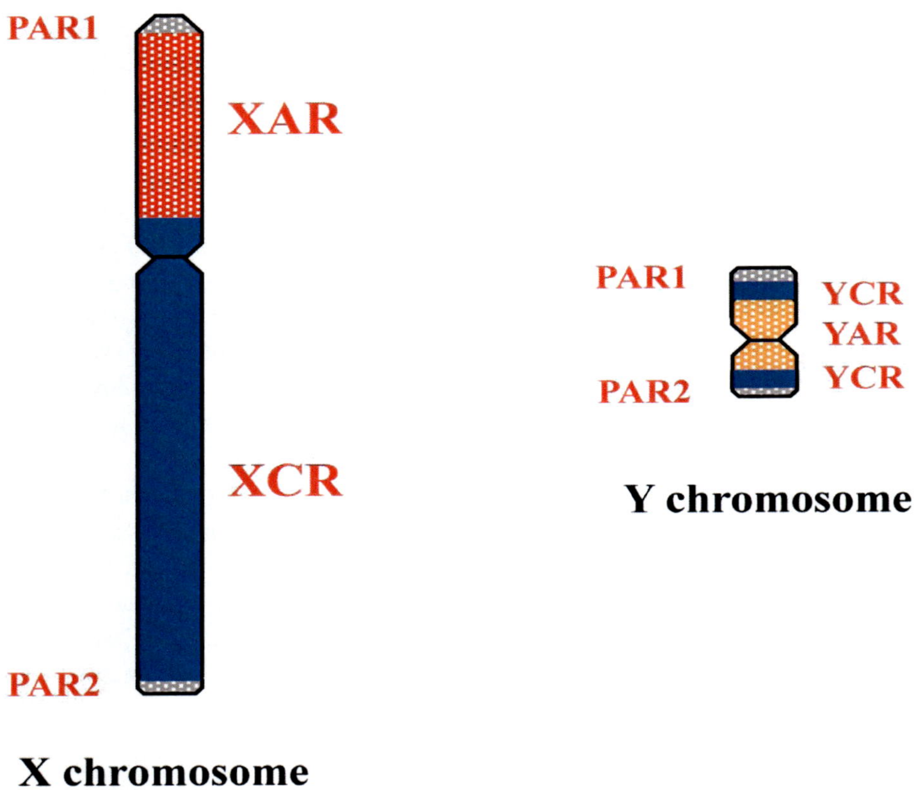

Figure 5. Comparative mapping of the orthologs of genes from other mammals on the human X detected an ancient region conserved on the X in all mammals (blue, XCR) and an added region (red, XAR) that is autosomal in marsupials and monotremes so was added to the X 180–100 MYA. The human Y chromosome, too, is subdivided into corresponding ancient and added regions. The ancient conserved region (YCR, blue) is tiny, and most of the Y derives from the added region (YAR).

Division of the human X and Y into these evolutionary layers on the basis of comparative gene mapping is supported by data from different sources. Nucleotide divergence between 19 different X-borne genes and their Y-borne homologues, showed that divergence times fell into clusters [43]. Synonymous nucleotide divergence in coding regions decreased in steps from the distal Xq to Xp in four "evolutionary strata". Some of these clusters are also evident in the mouse XY [44]. The oldest two strata defined by XY divergence mostly correspond to the two evolutionary layers conserved on the X in all mammals. The more recent strata 3 and 4, defined by XY divergence, correspond to the recently added XAR defined by comparative mapping. Correspondence with the building blocks defined by comparative gene mapping implies that they represent fusions between three independent evolutionary blocks. Thus, an original small mammal X, represented by human Xq, differentiated from the Y before 240 MYA. Then, a stratum 2 (corresponding to human Xq28 and pericentric region of Xp) was added before 210 MYA, and finally a stratum 3, the XAR, was added 100–180 MYA. Whole X sequencing subdivides the youngest block into three layers separated by inversions 38–44 MYA and 29–32 MYA [2].

Pseudoautosomal regions of sex chromosomes

Evolution of the sex chromosomes, particularly the progressive decay of the Y chromosome after its acquisition of SRY (Sex-determining region Y) as the male determining gene, has led to a unique genetic diversity of formerly homologous autosomes, owing to the establishment of a recombination barrier [45,46]. Meiotic chromosome pairing, however, necessitated the conservation on both chromosomes of specific regions that share sufficient homology to enable chromosome alignment during male meiosis, and this task is accomplished by the pseudoautosomal regions [47]. These regions represent short genomic stretches of sequence homology between the modern sex chromosomes. In humans, they represent about 2% of the X chromosome and 5% of the Y chromosome. PAR1 (pseudoautosomal region 1) is necessary for homologous X–Y chromosome- pairing during male meiosis and, undergoes one crossover event during this process [48-50]. Consequently, a loss of PAR1 is associated with male sterility [51]. Although PAR2 is not implicated in mediating male meiosis and recombines at a rate of only 2%, this still represents a sixfold higher recombination frequency when compared with the average of the remainder of the X chromosome [52]. Whereas the human PAR1 is homologous to the pseudoautosomal regions of several species, including great apes and Old World monkeys, the PAR2 sequence has a much shorter evolutionary history and is specific to humans [53-6]. Rodents seem to have lost the distal 9 Mb portion of the short arm of the X chromosome, and instead have a different, considerably shorter, PAR1, which in Mus musculus spans only 720 kb [54]. These attributes emphasize the extraordinary evolutionary forces imposed on the pseudoautosomal regions and explain their exceptional status within the genome.

Although the sequences of the entire PAR2 region and 99.3% of the euchromatic sequence of the X chromosome have been determined, only roughly 80% of the PAR1 sequence is available to date [2]. PAR1, very differently from PAR2, is characterized by a variety of structural properties that clearly distinguish it from the rest of the X chromosome. PAR1 exhibits a significantly higher GC content (> 48%) than the remainder of the X chromosome (39%) [55]. Although the PAR1 sequence has significantly less L1 family repeats, it exhibits a 4–5 times elevated Alu repeat content when compared with that of the remainder of the X chromosome. Analysis of PAR1 had suggested that recombination occurs fairly evenly [49]. It has recently been demonstrated that elevated recombination rates and recombination hotspots contribute to chromosome evolution. In fact, they break down genetic linkage between neighboring genes, enabling a faster rate of evolution by efficient multi-locus selection [56]. Moreover, given that the process of homologous recombination includes double-strand breaks and DNA synthesis by low-fidelity polymerases, it might lead to elevated mutation rates in the vicinity of the recombination event [57]. Finally, meiotic crossovers might contribute to the overall conservation between the X and Y chromosomal pseudoautosomal regions by enhancing gene conversion [58-59]. Therefore, these 'autosomal islands' at the tips of the otherwise divergent sex chromosomes represent an excellent model to investigate the influence of meiotic recombination on mutation rates and genetic stability.

With at least 29 genes (24 in PAR1 and 5 in PAR2) the pseudoautosomal regions display a higher gene-content (10 genes per Mb in PAR1 and 15 genes per Mb in PAR2) than the remainder of the X (7 genes per Mb) or the Y chromosome (3 genes per Mb) [1,2]. All characterized genes within PAR1 escape X inactivation, whereas the two most proximal PAR2 genes (SPRY3 and SYBL1) undergo X inactivation [1,60]. As a major genetic consequence, all PAR1 genes are present in two functional copies and follow an autosomal inheritance pattern [48,50]. Genes within PAR1 that exhibit dosage-sensitive functions might cause or contribute to the etiology of haploinsufficiency-related clinical conditions, including Turner syndrome (45, X). The previously suggested poor conservation of PAR1 genes among rodents has been confirmed recently by completing the sequences of several mammalian species. Only 3 of the 24 PAR1 genes have corresponding orthologs in mouse, and these have diverged considerably [61-63]. In fact, recently available sequence information has revealed that PAR1 resides within a 9 Mb block that has been removed from the X chromosome of a common murine ancestor of mouse and rat [2]. These data once more stress the evolutionary dynamics imposed on PAR1. Considerable interest has also been given to the analysis of the PAR1 boundary and the evolution of genes under its direct influence [64]. Comparative analysis of 270 kb from the PAR1 sequence flanking the pseudoautosomal boundary in humans and chimpanzees has unveiled a significantly higher sequence-divergence of pseudoautosomal sequence when compared with Y specific sequences [65]. In addition, data on other pseudoautosomal genes from different intervals have demonstrated that their substitution rates are significantly elevated in comparison with the individual genome averages for humans, chimps, gorillas and orangutans [66-68]. These results indicate that the substitution rate is high in distal and middle PAR1, and lower near the pseudoautosomal boundary, suggesting that this is indeed caused by a mutagenic effect of recombination. A striking difference was also seen between Y-specific and pseudoautosomal sequences when comparing different chimpanzee individuals [65]. If this holds true for humans, we should expect a great deal of sequence variation in PAR1 between human individuals of different geographical origins.

The X chromosome inactivation

One consequence of the degeneration of the Y chromosome is the unequal dosage between males and females for X linked genes with no Y partners. The necessity for dosage compensation of genes on the X has selected for a variety of dosage compensation mechanisms in invertebrates and vertebrates. In mammals, a system has evolved in which one of the two X chromosomes in a female becomes transcriptionally silent in the embryo and maintains this silence stably throughout life. Regions of the X are thought to have been recruited into the X chromosome inactivation system as their Y partner became degraded [69-71]. This hypothesis is supported by the observation that many genes on the human X escape inactivation, although they have no active Y partner [72,73]. On the human Xp, several genes escape X inactivation, suggesting that dosage compensation is not an urgent requirement of many genes as suggested by normal phenotype of heterozygotes for null mutations and deletions.

This epigenetic phenomenon involves noncoding RNAs (ncRNAs), antisense transcription, histone modifications, and DNA methylation to distinguish two genetically identical X chromosomes as active and silent entities within the same nucleus.

Mammalian X inactivation seems to come in two forms: imprinted and random X inactivation. Imprinted X activation occurs in early mammals, such as marsupials, which diverged from placental mammals approximately 180 million years ago. This form of dosage compensation is achieved by inactivating the paternally inherited X chromosome. Imprinted X inactivation can also be found in the extra-embryonic tissues of a subset of placental mammals. The cells forming the embryo in placental mammals undergo random X inactivation, through which either the paternal or the maternal X chromosome is inactivated. This novel mechanism is based on a noncoding RNA called Xist, which is unique to placental mammals and has not been found in marsupials or other vertebrate genomes [74]. The X Random inactivation poses an additional problem compared with other dosage-compensatory mechanisms because the two X chromosomes are present in different states (active and inactive) within the female nucleus. In imprinted X inactivation, the X chromosomes are distinguished by their parental origin.

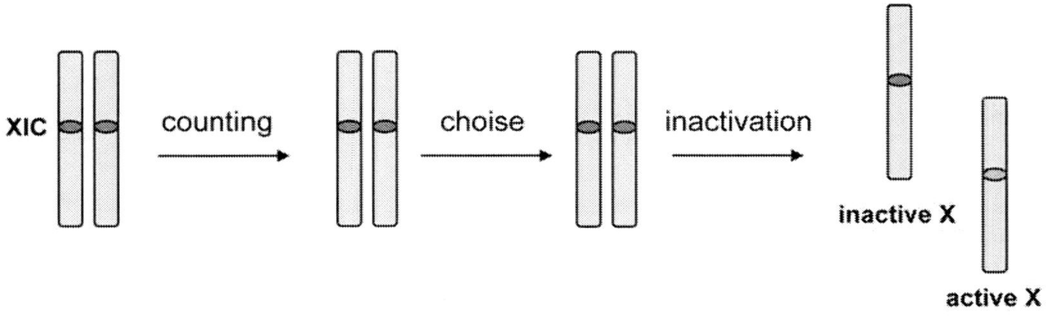

Figure 6. Scheme of counting and choice that involves pairing of the Xic loci at the initiation of X inactivation. The X-inactivation centre regulates Xist expression to ensure that one X chromosome remains active. Regulatory elements implicated in counting and choice in the Xic locus are indicated.

Random X chromosome inactivation (XCI) consists of an ordered series of processes. Each cell ensures, in a random manner, that only one X chromosome remains active and that the other X chromosome is inactivated. The differential treatment of the two X chromosomes results in an Xa and an inactive X (Xi), both of which are present in the same female nucleus.

The Xist RNA, which is expressed exclusively from, the Xi is used to establish the two functionally distinct forms of the X chromosomes. Random XCI is achieved in three genetically separable events (Figure 6). The X-chromosome-to-autosome ratio is counted to ensure that one X chromosome is inactivated per female diploid nucleus. One X chromosome is 'chosen' to be the future inactive X (Xi). Lastly, silencing is initiated by coating of the future Xi by the noncoding Xist RNA, recruitment of silencing factors, and condensation of the X-chromatin. In mammals, a region on the X chromosome named the 'X-inactivation center' (Xic) regulates the different steps of XCI. The Xic is comprised of several genetic elements that make long ncRNAs, including Xist, Tsix, Xite, DXPas34, and Jpx/Enox [75] (Figure 7). Xist encodes a 17 kb long, nuclear RNA expressed exclusively from the Xi [76]. Xist is essential for the silencing step of XCI, as a chromosome deleted for Xist cannot be silenced [77,78]. Xist RNA physically associates with the X-chromatin and nuclear matrix around the X and coats the Xi [79,80]. Xist is negatively regulated by its antisense gene partner, Tsix [81,82].

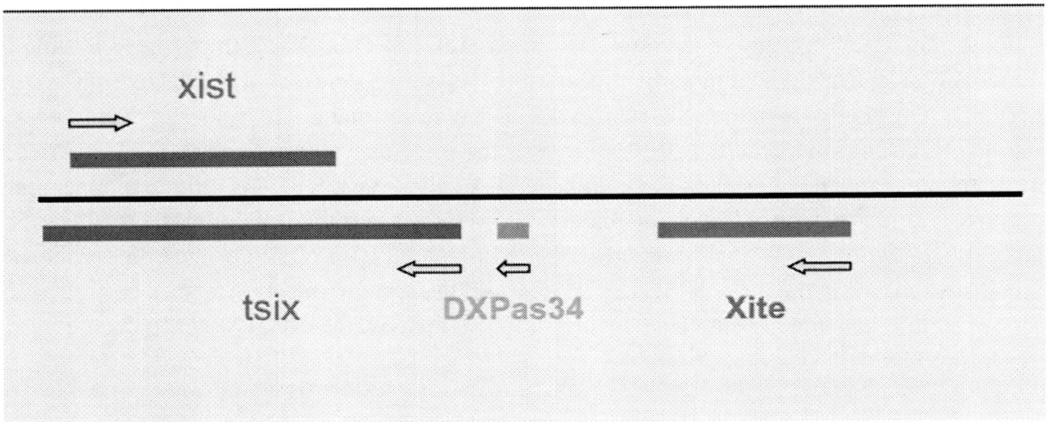

Figure 7. Map of the regulatory elements implicated in counting and choice in the mouse Xic locus. Are indicated the Xist gene, the antisense Tsix RNA, Xite and CCCTC-binding factor (CTCF)-binding sites at DXPas34.

On the future Xi, loss of Tsix permits upregulation of Xist and silencing of the cis chromosome. In parallel on the future active X (Xa), persistence of Tsix expression prevents upregulation of Xist and thereby prevents silencing on that chromosome.

Tsix is developmentally regulated by enhancers contained in the Xite and DXPas34 elements. Differential methylation of Xite and the CCCTC-binding factor (CTCF)-binding sites on DXPas34 correlate with X chromosome choice in mice [83]. The CTCF protein is associated with chromatin boundaries and has been proposed as a candidate factor involved in choice [84]. Consistent with this, deletion of DXPas34 results in ectopic X inactivation [85].

Xist RNA molecules then accumulate over the chromosome and initiate silencing. Gene targeting in mice has shown that Xist is required for both imprinted and random X inactivation, and that ectopic Xist expression in the absence of other Xic sequences can initiate chromosome-wide silencing [86-88]. However, initiation of silencing by Xist is restricted to the early stages of differentiation, implying that it is developmentally regulated. Initially, gene silencing is reversible and dependent on Xist; however, at a later stage in differentiation, X inactivation becomes independent of Xist and irreversible [89]. Hence, Xist is crucial for initiating silencing, but has a minor role in maintaining the Xi. Xist localization does not require X chromosome-specific sequences, and ectopic Xist expressed from autosomal transgenes can localize to autosomes and cause silencing, albeit to differing extents [88,90-91]. The observation that Xist spreading and the maintenance of silencing is less efficient outside the X chromosome instigated the idea of 'way stations' or 'boosters' at intervals along its length. Owing to the high density of long interspersed elements (LINEs) on the X chromosome, these regions have been proposed as candidate boosters [92]. Recent results showed that Xist spreading stops close to the X-autosomal translocation point and correlates with a severe drop in LINE density on chromosome 4 [93]. However, the mechanism underlying the function of distinct booster elements needs to be clarified. The analysis of functional motifs within the mouse Xist RNA can provide insights into how Xist causes gene silencing.

Xist is exceptional among cellular RNAs owing to its ability to encompass and silence an entire chromosome. Because of the numerous interesting epigenetic mechanisms involved, XCI remains an active field of research spanning over half a century. In conclusion, X inactivation is a model for the developmentally controlled formation of silent chromatin. A number of pathways act in a stepwise manner to establish a stable state of repression. Progress has been made in understanding the molecular details of X inactivation, but the method by which Xist initiates silencing remains unknown. It is also unclear how the process of X inactivation intertwines with cellular differentiation, and the molecular details of the timing of counting and the restriction of Xist in initiating silencing remain to be defined. Future studies of X inactivation might therefore also lead to a better understanding of stem cells and of the process of cellular differentiation.

Sex Chromosomes and Disease

Sex chromosome aneuploidy

1959 was a singularly important year in the emerging field of human cytogenetics. Almost simultaneously, two groups [94,95] reported the presence of an additional autosome in individuals with Down syndrome, and three of the four common sex chromosome abnormalities were identified: 47,XXY [96], 47,XXX [97], and 45,X [98] conditions. Cumulatively, these discoveries provided evidence for a new class of human genetic disorder and demonstrated that missing or additional chromosomes could lead to recognizable syndromes. Furthermore, the demonstration that 47,XXX and 45,X individuals were females and that 47,XXY individuals were males provided the first evidence of the sex-determining role of the human Y chromosome.

In the intervening half-century, a large body of information has accrued on the diagnosis, phenotypic consequences and clinical management of individuals with Down syndrome or sex chromosome aneuploidy.

With the advent of DNA polymorphism analysis in the late 1980s, it became possible to determine the parent and meiotic stage of origin of human aneuploid conditions. This approach has extensively been used to study the origin of autosomal trisomies, with one overarching conclusion: despite variation among individual chromosomes, the vast majority of autosomal trisomy derives from maternal nondisjunctional errors, typically occurring at meiosis I [99]. Surprisingly, however, this generalization does not extend to the sex chromosomes.

Human sex chromosome aneuploidies are remarkable in several different respects. First, the mechanisms of origin are unique: in contrast to all autosomal aneuploidies that have been examined, sex chromosome aneuploidies typically derive from loss or gain of a paternal chromosome. For example, the vast majority of paternally derived 47,XXYs derive from meioses in which the X and Y failed to recombine in the XpYp pseudoautosomal region [100].

Second, the phenotypic features are mild, at least by comparison with autosomal imbalance, and fertility is a possibility, at least for some individuals. Indeed, with the introduction of the assisted reproductive technologies (ART), this possibility has been extended to individuals who previously would have had no chance of becoming biological parents. Third, despite their abnormal chromosome constitutions, the vast majority of gametes produced by these individuals have the 'right' number of sex chromosomes. Although the mechanisms by which this occurs remain murky, at least two themes have emerged: loss of the extra sex chromosome prior to or during meiosis appears to occur with significant frequency.

This implies that a selective advantage is conferred to cells with proper X chromosome dosage (i.e. XX rather than XXX cells; XY rather than XXY cells) and to cells with a single Y chromosome (i.e. XY rather than XYY cells); in addition, it is clear that, in humans as well as in mice, stringent cell cycle control mechanisms operate to remove spermatocytes in situations wherein continued progression through meiosis would give rise to aneuploid gametes. Taken together, these mechanisms level the playing field for individuals with sex chromosome aneuploidies, so that the risk of producing aneuploid progeny is only a fraction of the predicted values.

Turner syndrome and partial monosomies

X chromosome monosomy or Turner syndrome (TS), is the only monosomy compatible with life and it is characterized by growth failure and infertility, often associated with a characteristic cognitive deficit and a range of anatomic abnormalities, highly variable form one patient to another [101]. The pathogenesis is complex. It might be due in part to the non-specific effects of the aneuploidy, but the prevalence of most Turner syndrome traits among 45,X patients is far in excess of what would be expected if only aneuploidy was involved. The finding that several X-linked genes escape X chromosome inactivation suggests that the most likely explanation for the common phenotypic manifestations of Turner syndrome is the effect of monosomy for distinct genes [101]. This was the case for short stature and most of the skeletal features of Turner syndrome that are caused by haploinsufficiency for the SHOX gene, in the pseudoautosomal region in Xp [102–105].

Infertility in 45,X patients is caused by oocyte-loss in the early stages of the meiotic prophase, before the pachytene meiotic stage, resulting in ovarian dysgenesis and streak ovaries [106]. Ogata and Matsuo [107] argued that ovarian failure in X monosomies could be caused by non-specific pairing errors at meiosis that increase the probability of germ cell atresia—the extent of ovarian failure correlates with the extent of pairing failure [107]. Partial Genotyping of X chromosome monosomies in women presenting the full Turner syndrome phenotype or only ovarian failure, highlighted specific regions that might be involved in ovarian function; furthermore, these studies seemed to favor haploinsufficiency for specific Xq or Xp genes as the cause of POF [108].

Deletion mapping seems to favor haploinsufficiency for several X-linked genes along the X chromosome as the mechanism responsible for ovarian failure. Given that both X chromosomes are active in the oocytes from the onset of meiosis, many loci for which a double dose is required for ovarian function could be present on the X chromosome [109]. Alternatively, it is intriguing to think that escape from X chromosome inactivation could have been maintained during X chromosome evolution in the somatic cells of the ovarian follicles, as a sex-specific mechanism important for follicular maturation and ovulation.

X-linked premature ovarian failure: a complex disease

The involvement of the X chromosome in premature ovarian failure was demonstrated by the relatively frequent chromosomal rearrangements in patients, but the requirement of two X chromosomes for ovarian function was quite unexplained until recently.

Disorders of ovulation are common in human females and account for a large proportion of infertility problems. Premature ovarian failure (POF; Online Mendelian Inheritance in Man, [OMIM] 311360 and 300511) comprises a group of disorders defined by hypergonadotropic ovarian failure before 40 years of age [110,111]. POF affects 5–10% of menopausal women and has a prevalence of about 1% in the general female population in Caucasian, Hispanic and African Americans [112]. The prevalence is lower in Japanese (0.14%) and Chinese (0.5%). POF results in infertility and lifelong sex steroid deficiency, and it is potentially associated with the severe health risks common to natural menopause, such as cardiovascular and neurological disorders, and osteoporosis. The etiology of POF is not known, but several mechanisms can be hypothesized, including a reduced oocyte or primordial follicle pool, accelerated follicular atresia, and alterations in follicular recruitment or maturation.

A genetic basis for POF has been well established by the report of numerous familial cases. The identification of genes responsible for autosomal recessive [113], X-linked dominant [114] or autosomal dominant syndromic forms [115,116] of the disease demonstrated a monogenic component. To date only two X-linked genes for ovarian failure were definitively demonstrated, *BMP15* (Bone Morphogenic Protein 15) and the premutated allele of the *FMR1* (Fragile X Mental Retardation 1) gene. *BMP15* is a member of the large super-family of the transforming growth factor beta (TGFbeta) proteins, involved in diverse cellular processes during embryonic development and tissue formation [117,118]. Many members of the family have been implicated in mammalian reproduction [119].

In humans, *BMP15* was reported to carry causative mutations in two sisters who were affected with primary amenorrhea and who carried a mutation inherited from the father [114].

The mutation caused substitution of a conserved amino acid in the preproregion of the protein [114]. The human mutation was shown to act as dominant negative and to decrease in vitro growth of granulosa cells after stimulation with wild type BMP15.

The *FMR1* premutation, which represents a risk factor for POF, is the most common of the known causes of POF. The *FMR1* gene, in Xq27, is responsible for Fragile X syndrome (120), a form of X-linked mental retardation associated with minor somatic traits [121–123]. Mental retardation in this case is caused a CGG trinucleotide in the 5' untranslated region (UTR) of the gene being expanded to more than 200 repeats (i.e. full mutation). Normal size alleles contain expansions ranging from 6 to 49 repeats. Alleles with expansions between 59 and 199 repeats might further expand to full mutation size within one generation [124]. Intermediate size alleles (50–59 repeats) are potentially unstable and might lead to full mutation in a few generations when transmitted; they are defined as premutations. As a consequence of the full mutation, the CpG island surrounding the 5' UTR of the *FMR1* gene becomes hypermethylated, the transcription is shut off and FMR1 protein is absent. In cells carrying the premutation alleles, the situation is quite different because *FMR1* mRNA levels are elevated, and FMR1 is present [125–127]. *FMR1* mRNA levels increase with increasing CGG repeat length within the premutation range, and FMRP levels decrease as a consequence of reduced translational efficiency of the premutated mRNA.

Premutation carriers, both females and males, were initially considered to be normal, and indeed they are not affected with mental retardation. Soon after identification of the FMR1 gene, it was realized that normal carriers of the FMR1 mutation had a significantly higher frequency of POF [128]. In fact, the expansion of *FMR1* CGG repeat from the normal size-range, represents the most common of the known risk-factors that might cause ovarian dysfunction and eventually POF. Part of the risk depends on the FMR1 allele involved. In some families — but not all — skewed X chromosome inactivation could be correlated with POF. More significant is the association found with the CGG repeat size. In all studies, the risk increased with increasing length of the repeat, up to the 80 repeats threshold, when the risk of ovarian dysfunction reached its highest [129-132]. This is in agreement with a toxic effect of the premutated mRNA itself, which might sequester CGG binding proteins that are important for RNA processing.

Moreover, a role for X chromosome genes was suggested by the frequent observation of X chromosomes anomalies in patients. Cytogenetic analysis of POF patients showed that balanced X–autosome translocations could be associated with POF and defined a 'critical region' for normal ovarian function on the long arm of the chromosome, corresponding to the Xq13.3–q27 interval, which is often divided into two portions, Xq13–q21 and Xq23–q27 [133,134]. A search for genes interrupted by the breakpoints in balanced X–autosome translocations identified five genes (*DIAPH2, XPNPEP2, POF1B, DACH2* and *CHM*) [135-138] interrupted by the translocations, out of >40 balanced translocations mapped. Mutation and expression analysis failed to demonstrate a role in POF for these five genes. Alternative explanations for the phenotype were proposed that could account for the size of the critical region: they ranged form the presence of loci along Xq that could be disrupted by the chromosomal rearrangements to a 'position effect' caused by the rearrangements on flanking genes. A direct effect of the rearrangements was also suggested, because the presence of unsynapsed regions might be recognized by meiotic check-points or other checkpoints that act during ovarian follicle maturation and that might increase apoptosis and reduce the number of ovarian follicles, thereby leading to POF [139,140].

Breakpoint mapping in POF patients and in normal women showed that the majority of the X chromosome breakpoints were mapped either to genomic regions free of transcribed sequences or to 'gene deserts', and confirmed that mechanisms different from gene interruption might be responsible for X-linked POF [141,142].

However, it is not unlikely that common variants might be associated to the disorder and might represent a relevant portion of the risk in patients. POF is, in fact, a disorder whose relevance has recently increased with the increase of the average female reproductive age. DNA sequence variants altering genes involved in ovarian function, could have been maintained during human evolution and could now be found associated with POF. If this is the case, studies of selected populations of the appropriate size should be carried out to show association and to identify the remaining risk factors. This approach might be more productive than the search for causative mutations in candidate genes in small populations, which, until recently, was the accepted method for most studies.

X-linked mental retardation: many genes for a complex disorder

X-linked mental retardation (XLMR) is a common cause of moderate to severe intellectual disability in males. With a prevalence of about 2%, mental retardation is the most common reason for referral to genetic services and one of the important unsolved problems in health care. XLMR is very heterogeneous, and about two-thirds of patients have clinically indistinguishable non-syndromic (NS-XLMR) forms, which has greatly hampered their molecular elucidation.

Mild forms of mental retardation (i.e. intelligence quotient [IQ] of 50 to 70) are thought to represent the lower end of the normal IQ distribution and to result from the interaction of many genes and non-genetic factors. In contrast, severe forms (i.e. IQ <50, incidence about 0.4%) might be caused by catastrophic events such as perinatal hypoxia or, more often, specific genetic factors such as chromosomal aberrations and defects of specific genes.

Mental retardation is significantly more frequent in males than in females [143–145]. This observation and the description of numerous large families with X-linked inheritance patterns [146–148] laid the foundations for the concept, now amply confirmed, that X-linked gene defects have important roles in the aetiology of mental retardation [149,150].

Clinical and genetic observations have shown that XLMR is very heterogeneous. The most common form is the fragile X mental retardation syndrome, characterized by a conspicuous cytogenetic abnormality, or fragile site, near the tip of Xq [151], which is caused by an expanded CCG triplet repeat sequence in the 5' untranslated region (UTR) of the *FMR1* gene, as shown by Verkerk et al. [152] and other groups. Ever since, fragile X syndrome has had a central role in XLMR research, and these studies have significantly broadened our insight into the molecular mechanisms underlying brain function in health and disease. For a long time, however, other forms of XLMR received relatively little attention, apart from amild form of XLMR, found to be caused by CCG repeat expansion in the 5' UTR of another gene, *FMR2* [153].

The cloning of the OLIGOPHRENIN (*OPHN1*) gene in 1998 [154] marked the beginning of systematic attempts to unravel the molecular causes of XLMR. Large cohorts of

XLMR families were the key to the success of these efforts, which, to date, have led to the identification of more than 80 XLMR genes [155]. Mutations in most of these give rise to clinically distinguishable syndromic forms of mental retardation, such as fragile X syndrome, Coffin-Lowry syndrome [156] or Rett syndrome [157]. Many of these genes have also been implicated in non-syndromic XLMR (NSXLMR).

NS-XLMR is characterized by 'pure' cognitive impairment without other recognizable clinical signs and is thought to be more common than syndromic forms of XLMR. The list of genes that have been implicated in nonsyndromic and syndromic forms of XLMR continues to grow; at present, it comprises 82 entries (Figure 8). XLMR genes are spread over the entire X chromosome with a higher density in two G-negative bands at Xp11.2 (12 genes) and at Xq28 region (12 genes).

However, a relevant portion of XLMR proteins is involved in different biological pathways, such as core metabolism (15%), DNA and RNA processing (6%), protein synthesis (3%), regulation of cell cycle and ubiquitin pathway (7%). These fundamental processes may disproportionately affect cognition because of a brain-specific expression pattern of the corresponding genes.

Approximately 4% of the more than 22.000 protein coding genes presently listed in ENSEMBL are located on the X-chromosome, and roughly 40% of these are currently known to be expressed in the brain. Therefore, several hundred genes could be involved in XLMR, and the 82 X-chromosomal genes that have been implicated in XLMR to date (Figure 8) might be just the tip of the iceberg.

In XLMR families, fragile X syndrome accounts approximately for 25% of the mutations [158]. Mutations of the *ARX* gene, considered to be the second most common cause of XLMR, give rise to a wide variety of distinguishable syndromes, including mental retardation with dystonic movements of the hands (Partington syndrome), X-linked infantile spasms or other forms of epilepsy, agenesis of the corpus callosum and abnormal genitalia, and severe brain malformations [159], and they are also an important cause of non-syndromic XLMR. *ARX* mutations have been found in 5-10% of the published (NS-XLMR) families [160]. Other relatively common causes of XLMR include mutations of *JARID1C* [161,162], *SLC6A8* [163], *MECP2* [164], *IL1RAPL1* and *PQBP1* [165,166].

To date, 34% of the large published 'MRX' have been found to carry mutations in known XLMR genes, but because systematic screening has not been performed, it is plausible that up to 50% of these families might be involved.

Recent studies suggest that the cumulative frequency of XLMR does not exceed 10%. This is much lower than should be expected by the well-documented excess of mentally retarded males; men are about 1.4-fold more likely to have severe mental retardation, and perhaps as much as 1.9-fold more likely to have mild mental retardation [167]. Therefore, this excess cannot be caused by X-linked gene defects alone [168]. A higher prevalence in males has also been observed in other disorders, including autism and Tourette's syndrome, but there are also diseases, such as congenital hypothyroidism, which are more common in females. Hormonal effects might be responsible for some of these differences, another explanation being different gene-dosage in the two sexes. Many genes on the human X-chromosome escape from X-inactivation, or at least partly. For most of these, there is no functionally equivalent homologue on the Y chromosome, which might explain gender-specific differences in the predisposition for certain diseases. Y-chromosomal genes might also be directly responsible for these sex differences, as documented by the SRY gene, which has a male-specific effect on brain function [169]; and the same might be true for other Y-chromosomal genes that acquired separate functions after having diverged from their respective, X-chromosomal counterparts.

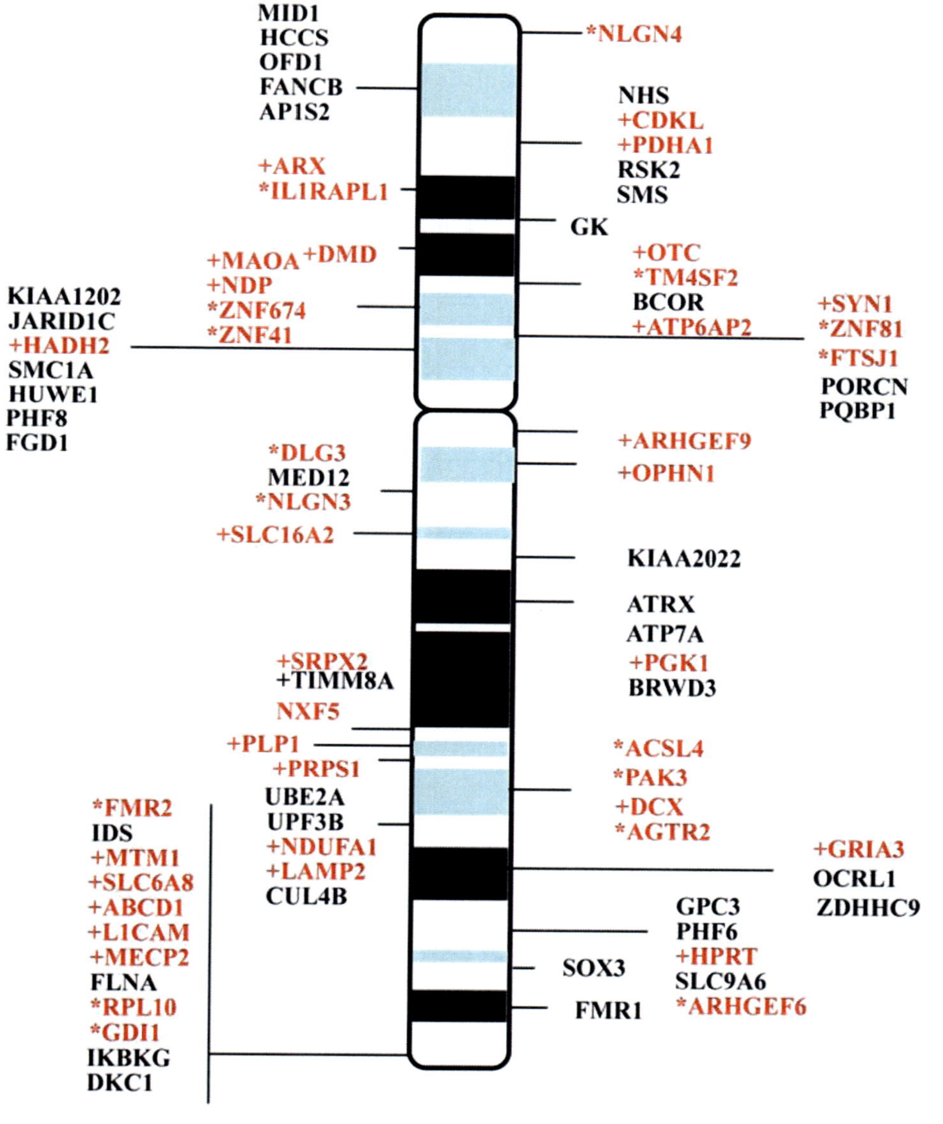

Figure 8. Ideogram of the X chromosome with the position of the 82 known XLMR genes. Genes written in black cause syndromes, those in red preceded by a + sign cause neuromuscular disorders, while those written in red preceded by an asterisk are involved in nonspecific (MRX) conditions..

A more specific - yet, to date, unproven – explanation, has been proposed by Mandel and Chelly [170], who speculated about the existence of X-linked risk factors for mental retardation (e.g. X-linked polymorphisms that lower the IQ, but not enough to cause mental retardation in the majority of males carrying these traits). Such polymorphic modulators of intelligence might be a relatively common cause of mental retardation in isolated males and small families, thereby contributing significantly to the observed excess of mentally retarded males. X-linked modifiers would be expected to have a stronger effect on the frequency of mild mental retardation, which is in keeping with the observation that the male excess is significantly higher in mild than in severe forms of mental retardation.

MicroRNAs (miRNAs) are known to regulate a wide variety of target genes, and they interact with FMRP, the product of the *FMR1* gene. Therefore, it is tempting to speculate that miRNAs might also be involved in the modulation of cognitive functions. Genetic variation of miRNA target sequences (e.g. variation in the 30 UTRs of known XLMR genes) could be also involved in the modulation of cognitive abilities. To date, most of the mutation screening programs have focused on coding exons, splice sites and promoters; therefore, functionally relevant genetic variation in target sequences of miRNAs might have been overlooked.

As previously proposed [171–173], genes that are important for brain development and function might be clustered on the human X-chromosome, because X-linked 'intelligence genes' would be expected to confer a selective advantage to males.The recent observation that more X-linked genes are highly expressed in the human brain than in the mouse brain [174] is compatible with this idea.

Since 1998, the number of genes implicated in NSXLMR has exponentially increased, owing to the generation of large cohorts of families, which has enabled systematic mutation-screening of positional and functional candidate genes. Ongoing efforts to sequence the vast majority of the >900 X-chromosomal genes in up to 300 XLMR families [175] are a logical extension of this strategy. Undoubtedly, these and even more ambitious plans, aiming at the sequencing of all X-chromosomal genes in 1000 families, will result in the identification of novel numerous XLMR genes.

Confocal time-lapse and two-photon imaging have revealed that an essential characteristic of the central nervous system is its conspicuous plasticity, involving dendrite formation, maintenance and synaptic interaction with upstream axons [168,175-176]. These findings have opened up the possibility that in patients with genetic defects causing mental retardation, drug-based therapeutic intervention might be possible even after birth. For example, antagonists of metabotropic glutamate receptors (mGluRs) were proposed for the treatment of fragile X syndrome, after it had been shown that loss of the FMR1 protein leads to overexpression of group I mGluRs, subsequently resulting in long-term depression, altered synaptic plasticity and dysgenesis of dendritic spines [168,177-178]. The recent spectacular finding that in a Drosophila model for fragile X syndrome, treatment of adult flies withmGluR antagonists can restore short-term memory and normal courtship behaviour [179], has substantiated this speculation and raised hopes that these studies will pave the way for drug treatment of fragile X syndrome in humans.

NEMO and incontinentia pigmenti disease

Few genes offer as many opportunities as the *NEMO* gene for explaining some of the more interesting principles of human genetics. Defined by human geneticists according to its role in the X-linked disorder incontinentia pigmenti, *NEMO* (officially *IKBKG* [inhibitor of kB kinase gamma] and also known as IKKg), is an X-linked gene with a number of fascinating features that make it a marvelous example for teaching. Among these features is the remarkable allelic heterogeneity that comprises a variety of forms: various distinctly different diseases; participation in an essential signal transduction pathway; lethality in males, and a female phenotype in which function is lost; X inactivation-related phenotypic variation; a high rate of new mutation due to repeat sequences that stimulate a deletion that is found in the majority of patients; and a complicated genomic architecture that provides insight into evolution.

Incontinentia pigmenti (IP; [OMIM], 308300) is an X-linked dominant disorder with a complex multisystemic and developmental phenotype [180]. It typically manifests as a male-lethal disorder, whereas most female patients survive because of selective elimination of cells expressing the mutant X chromosome [181]. The proliferation of normal cells rescues female patients from lethality and also yields the skewed X-inactivation observed in these subjects' blood cells. These observations are consistent with the notion that IP mutations are cell-lethal. IP is a completely penetrant disorder but shows significant variation in expression. Affected females in the same family often exhibit a widespread phenotypic severity; as with other X-linked disorders that affect females, this variability has been attributed to the pattern of X-inactivation during development.

The most conspicuous sign of the IP phenotype is a four-stage abnormality in skin integrity that commences at or before birth. Although the anomaly of skin pigmentation can be quite dramatic, the most significant medical problems in IP patients are blindness and neurological disturbances [180,182]. While virtually every IP patient exhibits the skin abnormality to some degree, the blindness and the central nervous system (CNS) anomalies are less frequent, occurring in about 40% and 30% of patients, respectively [182–184].

The perinatal incidence of IP has been estimated to be about 1 in 50,000 births, but it is probably higher because the complex phenotype is difficult to diagnose and the skin lesions are often mistaken for other conditions, such as viral (herpes simplex) or bacterial infections or toxic (erythema toxicum) reactions. Moreover, many affected infant females exhibit only the skin lesions in early life, whereas some of them seem to have experienced them during gestation, and thus show only the hyperpigmentation or scarred stages at birth. Sometimes, female patients do not shown any obvious phenotype despite carrying mutations; this has been attributed to early selection against the mutant X-chromosomes or to greater proliferation of normal cells in tissues affected from IP.

The differences in effect that are found among tissues in an individual, and among the same tissues in different individuals, point out the profound effect of selection for cells in which the normal X-chromosome escapes inactivation and becomes the active X-chromosome in this disease. This aspect of skewed X-inactivation with IP can lead to unusual examples of co-morbidity, the appearance of multiple unrelated disorders in the same individual.

The discovery in 2000 that IP results from mutations in *NEMO* (NFkB essential modulator) [185] was assisted by initial genomic sequencing efforts in Xq28 and by the placement of the *NEMO* gene into this region [186]. NEMO encodes the regulatory component of the IkB kinase (IKK) complex [187,188], which is responsible for activating the NFkB signaling pathway (Figure 9). The NFkB transcription factor complex can act to implement immune and inflammatory responses and to prevent apoptosis [189–192]. Thus, the absence of NEMO renders IKK non functional and consequently abolishes NFkB activity. Most IP patients carry loss-of-function mutations in *NEMO* and thus lack NFkB function in their mutant cells [185,193]. The loss of NEMO activity leaves mutant cells vulnerable to apoptosis when exposed to TNFα (tumor necrosis factor-alpha) [185]. The cell death causes lethality in male embryos and skewed X-inactivation in female patients, as a result of elimination of cells with an active mutant X-chromosome.

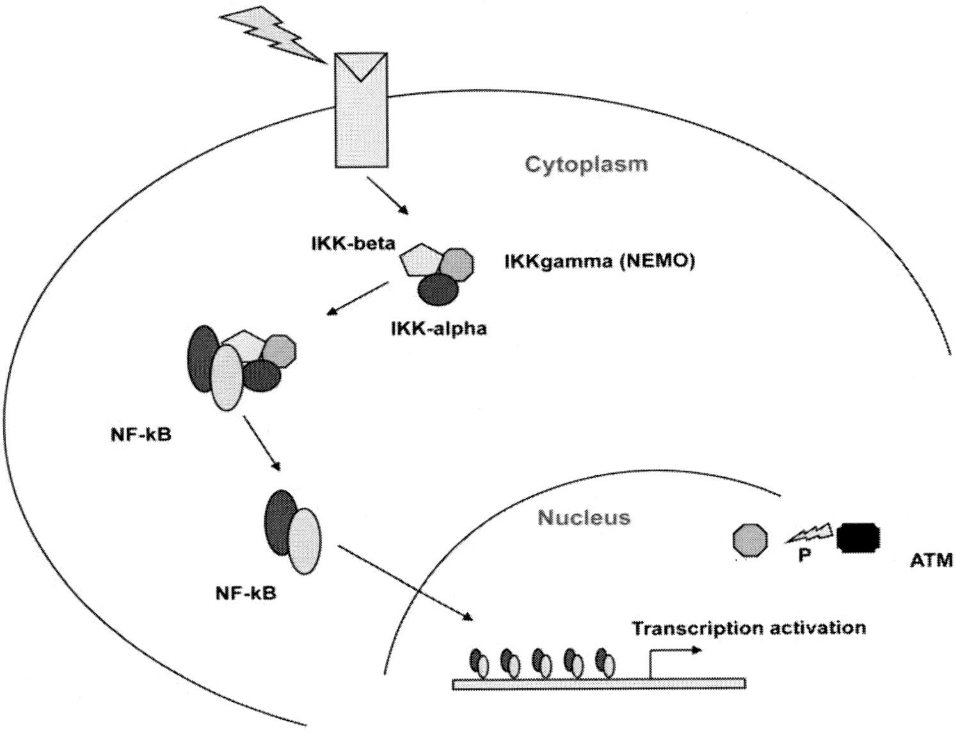

Figure 9. NEMO and the NFkB pathway. NEMO (IKKg) is the regulatory component of the IKK complex and receives upstream activating signals in the cytoplasm, and genotoxic damage sensing signals through ATM in the nucleus. When IKK becomes active, it phosphorylates IkB and targets it for degradation. NFkB travels to the nucleus and activates transcription of target genes.

Approximately 85% of IP patients exhibit an identical *NEMO* mutation. The presence of a common mutation among unrelated families - especially in the presence of a high rate of new mutation - was quite unexpected. The mechanism that leads to the common deletion mutation in IP is misalignment and unequal exchange between two highly identical repeat sequences located in intron 3 and about 4 kb 30 to the last exon [185,193] of *NEMO*. This mutation eliminates NEMO function and is lethal in males. Several other mutations have been observed in selected IP subjects [193-195], but the surprising revelation from the identification of IP *NEMO* mutations was that some of them were compatible with male survival as a result of residual NEMO and NFkB activity [196–197]. These mutations are not cell-lethal and consequently enable phenotypic expression to a greater degree, thereby better illuminating the functions of NFkB. IP males tend to have additional medical features not usually associated with IP, including immune dysfunction, osteopetrosis, and/or lymphedema. The atypical medical problems seen in male IP subjects implied that other disorders associated with similar problems might be related to IP and *NEMO* mutation. Thus, genetic analyses of some male patients with a similar disorder, ectodermal dysplasia (ED), revealed hypomorphic NEMO mutations [196-199]. The male IP phenotype also overlaps both familial expansile osteolysis (FEO) and primary lymphedema (PL). The genes mutated in all these IP-related disorders were discovered serendipitously at about the same time as *NEMO* [200–202]. All these genes function upstream of NEMO and thus implement their functions through NFkB signaling.

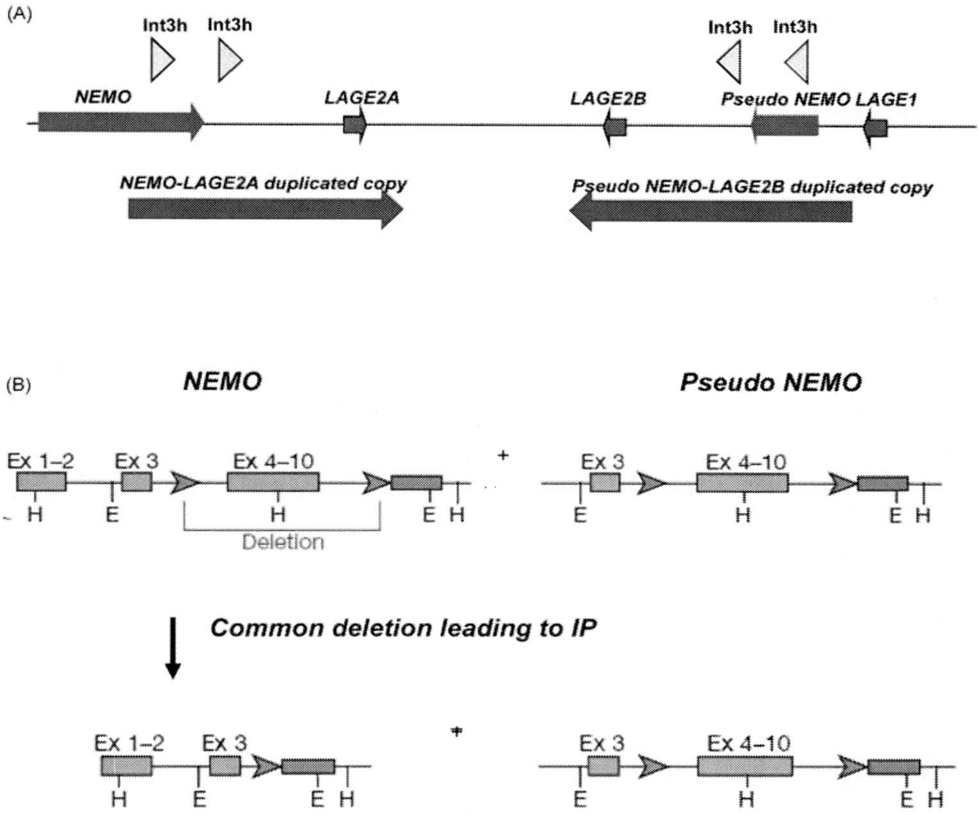

Figure 10. Genomic arrangement and rearrangement of NEMO and nearby sequences. (a) The NEMO–LAGE2 duplication. A section of NEMO, from intron 2, has been duplicated within a 35-kb sequence, along with the LAGE2 gene. The blue boxes below mark the duplicated regions. The red boxes represent the genes. A unique LAGE1 gene lies telomeric to the duplicated regions. The locations of the int3h repeats within each copy of NEMO are shown on top. (b) The NEMO–LAGE2 duplication and the common rearrangement in IP patients.

Therefore, IP might be viewed as a combination of ED, FEO and PL, often with a defect in immune function as well [203]. The medical signs associated with ED, FEO and PL are not seen in typical IP-patients, because of the male lethality and skewed X-inactivation in females.

In addition to explaining the genetic basis for the pathogenesis of IP, the NEMO gene itself has instructive structural properties that illustrate several genomic phenomena. A portion of *NEMO* (deltaNEMO) has been duplicated in an inverted nearby repeat, as has the LAGE2 (L-antigen family member) gene within a 35-kb sequence (Figure 10) [193,204]. Remarkably, the repeats are more than 99.9% identical in sequence, yet they are present in both the gorilla and the chimpanzee, and found in all copies of the human X-chromosome studied to date. Thus, a very highly conserved repeat appears to have been duplicated more than seven million years ago in an ancestor of modern humans, chimpanzees and gorillas. Combined with sequence data from chimpanzees and gorillas, this offers the opportunity to study recent events that maintain this high level of sequence identity. Other regions of the X chromosome, such as in the *Factor VIII* gene, have similar highly conserved repeat sequences that mediate disease-related rearrangements [204].

Both copies of *NEMO* contain identical int3h repeats, which are involved in mediating the common deletion in IP patients. This finding placed *NEMO* on a short list of genes involved in 'genomic disorders' that arise from large rearrangements, such as Charcot–Marie–Tooth disease, Hunter syndrome, and hemophilia A [205,206]. The two copies that arose from the *NEMO–LAGE2* duplication are virtually identical, and each copy contains repeated sequences within itself. A unique *LAGE1* gene lies telomeric to the duplicated regions [207]. The *NEMO–LAGE2* duplication is thus prone to various types of genomic rearrangements, including the lethal deletion seen in most IP patients. At least four types of genomic rearrangements have been discovered at the *NEMO–LAGE2* duplication (Figure 10), but several other types might be possible as a result of repeats located within each copy of the duplication. Both copies themselves constitute repeats because of their remarkable sequence homology and close proximity. It is increasingly evident that genomic rearrangements in the human genome are more common than once thought, and that polymorphic variation in human populations includes deletions, insertions and inversions mediated by repeated sequences [208]; and this locus provides an opportunity to define the range of both genotypic and phenotypic variation in the human population. Correlation of the various rearrangements with potential phenotypic consequences will be of significant interest.

X-linked Eye Disorders

A considerable fraction of the male retinal dystrophies in humans is thought to result from mutations in X-linked genes [209]. Indeed, different human X-linked retinal degenerations have been described, and the pursuit of disease-causing genes has been highly successful in the past decade [209]. These pathologies affect around 1 out of 3000 people under the age of 65 [209-211] and, although they comprise a wide group of blinding diseases, the X-linked degenerative forms fall into three main categories: Retinitis pigmentosa (RP), retinoschisis (RS), and color blindness (protanopia and deuteranopia).

Color blindness

Human color vision is trichromatic, based on three classes of photoreceptors, precisely of cones, in the retina with different light sensitivity. Short-wave-sensitive, S cones, have a maximum light sensitivity at 420 nm, middle-wave, M cones, at 530 nm and longwave, L cones, at 560 nm. The S, M and L cones are randomly distributed in the central retina. Unlike S cones, M and L cones widely vary in number within the central retina. Specific neurons, in some areas of the central nervous system, are able to compare the electric *stimuli* derived from these three categories of cones with different wavelength absorptions. The right combination of these *stimuli* gives the perception of red, yellow, green and blue colors, individually or in various combinations (Figure 11). Thus, it is clear that the synthesis of a single class of photopigment, with a distinct absorption spectrum in each photoreceptor cell, is crucial to color vision [212].

Figure 11. Advantages of Color Vision. The figure shows the same photograph of a summer landscape with and without color. Color vision permits distinctions between objects based on differences in the chromatic composition of reflected light.

At a molecular level, photopigments, G-protein-coupled receptors located on the surface of cone membranes, are responsible for the color vision. Photopigments are mainly composed of a large apoprotein, opsin, which forms a transmembrane bundle, covalently linked to a small conjugated chromophore, 11-cis retinaldehyde. Most of the differences between the wavelength of absorption of the L and M, are due to the interaction of amino acid side-chains in a key position of the opsin with the chromophore. One of the pigments is completely lacking in protanopia (no functional L-cones; OMIM: 303800) and deuteranopia (no functional M-cones; OMIM: 303900), whereas in protanomaly and deuteranomaly, cones show an altered absorption spectrum. In western Europeans, about 8% of males are colorblind [17]. Of these, about 75% have a defect in the deutan (green) series of cones, and about 25% have a defect in the protan (red) series. Males affected from protanopia or deuteranopia show a severe deficit with dichromatic color vision, based on functional S cones plus either M or L cones [213].

Waaler (1968) [214] distinguished 2 phenotypes associated to normal color vision according to "greenpoint", i.e., the point at which the subject sees pure green, and 2 phenotypes according to "bluepoint". He was the first to hypothesize that males can either be G1/B1, G1/B2, or G2/B2; females can be of 6 genotypes. In the next years, many studies based on the recombination fraction for the deutan (green), protan (red), *G6PD*, classic hemophilia and Xg *loci* [215-217] were performed. In Sardinia, studying *G6PD*, protan, deutan and Xg, Filippi et al. [217] found linkage disequilibrium between *G6PD* and protan colorblindness, but not between other pairs of these X-linked loci. They concluded that *G6PD* and protan were nearer one another than *G6PD* and deutan, hypothesizing that *G6PD locus* is between the deutan and protan loci. Purrello et al. [218] demonstrated on Sardinians that the *G6PD locus* is between the 2 colorblindness *loci*. Brennand et al. [219], examining one common X-linked DNA polymorphism - a BamHI polymorphism identified with a cDNA probe of the HPRT gene [219] - and the haplotypes in the sons of the phase-known penta-heterozygous mother concluded the probable order *HPRT*--deutan--*G6PD*--protan--Xqter. The colorblindness genes lie proximal to the F8C and EMD loci. Nathans et al. [220] were the first to isolate and sequence genomic and cDNA clones encoding each of the 3 visual pigments. Whereas there is a single red pigment gene, green pigment genes vary in number among color-vision-normal persons. The multiple green genes are arranged in a head-to-tail tandem array. The existence of multiple green pigment genes in tandem array may explain why deutan colorblindness is more frequent than protan colorblindness. Particularly, Nathans et al. [220] demonstrated that the genes encoding the L (*OPN1LW* [OPSIN 1 LONGWAVE]) and M (*OPNL1MW* [OPNLW MIDDLE-WAVE]) photopigments are arranged in a head-to-tail tandem array on the X chromosome at Xq28.

The cloning by Nathans and colleagues [220] of these genes has lead to the understanding of the molecular basis of the common red–green color vision deficiencies. The tandem array is composed of a single *OPN1LW* followed by one or more *OPN1MW* genes. The proximity and high sequence homology between the *OPN1LW* and *OPNL1MW* genes has predisposed this *locus* to unequal recombination and gene conversion events between the two X chromosomes during gametogenesis in females. These events determined the number variation of *OPNL1MW* genes [212].

Up to 1992, all red-green color vision defects had been associated with gross rearrangements within the red/green opsin gene array on Xq28. Winderickx et al. [221] first described a male with severe deuteranomaly without such a rearrangement. Conversely, they found that substitution of a highly conserved cysteine by arginine at position 203 (Cys203Arg) in the green opsins presumably accounted for his defect in color vision. Surprisingly, this mutation was found to be fairly common (2%) in the population but apparently was not always expressed. This relatively rare mutation, Cys203Arg, with detrimental effects on the stability of the encoded proteins, was found in approximately 1–2% of color vision-defective Caucasians from Northern Europe. Other few rare mutations in the opsin encoding genes were described. Ueyama et al. [222] found, in two Japanese subjects with deutan colorblindness, an Asn94Lys (C>A missense mutation) in the single green gene, and Arg330Gln (G>A) in both green genes. The mutant opsin showed no absorbance when expressed in cultured COS-7 cells [222]. Moreover, c.71A>C substitution in the promoter of the proximal *OPN1MW*, was found associated to congenital deutan-type color vision in about 14% of affected Japanese males [223].

X-linked retinitis pigmentosa (XLRP)

Retinitis Pigmentosa is a group of inherited disorders of the retina characterized by progressive peripheral degeneration of photoreceptor cells. Affected individuals first experience night blindness (nyctalopia), pigmentary alterations in the retina with the appearance of "bone-spicules", and progressive constriction of visual fields, that leads to blindness or severe visual disability in later life [224,225]. It is now known that RP comprises a very heterogeneous group of retinal and retinal pigment epithelium (RPE) dystrophies, both from the clinical and the genetics point of view [226].

RP is a typical rod-cone dystrophy, in which the defects in crucial genes cause cell death (apoptosis). The first histologic change found in the retina of RP patients is the shortening of the rod outer segments, less common, the genetic defects affect the RPE and cone photoreceptors. As the outer segments of rods progressively shorten in the mid periphery of the retina, peripheral and night vision loss occurs [17].

The estimated prevalence of various forms is believed to vary in different populations. Worldwide prevalence of RP is approximately 1 in 5000, and RP can be caused by molecular defects in more than 100 different genes, showing all patterns of inheritance: 20-25% is autosomal dominant (ADRP), 15-20% is autosomal recessive (ARRP), and 5-10% is X linked (XLRP), while the remaining 45-50% is sporadic, thus found in patients without any known affected relative [210]. Some cases of mitochondrial inherited RP have also been described. Moreover, although RP is most commonly found in isolation, it can be associated with systemic disease (i.e. hearing loss in the Usher syndrome).

Each phenotype can result from mutations in multiple genes and, in some cases, different mutations in the same gene can cause markedly diverse symptoms [227]. To date, 40 *loci* associated to RP have been mapped to chromosomal locations [210], and in about 70% of the cases the responsible gene has been identified using positional cloning and candidate gene approaches. Nevertheless, a relevant number of RP genes still remains to be identified, also supporting the evidence for the presence of additional loci. Most mutations lead to a similar phenotype, making genotype-phenotype correlations difficult. Nevertheless, mutations in XLRP genes are associated with a severe phenotype in terms of onset and progression of the disease [228].

XLRP has been recognized as a clinical entity in the medical literature since as early as the 1930s [17]. It is perhaps the most detrimental form of RP because of the severity and early onset of the disease [229]. Most males with XLRP show early onset of visual symptoms with night blindness before the age of 20 [230]. Moreover, it was first evident a wide variation in the clinical manifestations of heterozygous carrier females, which often display a "tapetal-like reflex", probably due to X-inactivation pattern [231].

Retinitis pigmentosa GTPase regulator (*RPGR)* gene and *RP2* are the two known RP-associated X-linked genes [232-234]. In the last years, 3 additional loci on the X-chromosome have been mapped, related genes have not yet been identified [235-237]. Indeed, five genetic *loci* for XLRP, including RP2 (MIM 312600) [238], RPGR (MIM 312610) [239], RP6 (MIM 312612) [240], RP23 (MIM 300424) [236], and RP24 (MIM 300155) [237], have been to date identified. It seems likely that most cases XLRP are accounted for these known *loci* [241-244]. Although, to date, some of the X-linked disease-causing genes still remain to be identified, most XLRP cases can be explained by mutations in *RPGR* and *RP2,* accounting for up to 80% and 20% of disease alleles, respectively [245]. Moreover, mutations in *RPGR* gene account, in some populations, for about 10–20% of all familial RP cases, a frequency higher than most other single RP *loci* [246].

Positional cloning was the strategy used for isolating XLRP-genes. This approach required the genetic analysis of affected families with polymorphic markers, in order to assess the chromosomal location of disease *locus*. Gross chromosomal rearrangements greatly facilitated this king of studies, allowing to narrow the specific genomic region to point to.

Moreover, a further simplification came from X-linked mode of inheritance (i.e., transmission of the disease from phenotypically normal female carriers to male children).

For instance, an interstitial deletion within the Xp21 region of a male patient (BB) with Duchenne muscular dystrophy (DMD), chronic granulomatous disease (CGD), McLeod phenotype, and retinitis pigmentosa, has been instrumental in cloning the genes for cytochrome b5 (CYBB) responsible for CGD, dystrophin for DMD and XK for the McLeod phenotype. The XLRP locus RP3 was believed to be in the proximal portion of the BB deletion because of its coincidence with the critical region delineated by linkage analysis [247]. Based on recombinations, identified by haplotype analysis and map positions of deletions in XLRP patients, RP3 was subsequently mapped to a relatively small region at Xp21.1 between the flanking markers OTC and DXS1110 [248]. The two genetic markers DXS1110 and OTC that flank the RP3 locus encompass a genomic region of <1 Mb; the proximal breakpoint of deletion in BB is 40 kb centomeric to DXS1110. Although, therefore, researchers have been searching for the transcribed sequences in this region surrounding the proximal breakpoint for several years, causative mutations were identified in any gene. To further refine the location of RP3, and to identify the disease locus in XLRP patients, genetic analysis of more than 30 pedigrees with several new polymorphic markers at both sides of the BB breakpoint, was performed. It was a surprise when a recombination event in one of our XLRP families indicated the location of the RP3 mutation outside the BB deletion, closer to the OTC locus [249]. The finding of the RPGR gene 400 kb centromeric to the BB proximal breakpoint confirmed the linkage data. The RP3 cloning was accomplished by the the groups led by Thomas Meitinger from the University of Munich, and Alan Wright from the MRC at Edinburgh. They re-evaluated the "establishment" of the contiguous deletions associated with syndromes in Xp21.1 and found evidence that contradicted the presence of RP3 in the BB deletion. A patient, SB, had XLRP as well as CGD and the McLeod phenotype, but the deletion in this patient extended 400 kb centromeric to the proximal breakpoint of the BB deletion. The deletion in another patient, NF, who had DMD and CGD but not RP, spanned the complete proximal part of the BB deletion and extended a few kb more towards OTC. Finally, the key to RPGR cloning was a patient, MO, who had RP and was reported as having a 75 kb deletion overlapping the proximal portion of SB [232]. RPGR, responsible for RP3 X−linked form of Retinitis pigmentosa, was identified by positional sequencing strategy gene [232].

This gene spans a region, in Xp21.1, of 172 kb that contains the sequence of the entire *RPGR* gene localized between SRPX and OTC genes. The *in vivo* function of RPGR is not fully understood. RPGR is localized in rod and cone photoreceptor connecting *cilia* [250], which link the biosynthetic inner segments and the light−sensing outer segments. RPGR localizes to the connecting cilia through binding, via its N−terminal domain. These region contains a domain homologous to the regulator of chromosome condensation 1 (RCC1), the guaninenucleotide- exchange factor for the GTPase Ran that has been shown to be relevant for the association with binding partners, such as RPGR−interacting protein (RPGRIP) [251,252]. RPGRIP itself is essential for photoreceptor function, as loss of RPGRIP causes another form of retinal degeneration, known as Leber‘s Congenital Amaurosis [253].

RPGR gene shows a great complexity in transcript splicing [242]. Earlier studies, sequencing a genomic region of about 172-kb containing the entire *RPGR* gene, identified a major transcript consisting of 19 exons, known as the constitutive, or default transcript, expressed in multiple tissues. The mutations in the domain homologous to RCC1, are present in only 10–15% of northern European RP3 families [254,255]. Moreover, Vervoort et al. [242] found 5 additional exons: 15b1, 15b2, 15a, ORF14, and a retina–enriched transcript, named ORF15, with a highly repetitive, purine–rich internal region (Figure 12). Exon ORF15 is a large 3' terminal exon consisting of exons from 1 to 15, and extending into part of intron 15. The predicted exon ORF15 protein domain contains an unusual region of low sequence complexity with high glutamic acid and glycine content (known as Plaid domain).

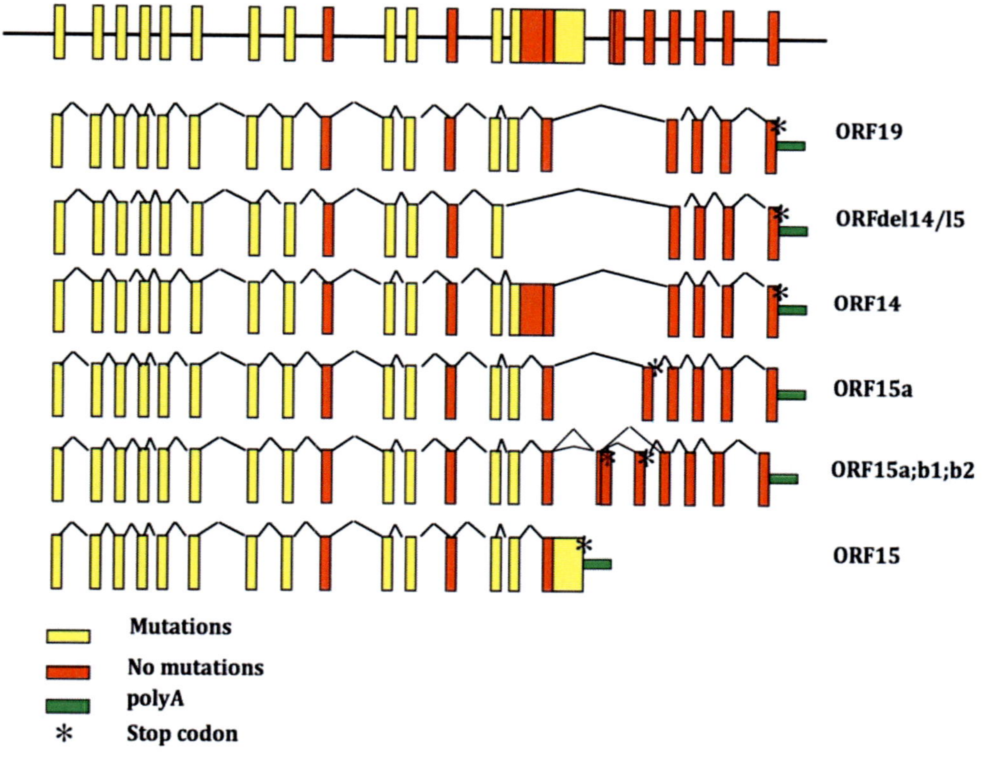

Figure 12. Alternative splicing of human *RPGR*. Schematic representation of *RPGR* gene structure and alternative transcripts. Here are illustrated 5 additional exons: 15b1, 15b2, 15a, ORF14 and ORF15. Exons 15b1 and 15b2 are two overlapping exons within intron 15, which use alternative acceptor splice sites and the same donor site. Their inclusion is predicted to result in premature termination of translation. Exon 15a, which is a third internal exon within intron 15 and also encodes a premature stop codon, is identical with the exon 15a reported previously11. Exon ORF14 corresponds to the mouse exon 14-14a-15; its inclusion does not disrupt the reading frame. Exon ORF15 is a large 3′ terminal exon consisting of exon 15 and extending into part of intron 15.

More recently, by RT-PCR analysis of human retina cDNA, Neidhardt et al. [256] identified a novel exon in intron 9, designated exon 9A, of the *RPGR* gene. The novel exon starts 418 bp downstream the splice donor site of exon 9 and contains 136 nucleotides. Exon 9A is present in a novel *RPGR* splice variant that truncates the RCC1 homologous protein domain. Predominant expression of this novel exon was found in the inner segment of human cones. Although the physiological significance of this complex splicing in *RPGR* gene remains unclear, in human genetic studies, a large number of disease–causing mutations in X-linked RP families, have been identified in the ORF15 exon, but none has been found in those exons specific for the constitutive transcript. Furthermore, at the protein level, the ORF15 variant is found only in photoreceptors, whereas the constitutive variant is expressed at higher levels outside of photoreceptors [257]. Hong et al. [257] showed that RPGR-ORF15 is the functionally significant variant in photoreceptors and that the length of its repetitive region can be reduced while preserving its function. Disease–causing mutations in the ORF15 exon cause a shift of the reading frame, whereas in–frame deletions and insertions are thought to be nonpathogenic [242]. The transcripts abundance, disclosed in *RPGR* gene, indicates that splicing is an important regulator of *RPGR* function [242,245,257].

From a molecular point of view, evidence shows that loss-of-function mutations within *RPGR* are responsible for RP3 type X-linked retinitis pigmentosa. Meindl et al. [232] identified 2 small intragenic deletions and 2 nonsense and 3 missense mutations in highly conserved residues in unrelated patients with X-linked RP. Roepman et al. [233] detected 3 different nucleotide exchanges and one 4-bp deletion in RP patients, but not in controls.

Miano et al. [255] found a total of 29 different *RPGR* mutations in northern European and United States patients. Seven different novel mutations, were identified, including 3 splice site mutations, 2 microdeletions, and 2 missense mutations. RNA analysis showed that the 3 splice site defects resulted in the generation of aberrant *RPGR* transcripts [255]. The same research group identified the exon ORF15 as a mutational hotspot, mutated in 60% of examined XLRP patients [242]. In addition, Vervoort and Wright [258] confirmed that exon ORF15 was a hotspot for mutations, at least in the British population, in which it accounts for 80% of the mutations found within a sample of 47 X-linked retinitis pigmentosa patients [258].

RP2 is the second major gene involved in the XLRP phenotype, localized to a chromosomal region by linkage to a polymorphic DNA marker [238]. This gene maps at Xp11.3. To date, mutations in *RP2* gene have been found in a further 10–20% of XLRP patients, as predicted from linkage studies and extensive mutation analysis in XLRP families [234,259]. *RP2* gene consists of five coding exons, encoding a widely expressed protein of 350 amino acids. The function of RP2 is not completely understood, but it shows homology to cofactor C, a protein involved in the ultimate step of gamma-tubulin folding, and interacts with ADP ribosylation factor-like 3 [260]. Post-translational acyl modifications at the N-terminus of RP2 mediate its targeting to the plasma membrane, and the disruption of the acylation site leads to RP. It has been recently shown that RP2 protein is an efficient GTPase-activating protein (GAP) for Arf-like 3, with structural features similar to other GAPs. Furthermore, the effect of mutations in XLRP patients correlated with their effect on catalysis [261]. The majority of pathogenic mutations identified so far in *RP2* coding region represent truncating mutations, whereas missense mutations often locate to the cofactor C-like domain of RP2. The remaining pathogenic mutations lead to truncated polypeptides by introducing premature stop codon or frameshifts [262].

X-linked retinoschisis (XLRS)

Retinoschisis (MIM 312700) is a severe disease of the nerve tissue in the eye. It affects the retinal cells in the macula (the central fixation point of vision at the back of the eye). Retinoschisis is commonly defined a form of macular degeneration (MD). However there are different types of MD, and many people diagnosed with MD do not have retinoschisis.

Particularly, X-linked retinoschisis (XLRS) is an inherited retinal dystrophy, which leads to schisis (splitting) of the neural retina and reduced visual acuity in affected men. It has been estimated that XLRS affects 1 in 15,000 to 25,000 individuals, primarily young males, even though the prevalence greatly varies among populations [263]. The condition accounts for almost all congenital retinoschisis, with occasional reports of autosomal dominant retinoschisis making up the remainder [264].

This condition frequently starts during childhood and is officially called Juvenile XLRS. Affected males show cystic degeneration leading to split in the retina, detachment of the retina, and finally complete retinal atrophy with sclerosis of the choroid [265]. They inherit this trait from their mothers, who carry the condition but retain normal vision [266]. Cystic maculopathy is sometimes the only finding in these patients. The basic lesion is cystic degeneration in the deep nerve layer. Visual handicap is usually mild up to age 40 to 50 years; thereafter, impairment of vision is slowly progressive [17].

Historically, Haas in the end of 19th century [267], was the first to describe the disease in 2 males with a beautiful drawing of the typical radiating cystic maculopathy [267]. This pathology was documented as an X linked inherited disease more than 10 years later, in 1913, when Pagenstecher published a pedigree that showed a clear X-linked pattern of inheritance [268]. Several similar cases were subsequently described, but were given alternative names including "neuroretinal disease" in men [269] and "congenital vascular veils" in the retina [270]. Jager was the first to use the term ''X linked retinoschisis'' in 1953 [271] and this, together with ''juvenile retinoschisis,'' is now the generally accepted term [211].

Although in the 20th century many clinical cases of typical X-linked inherited retinoschisis had been described [265,270,272-276], the underlying genetic cause was discovered in the 1997. By mapping and expression analysis of expressed sequence tags (ESTs), Sauer et al. [277] identified a novel retina-specific transcript, namely *XLRS1* (x-linked retinoschisin 1). The predicted protein was found to contain a highly conserved motif implicated in cell-cell interaction and thus may be active in cell adhesion processes during retinal development. The first mutation analyses of *XLRS1* in affected families, reveled nine mutations segregating with the disease phenotype in the respective families [277]. Since then, numerous inactivating mutations in the *RS1* gene have been described [278].

The *RS1* gene maps to Xp22, and consists of 6 exons, encoding a 224-amino acid protein. This protein is processed by N-terminal cleavage into a mature protein (201 amino acids) of about 23 kD. Investigation of the protein retinoschisin, has revealed it to be a secretory protein, containing a a highly conserved discoidin domain, shared with a number of other proteins [277], and functioning as an octamer [279,280]. This domain is involved in cell-cell adhesion, in agreement with the phenotype (i. e. splitting of the retina) observed in affected males with deleterious mutations in *RS1* gene. There is a great variability in disease severity, even among individuals who have the same mutation in *RS1* [281-283], and no correlation has been identified between mutation and disease severity or progression [281].

Numerous *RS1* mutations, associated to XLRS, have been described [278,283]. Described mutations are predominantly missense, clustered in the exons 4 to 6, encoding the discoidin domain. Furthermore, deletions, insertions and also splice site mutations have been described. Studies of mutant retinoschisin indicate that missense mutations can be involved in the disease etiology by at least three mechanisms [279,284]: 1) interfering with retinoschisin protein secretion; 2) allowing secretion but interfering with octamerisation; or 3) allowing secretion and octamerisation but interfere with protein function [211]. Since it has not been described any correlation between the molecular mechanism of disease and its severity [281], there may be other influencing factors in the disease severity, such as genetic modifiers or environmental influences [285].

Although little is known about RS pathogenesis, the lesions are, thought to be related to a defect in retinal Mueller cells. Sauer et al. [277] noted that Mueller cells, the principal glia cells in the retina, have a variety of crucial functions, such as provide nutrients to neurons and photoreceptor cells, mechanical stability, synthesis and renewal of visual pigment, phagocytosis, neuronal signaling, and clearance of potassium ions from the extracellular space. Particularly, a defect in the K+ clearance is thought to contribute to the characteristic electroretinography abnormalities in XLRS patients. Moreover, Mueller cells have been shown to aid neurite outgrowth and neuronal connections. Failure to establish a proper neuronal interaction may be the indirect result of a Mueller-cell defect. Consequently, it has been suggested by Arden et al. [266] that RS may be considered a disorder of retinal development rather than a dystrophic process [266].

To understand the XLRS underlying pathophysiology, three retinoschisis mouse models – with retinal pathology similar to the human disease - have been developed [286-288]. Fundus examination revealed the presence of small cyst-like structures in the inner retina and intraretinal flecks in all three XLRS1h knockout models mouse. Similarly, ERG analysis of mouse models showed a characteristic reduced dark-adapted b-wave. Histologically, the mice displayed disorganisation of retinal layers due to mislocalisation of cells within the inner plexiform, inner nuclear and outer plexiform layers; focal areas of retinal splitting or ''schisis'' were also evident within the inner nuclear layer and structural abnormalities of synapses occur within the outer plexiform layer [286-288]. The mouse retina was also used as developmental model, allowing to understand the involvement of retinoschisin in both retinal development and maintenance. This suggests that XLRS is, at least initially, a developmental abnormality of retina. These evidences strongly suggest that gene replacement might therefore be a therapeutic possibility for this X-linked pathology. This has been used with some success on both knock-out mouse XLRS models [287,289]. For *RS1* gene therapy adeno-associated viral vector and intraocular injection were used. Subsequent investigation of the retina in these animals (males) indicated successful expression of retinoschisin in all retinal layers [287,289]. However, these data have to be interpreted with caution since the mouse models were both null mutants, expressing no retinoschisin protein, which is a different *scenario* with respect to that observed in most patients with missense mutations.

Choroideremia

Choroideremia (CHM) is a rare X-linked eye pathology characterized by progressive degeneration of the choriocapillaris, the retinal pigment epithelium, and the photoreceptor cells (MIM 303100) [290-292]. Its incidence is estimated as 1 in 100,000 [17].

Although the clinical manifestation of CHM shows considerable interfamilial as well as intrafamilial variability, affected males display progressive loss of vision, with reduced central vision, constriction of visual fields and night blindness, beginning at an early age. Furthermore, the choroid and retina undergo complete atrophy. Heterozygous females generally show no serious visual impairment, but often show patchy pigmentation and degeneration of the RPE and mild atrophy around the optic disc, suggesting clonal areas of disease attributable to random X-inactivation [293-296]. Moreover, fully affected females have been reported [297]. Occasionally, female carriers show clinical signs of CHM that can be ascribed to skewed X-inactivation [297-298]. Manifestation of the disease in females can also result from disruption of the CHM gene by X-autosomal translocations [299-300]. In these cases, X-inactivation is nonrandom; the normal X is preferentially inactivated, while both translocation fragments remain active.

The gene for *CHM* was localized to the Xq13–q22 region by linkage analysis [296,301-303]. Nussbaum et al. [302] found that, with 90% probability, choroideremia mapped within 9 cM of DXYS1, a polymorphic DNA probe. Lesko et al. [296] had a high lod score in linkage analysis with DXYS1. Gal et al. [304] suggested the following order: Xcen—DXYS1--DXS3--TCD--DXS11--Xqter. This assignment was further refined by the characterization of cytogenetically visible deletions in patients with CHM, mental retardation (MR), and deafness type 3 (DFN3) [294,305,306]. Molecular analysis of deletions in patients with classic CHM [307-310] formed the basis for the positional cloning of a part of the CHM gene from Xq21.2 [292,311]. Only recently, the complete protein coding region of the gene has been isolated [312]. The full-length *CHM* mRNA is of approximately 5.6 kb and is expressed in various tissue types, including choroid, retina, and RPE [292]. The entire gene, composed of 15 exons, spans a genomic sequence of at least 150 kb [312]. Insight into the function of the *CHM* encoded protein came with the biochemical purification of Rab geranylgeranyl transferase (Rab GGTase) from rat brain. This multi-subunit enzyme attaches geranylgeranyl (GG) groups to Rab proteins. This modification is crucial for their action in intracellular vesicular transport. Rab GGTase is made of a catalytic α/β heterodimer and an accessory component, the Rab escort protein (REP), shown to correspond to the protein encoded by the *CHM* gene [313-315]. REP-1 binds to newly synthesized Rab proteins, presents them to the catalytic part of the enzyme, and delivers the prenylated Rabs to their target membranes [315,316]. In yeast, both the α/β heterodimer of Rab GGTase (encoded by the BET4 and BET2 genes) and REP (encoded by the MRS6/MSI4 gene) are essential for cell viability [317-322]. A complete loss of REP activity would likely be lethal in humans too. An autosomal homologue of the *REP-1* gene has been identified, and designated CHML for choroideremia-like. The CHML protein, also named REP-2, was demonstrated to perform a function similar to that of REP-1, and it was therefore hypothesized that REP-2 substitutes for the loss of REP-1 activity in choroideremia patients, thereby preventing symptoms in tissues and organs other than the eye [322].

Classic choroideremia, above all in European patients, is associated to large deletions in *CHM* gene. By contrast, no deletions have yet been found in the North American population [311]. The size of the deletions varies widely, ranging from a few kilobases to 5–15 Mb spanning the complete *CHM* gene. Only 3 out of 18 deletions are located entirely within the *CHM* gene. Deletion breakpoints show a random distribution except for 5 deletions that all end distally between exon 1 and pJ7.6B. These clustered endpoints have not been characterized in detail and the physical size of the chromosomal segment between exon 1 and pJ7.6B is not known. Apart from this region there are no obvious hotspots for the induction of chromosomal breakpoints near or within the *CHM* gene.

X-autosome translocations, involving the *CHM* gene have also been reported in two females suffering from choroideremia. One female carried an (X;13)(q21.2;p12) translocation and the other an (X;7)(q21.2;p12) translocation. Both females also showed ovarian dysgenesis, a common feature in X-autosomal translocations involving the proximal part of Xq [324]. The Xchromosomal breakpoint of the X; 13 translocation was mapped to *CHM* intron 12 [292]. The X;7 translocation breakpoint is located either in intron 3 or in exon 4 of the REP- 1 gene [312]. So far, nonsense mutations have been detected in the *CHM* gene. The majority (about 70%) of small alterations in the *CHM* gene are nonsense and frameshift mutations that result in premature translation termination codons. The remaining 30% of small alterations in the *CHM* gene are mutations that affect splice sites. The vast majority of subtle mutations appear to completely inactivate the REP-1 protein product. This finding explains why there is no apparent correlation between the clinical severity of the disorder and the underlying mutations. The absence of missense mutations may indicate that these have no clinical effect. A more likely possibility would be that missense mutations give rise to a different disease phenotype [325].

Conclusion

The value of complete and accurate reference sequences for the sex chromosomes cannot be over-estimated. Although sequencing and assembly - even of difficult repeat-rich regions of euchromatin - are now reliable, the identification of protein-coding genes remains incomplete, and that of other functional elements is rudimentary. More complete annotation is urgently needed, together with a fuller understanding of the extent of variation of both the DNA sequence and its epigenetic states. With these in place, we will be in a position to assess the genetic influences of the sex chromosomes on health and disease.

Although in this review we have described a quite comprehensive landscape of many fascinating examples of genetic variation and disease-causing mutations among the sex chromosomes, a complete description of all sex chromosomes' features is well beyond the size limitations of this issue. The overviews of broader topics, along with specific vignettes drawn from individual genes and diseases are meant to illustrate the range of subjects that have intrigued human geneticists since X-linkage was first implicated in the etiology of common disorders such as hemophilia, and since heteromorphism for sex chromosomes was understood. Numerous other genes and diseases were not included here, but they have great interest in their own right. Perhaps a future issue can focus on those other stories of the remarkable sex chromosomes.

References

[1] Skaletsky H, Kuroda-Kawaguchi T, Minx PJ, Cordum HS, Hillier L, Brown LG, Repping S, Pyntikova T, Ali J, Bieri T et al. (2003): The malespecific region of the human Y chromosome is a mosaic of discrete sequence classes. *Nature, 423*:825-837.

[2] Ross MT, Grafham DV, Coffey AJ, Scherer S, McLay K, Muzny D, Platzer M, Howell GR, Burrows C, Bird CP et al. (2005): The DNA sequence of the human X chromosome. *Nature, 434*:325-337.

[3] Bentley, D. R. et al. (2001). The physical maps for sequencing human chromosomes 1, 6, 9, 10, 13, 20 and *X. Nature 409*, 942-943.

[4] International Human Genome Sequencing Consortium. Initial sequencing and analysis of the human genome. *Nature 409*, 860-921 (2001).

[5] Schmutz, J. et al. (2004). Quality assessment of the human genome sequence. *Nature 429*, 365-368.

[6] Ciccodicola A, D'Esposito M, Esposito T, Gianfrancesco F, Migliaccio C, Miano MG, Matarazzo MR, Vacca M, Franze A, Cuccurese M et al. (2000). Differentially regulated and evolved genes in the fully sequenced Xq/Yq pseudoautosomal region. *Hum Mol Genet, 9*:395-401.

[7] Mahtani, M. M. & Willard (1998), H. F. Physical and genetic mapping of the human X chromosome centromere: repression of recombination. *Genome Res. 8*, 100-110.

[8] Griffiths-Jones, S (2004). The microRNA Registry. Nucleic Acids Res. 32 (Database issue), D109-111.

[9] Brown, C. J. et al. (1991). A gene from the region of the human X inactivation centre is expressed exclusively from the inactive X chromosome. *Nature 349*, 38-44.

[10] Brown, C. J. et al. (1992). The human XIST gene: analysis of a 17 kb inactive X-specific RNA that contains conserved repeats and is highly localized within the nucleus. *Cell 71*, 527-542.

[11] Lee, J. T., Davidow, L. S. & Warshawsky, D. (1999). Tsix, a gene antisense to Xist at the X-inactivation centre. *Nature Genet. 21*, 400-404.

[12] Migeon, B. R., Chowdhury, A. K., Dunston, J. A. & McIntosh, I. (2001). Identification of TSIX, encoding an RNA antisense to human XIST, reveals differences from its murine counterpart: implications for X inactivation. Am. J. Hum. *Genet. 69*, 951-960.

[13] Chow, J. C., Hall, L. L., Clemson, C. M., Lawrence, J. B. & Brown, C. J. (2003). Characterization of expression at the human XIST locus in somatic, embryonal carcinoma, and transgenic cell lines. *Genomics 82*, 309-322.

[14] Chureau, C. et al. (2002). Comparative sequence analysis of the X-inactivation center region in mouse, human, and bovine. *Genome Res. 12*, 894-908.

[15] Cheung, J. et al. (2003). Genome-wide detection of segmental duplications and potential assembly errors in the human genome sequence. *Genome Biol. 4*, R25.

[16] Stankiewicz, P. & Lupski, J. R. (2002). Genome architecture, rearrangements and genomic disorders. *Trends Genet. 18*, 74-82.

[17] McKusick-Nathans Institute for Genetic Medicine, Johns Hopkins University and National Center for Biotechnology Information, National Library of Medicine. OMIM: Online Mendelian Inheritance in Man. http://www.ncbi.nlm.nih.gov /omim/ (2000).

[18] Schueler, M. G., Higgins, A. W., Rudd, M. K., Gustashaw, K. & Willard, H. F. (2001). Genomic and genetic definition of a functional human centromere. *Science 294*, 109-115.

[19] Spence, J. M. et al. (2002). Co-localization of centromere activity, proteins and topoisomerase II within a subdomain of the major human X alpha-satellite array. *EMBO J. 21*, 5269-5280.

[20] Rudd, M. K. & Willard, H. F. (2004). Analysis of the centromeric regions of the human genome assembly. *Trends Genet. 20*, 529-533.

[21] She, X. et al. (2004). The structure and evolution of centromeric transition regions within the human genome. *Nature 430*, 857-864.

[22] The International SNP MapWorking Group. A map of human genome sequence variation containing 1.42 million single nucleotide polymorphisms. *Nature 409*, 928-933 (2001).

[23] Venter, J. C. et al. (2001). The sequence of the human genome. *Science 291*, 1304-1351.

[24] Schueler, M. G., Higgins, A. W., Rudd, M. K., Gustashaw, K. & Willard, H. F. (2001). Genomic and genetic definition of a functional human centromere. *Science 294*, 109-115.

[25] Tyler-Smith, C. et al. (1993). Localization of DNA sequences required for human centromere function through an analysis of rearranged Y chromosomes. *Nature Genet. 5*, 368-375.

[26] Lahn, B. T. & Page, D. C. (1997). Functional coherence of the human Y chromosome. *Science 278*, 675-680.

[27] Page, D. C., Harper, M. E., Love, J. & Botstein, D. (1984). Occurrence of a transposition from the Xchromosome long arm to the Y-chromosome short arm during human evolution. *Nature 311*, 119-123.

[28] Mumm, S., Molini, B., Terrell, J., Srivastava, A. & Schlessinger, D. (1997). Evolutionary features of the 4-Mb Xq21.3 XY homology region revealed by a map at 60-kb resolution. *Genome Res. 7*, 307-314.

[29] Schwartz, A. et al. (1998). Reconstructing hominid Y evolution: X-homologous block, created by X-Y transposition, was disrupted by Yp inversion through LINE-LINE recombination. Hum. Mol. *Genet. 7*, 1-11.

[30] Tilford, C. et al. (2001). A physical map of the human Y chromosome. *Nature 409*, 943-945.

[31] Tyler-Smith, C., Taylor, L. & Muller, U. (1988). Structure of a hypervariable tandemly repeated DNA sequence on the short arm of the human Y chromosome. J. Mol. *Biol. 203*, 837-848.

[32] Rice, WR. Evolution of the Y sex chromosome in animals. *Bioscience 1996*, *46:*331-343.

[33] Charlesworth, B. (1996). The evolution of chromosomal sex determination and dosage compensation. *Curr Biol, 6*:149-162.

[34] Bachtrog, D. (2006). A dynamic view of sex chromosome evolution. *Curr Opin Genet Dev, 16:*578-585.

[35] Vicoso, B, Charlesworth, B. (2006). Evolution on the X chromosome: unusual patterns and processes. *Nat Rev Genet, 7:*645-653.

[36] Aitken, RJ, & Graves, JAM. (2002). The future of sex. *Nature, 415*, 963.

[37] Raudsepp, T, Lee, EJ, Kata, SR, Brinkmeyer, C, Mickelson, JR, Skow, LC, et al. (2004). Exceptional conservation of horse-human gene order on X chromosome revealed by high-resolution radiation hybrid mapping. *Proc Natl Acad Sci USA, 101:*2386-91.

[38] Bourque, G, Pevzner, PA & Tesler, G. (2004). Reconstructing the genomic architecture of ancestral mammals: lessons from human, mouse, and rat genomes. *Genome Res, 14:*507-16.

[39] Saifi, GM & Chandra, HS. (1999). An apparent excess of sex- and reproduction related genes on the human X chromosome. *Proc R Soc Lond B Biol Sci, 266:*203-9.

[40] Rozen, S, Skaletsky, H, Marzalek, JD, Minx, PJ, Cordum, HS, Waterston, RH, Wilson, RK, and Page, DC. Abundant gene conversion between arms of palindromes in human and ape Y chromosomes. *Nature 2003*, *423*: 873-876.

[41] Graves JAM. (1995). The origin and function of the mammalian Y chromosome and Y-borne genes - an evolving understanding. *Bioessays, 17:*311-320.

[42] Waters, P, Duffy, B, Frost, CJ, Delbridge, ML & Graves, JAM. (2001). The human Y chromosome derives largely from a single autosomal region added 80-130 million years ago. *Cytogenet Cell Genet, 92:*74-79.

[43] Lahn, BT, Page, DC. (1999). Four evolutionary strata on the human X chromosome. *Science, 286:*964-967.

[44] Sandstedt, SA, Tucker & PK. (2004),. Evolutionary strata on the mouse X chromosome correspond to strata on the human X chromosome. *Genome Res 14:* 267-272.

[45] Charlesworth, D, Charlesworth, B, Marais, G. (2005). Steps in the evolution of heteromorphic sex chromosomes. *Heredity, 95:*118-128.

[46] Charlesworth, B, Charlesworth, D. (2000). The degeneration of Y chromosomes. *Philos Trans R Soc Lond B Biol Sci, 355:*1563-1572.

[47] Graves, JA, Wakefield, MJ & Toder R. (1998). The origin and evolution of the pseudoautosomal regions of human sex chromosomes. *Hum Mol Genet, 7:*1991-1996.

[48] Cooke, HJ, Brown WR & Rappold GA. (1985). Hypervariable telomeric sequences from the human sex chromosomes are pseudoautosomal. *Nature, 317:*687-692.

[49] Lien S, Szyda J, Schechinger B, Rappold G & Arnheim N. (2000). Evidence for heterogeneity in recombination in the human pseudoautosomal region: high resolution analysis by sperm typing and radiation-hybrid mapping. *Am J Hum Genet, 66:*557-566.

[50] Rouyer F, Simmler MC, Johnsson C, Vergnaud G, Cooke HJ & Weissenbach J. (1986). A gradient of sex linkage in the pseudoautosomal region of the human sex chromosomes. *Nature, 319:*291-295.

[51] Mohandas TK, Speed RM, Passage MB, Yen PH, Chandley AC & Shapiro LJ. (1992). Role of the pseudoautosomal region in sexchromosome pairing during male meiosis: meiotic studies in a man with a deletion of distal Xp. *Am J Hum Genet, 51:*526-533.

[52] Li L, Hamer DH (1995): Recombination and allelic association in the Xq/Yq homology region. *Hum Mol Genet, 4:*2013-2016.

[53] Charchar FJ, Svartman M, El-Mogharbel N, Ventura M, Kirby P, Matarazzo MR, Ciccodicola A, Rocchi M, D'Esposito M & Graves JAM. (2003). Complex events in the evolution of the human pseudoautosomal region 2 (PAR2). *Genome Res, 13:*281-286.

[54] Perry J, Palmer S, Gabriel A, Ashworth A. (2001). A short pseudoautosomal region in laboratory mice. *Genome Res, 11:*1826-1832.

[55] Humphray SJ, Oliver K, Hunt AR, Plumb RW, Loveland JE, Howe KL, Andrews TD, Searle S, Hunt SE & Scott CE et al. (2004). DNA sequence and analysis of human chromosome 9. *Nature, 429:*369-374.

[56] Marais G & Charlesworth B. Genome evolution (2003): recombination speeds up adaptive evolution. *Curr Biol, 13:*R68-R70.

[57] Lercher MJ & Hurst LD. (2002). Human SNP variability and mutation rate are higher in regions of high recombination. *Trends Genet, 18:*337-340.

[58] May CA, Shone AC, Kalaydjieva L, Sajantila A & Jeffreys AJ. (2002). Crossover clustering and rapid decay of linkage disequilibrium in the Xp/Yp pseudoautosomal gene SHOX. *Nat Genet, 31:*272-275.

[59] Jeffreys AJ & May CA. (2004). Intense and highly localized gene conversion activity in human meiotic crossover hot spots. *Nat Genet, 36:*151-156.

[60] D'Esposito M, Ciccodicola A, Gianfrancesco F, Esposito T, Flagiello L, Mazzarella R, Schlessinger D & D'Urso M. (1996). A synaptobrevin-like gene in the Xq28 pseudoautosomal region undergoes X inactivation. *Nat Genet 13:*227-9.

[61] Gianfrancesco F, Esposito T, Montanini L, Ciccodicola A, Mumm S, Mazzarella R, Rao E, Giglio S, Rappold G & Forabosco A. (1998). A novel pseudoautosomal gene encoding a putative GTP-binding protein resides in the vicinity of the Xp/Yp telomere. *Hum Mol Genet, 7:*407-14.

[62] Esposito T, Gianfrancesco F, Ciccodicola A, Montanini L, Mumm S, D'Urso M & Forabosco A. (1999). A novel pseudoautosomal human gene encodes a putative protein similar to Ac-like transposases. *Hum Mol Genet, 8:*61-7.

[63] Gianfrancesco F, Sanges R, Esposito T, Tempesta S, Rao E, Rappold G, Archidiacono N, Graves JA, Forabosco A & D'Urso M. (2001). Differential divergence of three human pseudoautosomal genes and their mouse homologs: implications for sex chromosome evolution. *Genome Res, 11:*2095-2100.

[64] Galtier N. Recombination (2004), GC-content and the human pseudoautosomal boundary paradox. *Trends Genet, 20:*347-349.

[65] Kuroki Y, Toyoda A, Noguchi H, Taylor TD, Itoh T, Kim DS, Kim DW, Choi SH, Kim IC & Choi HH et al. (2006). Comparative analysis of chimpanzee and human Y chromosomes unveils complex evolutionary pathway. *Nat Genet, 38:*158-167.

[66] Bussell JJ, Pearson NM, Kanda R, Filatov DA & Lahn BT (2006): Human polymorphism and human–chimpanzee divergence in pseudoautosomal region correlate with local recombination rate. *Gene, 368:*94-100.

[67] Filatov DA. (2004). A gradient of silent substitution rate in the human pseudoautosomal region. *Mol Biol Evol, 21:*410-417.

[68] Filatov DA. (2003), Gerrard DT: High mutation rates in human and ape pseudoautosomal genes. *Gene, 317:*67-77.

[69] Graves, JAM, & Schmidt, MM. (1992). Mammalian sex chromosomes - design or accident? *Curr Opin Genet Dev, 2:*890-901.

[70] Graves, JA, Disteche, CM, and Toder & R. (1998). Gene dosage in the evolution and function of mammalian sex chromosomes. *Cytogenet Cell Genet, 80:*94-103.

[71] Jegalian, K, & Page, DC. (1998). A proposed path by which genes common to mammalian x and Y chromosomes evolve to become X inactivated. *Nature, 394:*776-780.

[72] Esposito, T, Gianfrancesco, F, Ciccodicola, A, D'Esposito, M, Nagaraja, R, Mazzarella, R, D'Urso, M & Forabosco, A. (1997). Escape from X inactivation of two new genes associated with DXS6974E and DXS7020E. *Genomics, 43:* 183-90.

[73] Carrel, L, & Willard, HF. (2005). X-inactivation profile reveals extensive variability in X-linked gene expression in females. *Nature, 434:*400-404.

[74] Duret, L, Chureau, C, Samain, S, Weissenbach, J & Avner, P. (2006). The xist RNA gene evolved in eutherians by pseudogenization of a protein-coding gene. *Science, 312:*1653-1655.

[75] Avner, P & Heard, E. (2001). X-chromosome inactivation: counting, choice and initiation. *Nat Rev Genet, 2:*59-67.

[76] Brown, CJ, Ballabio, A, Rupert, JL, Lafreniere, RG, Grompe, M, Tonlorenzi, R & Willard, HF. (1991). A gene from the region of the human X inactivation centre is expressed exclusively from the inactive X chromosome. *Nature, 349:*38-44.

[77] Marahrens, Y, Panning, B, Dausman, J, Strauss, W & Jaenisch, R. (1997). Xist deficient mice are defective in dosage compensation but not spermatogenesis. *Genes Dev, 11:*156-166.

[78] Penny, GDKG, Sheardown, SA, Rastan, S & Brockdorff, N. (1996). Requirement for Xist in X chromosome inactivation. *Nature, 379:*131-137.

[79] Brown, CJ, Hendrich, BD, Rupert, JL, Lafreniere, RG, Xing, Y, Lawrence, J & Willard, HG. (1992). The human XIST gene: analysis of a 17kb inactive X-specific RNA that contains conserved repeats and is highly localized within the nucleus. *Cell, 71:*527-542.

[80] Clemson, C, McNeil, JA, Willard, HF & Lawrence, JB. (1996). XIST RNA paints the inactive X chromosome at interphase: evidence for a novel RNA involved in nuclear/chromosome structure. *J Cell Biol, 132:*259-275.

[81] Lee, JT, Davidow, LS & Warshawsky, D. (1999). Tsix, a gene antisense to Xist at the X-inactivation centre. *Nat Genet, 21:*400-404.

[82] Lee, JT & Lu, N. (1999). Targeted mutagenesis of Tsix leads to nonrandom X inactivation. *Cell, 99:*47-57.

[83] Boumil, RM, Ogawa, Y, Sun, BK, Huynh, KD & Lee, JT. (2006). Differential methylation of Xite and CTCF sites in Tsix mirrors the pattern of X-inactivation choice in mice. *Mol Cell Biol, 26:* 2109-2117.

[84] Chao, W, Huynh, KD, Spencer, RJ, Davidow, LS & Lee, JT. (2002). CTCF, a candidate trans-acting factor for X-inactivation choice. *Science, 295:* 345-347.

[85] Vigneau, S, Augui, S, Navarro, P, Avner, P & Clerc, P. (2006). An essential role for the DXPas34 tandem repeat and Tsix transcription in the counting process of X chromosome inactivation. *Proc Natl Acad Sci USA, 103:*7390-7395.

[86] Marahrens, Y, Panning, B, Dausman, J, Strauss, W & Jaenisch, R. (1997). Xist-deficient mice are defective in dosage compensation but not spermatogenesis. *Genes Dev, 11:*156-166.

[87] Penny, GD, Kay, GF, Sheardown, SA, Rastan, S & Brockdorff, N. (1996). Requirement for Xist in X chromosome inactivation. *Nature, 379:*131-137.

[88] Wutz, A & Jaenisch, R. (2000). A shift from reversible to irreversible X inactivation is triggered during ES cell differentiation. *Mol Cell, 5:* 695-705.

[89] Csankovszki, G, Panning, B, Bates, B, Pehrson, JR & Jaenisch, R. (1999). Conditional deletion of Xist disrupts histone macroH2A localization but not maintenance of X inactivation. *Nat Genet, 22:* 323-324.

[90] Herzing, LB, Romer, JT, Horn, JM & Ashworth, A. (1997). Xist has properties of the X-chromosome inactivation centre. *Nature, 386:* 272-275.

[91] Lee, JT, Strauss, WM, Dausman, JA & Jaenisch, R. (1996). A 450 kb transgene displays properties of the mammalian X-inactivation center. *Cell, 86:*83-94.

[92] Lyon, MF. (2003). The Lyon and the LINE hypothesis. *Semin Cell Dev Biol, 14:* 313-318.

[93] Popova, BC, Tada, T, Takagi, N, Brockdorff, N & Nesterova, TB. (2006). Attenuated spread of X-inactivation in an X; autosome translocation. *Proc Natl Acad Sci USA, 103:*7706-7711.

[94] Jacobs PA, Baikie AG & Court Brown WM. (1959). Strong JA: The somatic chromosomes in mongolism. *Lancet, 1:*710.

[95] Lejeune J, Turpin R & Gautier M. (1959). Mongolism; a chromosomal disease (trisomy). *Bull Acad Natl Med, 143:*256-265.

[96] Jacobs PA, Strong JA. (1959). A case of human intersexuality having a possible XXY sex-determining mechanism. *Nature, 183:*302-303.

[97] Jacobs PA, Baikie AG, Brown WM, Macgregor TN, Maclean N & Harnden DG. (1959). Evidence for the existence of the human ''super female''. *Lancet, 2:*423-425.

[98] Ford CE, Jones KW, Polani PE, De Almeida JC & Briggs JH. (1959). A sex-chromosome anomaly in a case of gonadal dysgenesis (Turner's syndrome). *Lancet, 1:*711-713.

[99] Hassold T & Hunt P. (2001). To err (meiotically) is human: the genesis of human aneuploidy. *Nat Rev Genet, 2:*280-291.

[100] Thomas NS & Hassold TJ. (2003). Aberrant recombination and the origin of Klinefelter syndrome. *Hum Reprod Update, 9:*309-317.

[101] Zinn AR & Ross JL. (1998). Turner syndrome and haploinsufficiency. *Curr Opin Genet Dev, 8:*322-327.

[102] Rao E, Weiss B, Fukami M, Rump A, Niesler B, Mertz A, Muroya K, Binder G, Kirsch S, Winkelmann M et al. (1997). Pseudoautosomal deletions encompassing a novel homeobox gene cause growth failure in idiopathic short stature and Turner syndrome. *Nat Genet, 16:*54-63.

[103] Belin V, Cusin V, Viot G, Girlich D, Toutain A, Moncla A, Vekemans M, Le Merrer M & Munnich A. (1998). Cormier-Daire V: SHOX mutations in dyschondrosteosis (Leri-Weill syndrome). *Nat Genet, 19:*67-69.

[104] Shears DJ, Vassal HJ, Goodman FR, Palmer RW, Reardon W, Superti-Furga A & Scambler PJ. (1998). Winter RM: Mutation and deletion of the pseudoautosomal gene SHOX cause leri-weill dyschondrosteosis. *Nat Genet, 19:*70-73.

[105] Clement-Jones M, Schiller S, Rao E, Blaschke RJ, Zuniga A, Zeller R, Robson SC, Binder G, Glass I, Strachan T et al. (2000). The short stature homeobox gene SHOX is involved in skeletal abnormalities in Turner syndrome. *Hum Mol Genet, 9:*695-702.

[106] Speed RM. (1988). The possible role of meiotic pairing anomalies in the atresia of human fetal oocytes. *Hum Genet, 78:*260-266.

[107] Ogata T & Matsuo N. (1995). Turner syndrome and female sex chromosome aberrations: deduction of the principal factors involved in the development of clinical features. *Hum Genet, 95:*607-629.

[108] Simpson JL & Rajkovic A. (1999). Ovarian differentiation and gonadal failure. *Am J Med Genet, 89:*186-200.

[109] Gartler SM, Andina R & Gant N. (1975). Ontogeny of X-chromosome inactivation in the female germ line. *Exp Cell Res, 91:*454-457.

[110] Coulam CB, Adamson SC & Annegers JF. (1986). Incidence of premature ovarian failure. *Obstet Gynecol, 67:*604-606.

[111] Goswami D & Conway G. (2005). Premature ovarian failure. *Hum Reprod Update, 11:*391-410.

[112] Luborsky JL, Meyer P, Sowers MF, Gold EB & Santoro N. (2003). Premature menopause in a multi-ethnic population study of the menopause transition. *Hum Reprod, 18:*199-206.

[113] Aittomaki K, Lucena JL, Pakarinen P, Sistonen P, Tapanainen J, Gromoll J, Kaskikari R, Sankila EM, Lehvaslaiho H, Engel AR et al. (1995). Mutation in the follicle-stimulating hormone receptor gene causes hereditary hypergonadotropic ovarian failure. *Cell, 82:*959-968.

[114] Di Pasquale E, Beck-Peccoz P & Persani L. (2004). Hypergonadotropic ovarian failure associated with an inherited mutation of human bone morphogenetic protein-15 (BMP15) gene. *Am J Hum Genet, 75:*106-111.

[115] Crisponi L, Deiana M, Loi A, Chiappe F, Uda M, Amati P, Bisceglia L, Zelante L, Nagaraja R, Porcu S et al. (2001). The putative forkhead transcription factor FOXL2 is mutated in blepharophimosis/ptosis/epicanthus inversus syndrome. *Nat Genet, 27:*159-166.

[116] Fogli A, Rodriguez D, Eymard-Pierre E, Bouhour F, Labauge P, Meaney BF, Zeesman S, Kaneski CR, Schiffmann R & Boespflug-Tanguy O. (2003). Ovarian failure related to eukaryotic initiation factor 2B mutations. *Am J Hum Genet, 72:*1544-1550.

[117] Kingsley DM. (1994). The TGF-b superfamily: new members, new receptors, and new genetic tests of function in different organisms. *Genes Dev, 8:*133-146.

[118] Hogan BL. (1996). Bmps: multifunctional regulators of mammalian embryonic development. *Harvey Lect, 92:*83-98.

[119] Drummond AE. (2005). TGFb signalling in the development of ovarian function. *Cell Tissue Res, 322:*107-115.

[120] Garber K, Smith KT, Reines D & Warren ST. (2006). Transcription, translation, and fragile X syndrome: Curr Opin Genet Dev.

[121] Verkerk AJ, Pieretti M, Sutcliffe JS, Fu YH, Kuhl DP, Pizzuti A, Reiner O, Richards S, Victoria MF, Zhang FP et al. (1991). Identification of a gene (FMR-1) containing a CGG repeat coincident with a breakpoint cluster region exhibiting length variation in fragile X syndrome. *Cell, 65:*905-914.

[122] Oberle I, Rousseau F, Heitz D, Kretz C, Devys D, Hanauer A, Boue J, Bertheas MF & Mandel JL. (1991). Instability of a 550-base pair DNA segment and abnormal methylation in fragile X syndrome. *Science, 252:*1097-1102.

[123] Fu YH, Kuhl DP, Pizzuti A, Pieretti M, Sutcliffe JS, Richards S, Verkerk AJ, Holden JJ, Fenwick RG Jr, Warren ST et al. (1991). Variation of the CGG repeat at the fragile X site results in genetic instability: resolution of the Sherman paradox. *Cell, 67:*1047-1058.

[124] Nolin SL, Brown WT, Glicksman A, Houck GE Jr, Gargano AD, Sullivan A, Biancalana V, Brondum-Nielsen K, Hjalgrim H, Holinski-Feder E et al. (2003). Expansion of the fragile X CGG repeat in females with premutation or intermediate alleles. *Am J Hum Genet, 72:*454-464.

[125] Tassone F, Hagerman RJ, Taylor AK, Gane LW, Godfrey TE & Hagerman PJ. (2000). Elevated levels of FMR1 mRNA in carrier males: a new mechanism of involvement in the fragile-X syndrome. *Am J Hum Genet, 66:*6-15.

[126] Kenneson A, Zhang F, Hagedorn CH & Warren ST. (2001). Reduced FMRP and increased FMR1 transcription is proportionally associated with CGG repeat number in intermediate-length and premutation carriers. *Hum Mol Genet, 10:*1449-1454.

[127] Tassone F & Hagerman PJ. (2003). Expression of the FMR1 gene. *Cytogenet Genome Res, 100:*124-128.

[128] Cronister A, Hagerman RJ, Wittenberger M & Amiri K. (1991). Mental impairment in cytogenetically positive fragile X females. *Am J Med Genet, 38:*503-504.

[129] Bretherick KL, Fluker MR & Robinson WP. (2005). FMR1 repeat sizes in the gray zone and high end of the normal range are associated with premature ovarian failure. *Hum Genet, 117:*376-382.

[130] Bodega B, Bione S, Dalpra L, Toniolo D, Ornaghi F, Vegetti W, Ginelli E & Marozzi A. (2005). Influence of intermediate and uninterrupted FMR1 CGG expansions in premature ovarian failure manifestation. *Hum Reprod 10.*1093/humrep/ dei432.

[131] Sullivan AK, MarcusM, Epstein MP, Allen EG, Anido AE, Paquin JJ, Yadav-Shah M & Sherman SL. (2005). Association of FMR1 repeat size with ovarian dysfunction. *Hum Reprod, 20:*402-412.

[132] Ennis S, Ward D & Murray A. (2006). Nonlinear association between CGG repeat number and age of menopause in FMR1 premutation carriers. *Eur J Hum Genet, 14:*253-255.

[133] Therman E & Susman B. (1990). The similarity of phenotypic effects caused by Xp and Xq deletions in the human female: a hypothesis. *Hum Genet, 85:*175-183.

[134] Therman E, Laxova R & Susman B. (1990). The critical region on the human Xq. *Hum Genet, 85:*455-461.

[135] Bione S, Sala C, Manzini C, Arrigo G, Zuffardi O, Banfi S, Borsani G, Jonveaux P, Philippe C, Zuccotti M et al. (1998). A human homologue of the Drosophila melanogaster diaphanous gene is disrupted in a patient with premature ovarian failure: evidence for conserved function in oogenesis and implications for human sterility. *Am J Hum Genet, 62:*533-541.

[136] Prueitt RL, Ross JL & Zinn AR. (2000). Physical mapping of nine Xq translocation breakpoints and identification of XPNPEP2 as a premature ovarian failure candidate gene. *Cytogenet Cell Genet, 89*:44-50.

[137] Bione S, Rizzolio F, Sala C, Ricotti R, Goegan M, Manzini MC, Battaglia R, Marozzi A, Vegetti W, Dalpra L et al. (2004). Mutation analysis of two candidate genes for premature ovarian failure, DACH2 and POF1B. *Hum Reprod, 19:*2759-2766.

[138] van Bokhoven H, Schwartz M, Andreasson S, van den Hurk JA, Bogerd L, Jay M, Ruther K, Jay B, Pawlowitzki IH, Sankila EM et al. (1994). Mutation spectrum in the CHM gene of Danish and Swedish choroideremia patients. *Hum Mol Genet, 3:*1047-1051.

[139] Schlessinger D, Herrera L, Crisponi L, Mumm S, Percesepe A, Pellegrini M, Pilia G & Forabosco A. (2002). Genes and translocations involved in POF. *Am J Med Genet, 111:*328-333.

[140] Turner JM, Mahadevaiah SK, Fernandez-Capetillo O, Nussenzweig A, Xu X, Deng CX & Burgoyne PS. (2005). Silencing of unsynapsed meiotic chromosomes in the mouse. *Nat Genet*, *37*:41-47.

[141] Rizzolio F, Bione S, Sala C, Goegan M, Gentile M, Gregato G, Rossi E, Pramparo T, Zuffardi O & Toniolo D. (2006). Chromosomal rearrangements in Xq and premature ovarian failure: mapping of 25 new cases and review of the literature. *Hum Reprod*, DOI:10.1093/humrep/dei495.

[142] Mumm S, Herrera L, Waeltz PW, Scardovi A, Nagaraja R, Esposito T, Schlessinger D, Rocchi M & Forabosco A. (2001). X/autosomal translocations in the Xq critical region associated with premature ovarian failure fall within and outside genes. *Genomics*, *76*:30-36.

[143] Penrose LS. (1938). A clinical and genetic study of 1280 cases of mental defect. London: Medical Research Council. *Special reports series*, p. 229.

[144] Wright SW, Tarjan G, Eyer L. (1959). Investigation of families with two or more mentally defective siblings: clinical observations. *Am J Dis Child*, *97*:445-456.

[145] Priest JH, Thuline HC, Laveck GD & Jarvis DB. (1961). An approach to genetic factors in mental retardation. Studies of families containing at least two siblings admitted to a state institution for the retarded. *Am J Ment Defic*, *66*:42-50.

[146] Martin JP & Bell J. (1943). A pedigree of mental defect showing sex-linkage. *J Neurol Neurosurg Psychiatry*, *6*:154-156.

[147] Allan W, Herndon CN & Dudley FC. (1944). Some examples of the inheritance of mental deficiency: apparently sex-linked idiocy and microcephaly. *Am J Ment Defic*, *XLVIII*:325-334.

[148] Renpenning H, Gerrard JW, Zaleski WA & Tabata T. (1962).: Familial sex-linked mental retardation. *Can Med Assoc J*, *87*:954-956.

[149] Lehrke R. (1972). A theory of X-linkage of major intellectual traits. *Am J Ment Defic*, *76*:611-619.

[150] Lehrke RG. (1974). X-linked mental retardation and verbal disability. *Birth Defects Orig Artic Ser*, *10*:1-100.

[151] Lubs HA. (1969). A marker X chromosome. *Am J Hum Genet*, *21*:231-244.

[152] Verkerk AJMH, Pieretti M, Sutcliffe JS, Fu Y-H, Kuhl DPA, Pizzuti A, Reiner O, Richards S, Victoria MF, Zhang F et al. (1991). Identification of a gene (FMR-1) containing a CGG repeat coincident with a breakpoint cluster region exhibiting length variation in fragile X syndrome. *Cell*, *65*:905-914.

[153] Gecz J, Gedeon AK, Sutherland GR & Mulley JC. (1996). Identification of the gene FMR2, associated with FRAXE mental retardation. *Nat Genet*, 13:105-108.

[154] Billuart P, Bienvenu T, Ronce N, des Portes V, Vinet MC, Zemni R, Roest Crollius H, Carrie A, Fauchereau F, Cherry M et al. (1998). Oligophrenin-1 encodes a rhoGAP protein involved in X-linked mental retardation. *Nature*, *392*:923-926.

[155] Chiurazzi P, Schwartz CE, Gecz J, Neri G. (2008). XLMR genes: update 2007. *Eur J Hum Genet*. *Apr*;*16*(4):422-34.

[156] Trivier E, De Cesare D, Jacquot S, Pannetier S, Zackai E, Young I, Mandel JL, Sassone-Corsi P & Hanauer A. (1996). Mutations in the kinase Rsk-2 associated with coffin-lowry syndrome. *Nature*, 384:567-570.

[157] Amir RE, Van den Veyver IB, Wan M, Tran CQ, Francke U & Zoghbi HY. (1999). Rett syndrome is caused by mutations in X-linked MECP2, encoding methyl-CpG-binding protein 2. *Nat Genet*, *23*:185-188.

[158] Fishburn J, Turner G, Daniel A & Brookwell R. (1983). The diagnosis and frequency of X-linked conditions in a cohort of moderately retarded males with affected brothers. *Am J Med Genet*, *14*:713-724.

[159] Friocourt G, Poirier K, Rakic S, Parnavelas JG & Chelly J. (2006). The role of ARX in cortical development. *Eur J Neurosci*, *23*:869-876.

[160] Nawara M, Szczaluba K, Poirier K, Chrzanowska K, Pilch J, Bal J, Chelly J & Mazurczak T. (2006). The ARX mutations: A frequent cause of X-linked mental retardation. *Am J Med Genet*, *140*:727-732.

[161] Jensen LR, AmendeM, Gurok U, Tzschach A, Moser B, Janecke A, Tariverdian G, Chelly J, Fryns J-P, van Esch H et al. (2005). Mutations in the JARID1C gene, encoding a protein involved in transcriptional regulation and chromatin remodeling, cause Xlinked mental retardation. *Am J Hum Genet*, *76*:227-236.

[162] Tzschach A, Lenzner S, Moser B, Reinhardt R, Chelly J, Fryns J-P, Kleefstra T, Raynaud M, Turner G, Ropers HH et al. (2006). Novel JARID1C/SMCX mutations in patients with X-linked mental retardation. *Hum Mutat*, *27*:389.

[163] Rosenberg EH, Almeida LS, Kleefstra T, deGrauw RS, Yntema HG, Bahi N, Moraine C, Ropers HH, Fryns JP, deGrauw TJ et al. (2004). High prevalence of SLC6A8 deficiency in X-linked mental retardation. *Am J Hum Genet*, *75*:97-105.

[164] Van Esch H, Bauters M, Ignatius J, Jansen M, Raynaud M, Hollanders K, Lugtenberg D, Bienvenu T, Jensen LR, Gecz J et al. (2005). Duplication of the MECP2 region is a frequent cause of severe mental retardation and progressive neurological symptoms in males. *Am J Hum Genet*, *77*:442-453.

[165] Kalscheuer VM, Freude K, Musante L, Jensen LR, Yntema HG, Gecz J, Sefiani A, Hoffmann K, Moser B, Haas S et al. (2003). Mutations in the polyglutamine binding protein 1 gene cause X-linked mental retardation. *Nat Genet*, *35*:313-315.

[166] Stevenson RE, Bennetts B & Abidi F. (2005). Renpenning syndrome comes into focus. *Am J Med Genet*, *134A*:415-421.

[167] Leonard H & Wen X. (2002). The epidemiology of mental retardation: challenges and opportunities in the new millennium. *Ment Retard Dev Disabil Res Rev*, *8*:117-134.

[168] Ropers HH & Hamel BC. (2005). X-linked mental retardation. *Nat Rev Genet*, *6*:46-57.

[169] Dewing P, Chiang CW, Sinchak K, Sim H, Fernagut PO, Kelly S, Chesselet MF, Micevych PE, Albrecht KH, Harley VR et al. (2006). Direct regulation of adult brain function by the male-specific factor SRY. *Curr Biol*, *16*:415-420.

[170] Mandel JL & Chelly J. (2004). Monogenic X-linked mental retardation: Is it as frequent as currently estimated? The paradox of the ARX (aristaless X) mutations. *Eur J Hum Genet*, *12*:689-693.

[171] Turner G. (1996). Finding genes on the X chromosome by which Homo may have become sapiens. *Am J Hum Genet*, *58*:1109-1110.

[172] Zechner U, Shi W, Hemberger M, Himmelbauer H, Otto S, Orth A, Kalscheuer V, Fischer U, Elango R, Reis A et al. (2004). Divergent genetic and epigenetic post-zygotic isolation mechanisms in Mus and Peromyscus. *J Evol Biol*, *17*:453-460.

[173] Skuse DH. (2005). X-linked genes and mental functioning. *Hum Mol Genet*, *14*:R27-R32.

[174] Nguyen DK & Disteche CM. (2006). Dosage compensation of the active X chromosome in mammals. *Nat Genet*, *38*:47-53.

[175] Tarpey P, Fryer A, Gecz J, Goodship JA, Partington M, Price S, Schwartz C, Stevenson R, Tolmie JL, Turner G et al. (2005). Prevalence of mutations in X-linked mental retardation genes in familial mental retardation. *In 12th International Workshop on Fragile X and X-linked Mental Retardation; Williamsburg, USA*: 46.

[176] Bonhoeffer T & Yuste R. (2002). Spine motility. Phenomenology, mechanisms, and function. *Neuron*, *35*:1019-1027.

[177] Yuste R & Bonhoeffer T. (2004). Genesis of dendritic spines: insights from ultrastructural and imaging studies. *Nat Rev Neurosci*, *5*:24-34.

[178] Jin P & Warren ST. (2003). New insights into fragile X syndrome: from molecules to neurobehaviors. *Trends Biochem Sci*, *28*:152-158.

[179] McBride SMJ, Choi CH, Wang Y, Liebelt D, Braunstein E, Ferreiro D, Sehgal A, Siwicki KK, Dockendorff TC, Nguyen HT et al. (2005). Pharmacological rescue of

synaptic plasticity, courtship behavior, and mushroom body defects in a Drosophila model of fragile X syndrome. *Neuron*, *45*:753-764.

[180] Landy SJ & Donnai D. (1993). Incontinentia pigmenti (Bloch-Sulzberger syndrome). *J Med Genet*, 30:53-59.

[181] Parrish JE, Scheuerle AE, Lewis RA, Levy ML & Nelson DL. (1996). Selection against mutant alleles in blood leukocytes is a consistent feature in incontinentia pigmenti type 2. *Hum Mol Genet*, *5*:1777-1783.

[182] Cohen PR. (1994). Incontinentia pigmenti: clinicopathologic characteristics and differential diagnosis. *Cutis*, *54*:161-166.

[183] Cohen PR & Kurzrock R. (1995). Miscellaneous genodermatoses: Beckwith-Wiedemann syndrome, Birt-Hogg-Dube syndrome, familial atypical multiple mole melanoma syndrome, hereditary tylosis, incontinentia pigmenti, and supernumerary nipples. *Dermatol Clin*, *13*:211-229.

[184] Goldberg MF & Custis PH. (1993). Retinal and other manifestations of incontinentia pigmenti (Bloch-Sulzberger syndrome). *Ophthalmology*, *100*:1645-1654.

[185] Smahi A et al. (2000). Genomic rearrangement in NEMO impairs NF-kB activation and is a cause of incontinentia pigmenti. The international incontinentia pigmenti (IP) consortium. *Nature*, *405*:466-472.

[186] Jin DY & Jeang KT. (1999) Isolation of full-length cDNA and chromosomal localization of human NF-kB modulator NEMO to Xq28. *J Biomed Sci*, *6*:115-120.

[187] Yamaoka S, Courtois G, Bessia C, Whiteside ST, Weil R, Agou F, Kirk HE, Kay RJ & Israel A. (1998). Complementation cloning of NEMO, a component of the IkB kinase complex essential for NF-kB activation. *Cell*, *93*:1231-1240.

[188] Rothwarf DM, Zandi E, Natoli G & Karin M. (1998). IKK-g is an essential regulatory subunit of the IkB kinase complex. *Nature*, *395*:297-300.

[189] Tak PP & Firestein GS. (2001). NF-kB: a key role in inflammatory diseases. *J Clin Invest*, 107:7-11.

[190] Karin M & Delhase M. (2000). The IkB kinase (IKK) and NF-kB: key elements of proinflammatory signalling. *Semin Immunol*, *12*:85-98.

[191] Baeuerle PA & Baltimore D. (1996). NF-kB: ten years after. *Cell*, *87*:13-20.

[192] Baeuerle PA & Henkel T. (1994). Function and activation of NF-kB in the immune system. *Annu Rev Immunol*, *12*:141-179.

[193] Aradhya S, Woffendin H, Bonnen P, Yamagata T, Heiss NS, Esposito T, Bardaro T, Poustka A, D'Urso M, Kenwrick S et al. (2002). Physical and genetic characterization reveals a pseudogene, an evolutionary junction, and unstable loci in distal Xq28. *Genomics*, *79*:31-40.

[194] Li XH, Fang X & Gaynor RB. (2000). Role of IKKg/NEMO in assembly of the IKK complex. *J Biol Chem*, *276*:4494-4500.

[195] Tarassishin L & Horwitz MS. (2001). Sites on FIP-3 (NEMO/IKKg) essential for its phosphorylation and NF-kB modulating activity. *Biochem Biophys Res Commun*, *285*:555-560.

[196] Jain A, Ma CA, Liu S, Brown M, Cohen J & Strober W. (2001). Specific missense mutations in NEMO result in hyper-IgM syndrome with hypohydrotic ectodermal dysplasia. *Nat Immunol*, *2*:223-228.

[197] Aradhya S, Courtois G, Rajkovic A, Lewis RA, Levy M, Israel A & Nelson DL. (2001). Atypical forms of incontinentia pigmenti in male individuals result from mutations of a cytosine tract in exon 10 of NEMO (IKK-g). *Am J Hum Genet*, *68*: 765-771.

[198] Zonana J, Elder ME, Schneider LC, Orlow SJ, Moss C, Golabi M, Shapira SK, Farndon PA, Wara DW, Emmal SA et al. (2000). A novel X-linked disorder of immune deficiency and hypohidrotic ectodermal dysplasia is allelic to incontinentia pigmenti and due to mutations in IKK-g (NEMO). *Am J Hum Genet*, *67*:1555-1562.

[199] Doffinger R, Smahi A, Bessia C, Geissmann F, Feinberg J, Durandy A, Bodemer C, Kenwrick S, Dupuis-Girod S, Blanche S et al. (2001). X-linked anhidrotic ectodermal dysplasia with immunodeficiency is caused by impaired NF-kB signaling. *Nat Genet*, *27*:277-285.

[200] Kere J, Srivastava AK, Montonen O, Zonana J, Thomas N, Ferguson B, Munoz F, Morgan D, Clarke A, Baybayan P et al. (1996). X-linked anhidrotic (hypohidrotic) ectodermal dysplasia is caused by mutation in a novel transmembrane protein. *Nat Genet*, *13*:409-416.

[201] Karkkainen MJ, Ferrell RE, Lawrence EC, Kimak MA, Levinson KL, McTigue MA, Alitalo K & Finegold DN. (2000). Missense mutations interfere with VEGFR-3 signalling in primary lymphoedema. *Nat Genet*, *25*:153-159.

[202] Hughes AE, Ralston SH, Marken J, Bell C, MacPherson H, Wallace RG, van Hul W, Whyte MP, Nakatsuka K, Hovy L et al. (2000). Mutations in TNFRSF11A, affecting the signal peptide of RANK, cause familial expansile osteolysis. *Nat Genet*, *24*:45-48.

[203] Aradhya S & Nelson DL. (2001). NF-kB signaling and human disease. *Curr Opin Genet Dev*, *11*:300-306.

[204] Aradhya S, Bardaro T, Galgóczy P, Yamagata T, Esposito T, Patlan H, Ciccodicola A, Munnich A, Kenwrick S, Platzer M, D'Urso M & Nelson DL. (2001). Multiple pathogenic and benign rearrangements occur at a 35-kb duplication involving the NEMO and LAGE2 genes. *Hum Mol Genet*, *10*:2557-2567.

[205] Lakich D, Kazazian HH Jr, Antonarakis SE & Gitschier J. (1993). Inversions disrupting the factor VIII gene are a common cause of severe haemophilia A. *Nat Genet*, *5*:236-241.

[206] Lupski JR: Genomic disorders: structural features of the genome can lead to DNA rearrangements and human disease traits. *Trends Genet* 1998, 14:417-422.

[207] Chen YT, Boyer AD, Viars CS, Tsang S, Old LJ & Arden KC. (1997). Genomic cloning and localization of CTAG, a gene encoding an autoimmunogenic cancer-testis antigen NY-ESO-1, to human chromosome Xq28. *Cytogenet Cell Genet*, *79*:237-240.

[208] Eichler EE. (2006). Widening the spectrum of human genetic variation. *Nat Genet*, *38*:9-11.

[209] Wright AF & Shu X. (2007). Focus on Molecules: RPGR. *Eye Res. Jul*;*85(1)*: 1-2.

[210] RetNet (http://www.sph.uth.tmc.edu/Retnet)

[211] Sikkink SK, Biswas S, Parry NR, Stanga PE & Trump D. (2007). X-linked retinoschisis: an update. *J Med Genet. Apr;44(4)*:225-32.

[212] Deeb SS. (2006). Genetics of variation in human color vision and the retinal cone mosaic. *Curr Opin Genet Dev. Jun;16(3)*:301-7.

[213] Carroll J. (2008). Focus on molecules: the cone opsins. *Exp Eye Res. Jun;86(6)*: 865-6.

[214] Waaler GH. (1968). The heredity of normal and defective colour vision. *Acta Ophthalmol (Copenh). 46(1)*:33-40.

[215] Arias S & Rodríguez A. (1972). New families, one with two recombinants for estimation of recombination between the deutan and protan loci. *Humangenetik, 14(4)*: 264-8.

[216] Zatz M, Itskan SB, Sanger R, Frota-Pessoa O & Saldanha PH. (1974). New linkage data for the X-linked types of muscular dystrophy and G6PD variants, colour blindness, and Xg blood groups. *J Med Genet. Dec;11(4)*: 321-7.

[217] Filippi G, Rinaldi A, Palmarino R, Seravalli E & Siniscalco M. (1977). Linkage disequilibrium for two X-linked genes in Sardinia and its bearing on the statistical mapping of the human X chromosome. *Genetics. May;86(1)*: 199-212.

[218] Purrello M, Nussbaum R, Rinaldi A, Filippi G, Traccis S, Latte B & Siniscalco M. (1984). Old and new genetics help ordering loci at the telomere of the human X-chromosome long arm. *Hum Genet., 65(3)*:295-9.

[219] Brennand J, Chinault AC, Konecki DS, Melton DW & Caskey CT. (1982). Cloned cDNA sequences of the hypoxanthine/guanine phosphoribosyltransferase gene from a mouse neuroblastoma cell line found to have amplified genomic sequences. *Proc Natl Acad Sci U S A. Mar;79(6)*:1950-4

[220] Nathans J, Thomas D & Hogness DS. (1986). Molecular genetics of human color vision: the genes encoding blue, green, and red pigments. *Science. Apr 11;232(4747)*: 193-202.

[221] Winderickx J, Sanocki E, Lindsey DT, Teller DY, Motulsky AG & Deeb SS. (1992). Defective colour vision associated with a missense mutation in the human green visual pigment gene. *Nature Genet. 1*: 251-256.

[222] Ueyama H, Kuwayama S, Imai H, Tanabe S, Oda S, Nishida Y, Wada A, Shichida Y & Yamade S. (2002). Novel missense mutations in red/green opsin genes in congenital color-vision deficiencies. *Biochem Biophys Res Commun. Jun 7;294(2)*: 205-9.

[223] Ueyama H, Li YH, Fu GL, Lertrit P, Atchaneeyasakul LO, Oda S, Tanabe S, Nishida Y, Yamade S & Ohkubo I. (2003). An A-71C substitution in a green gene at the second position in the red/green visualpigment gene array is associated with deutan color-vision deficiency. *Proc Natl Acad Sci USA, 100*:3357-3362.

[224] Grover S, Fishman GA, Anderson RJ, Tozatti MS, Heckenlively JR, Weleber RG, Edwards AO & Brown J Jr. (1999). Visual acuity impairment in patients with retinitis pigmentosa at age 45 years or older. *Ophthalmology. 106(9)*:1780-5.

[225] Bird AC. (1987). Clinical investigation of retinitis pigmentosa. *Prog Clin Biol Res., 247*:3-20.

[226] Wang MX, Sando RS Jr, Crandall AS & Donoso LA. (1995). Recent advances in the molecular genetics of retinitis pigmentosa. *Curr Opin Ophthalmol., 6(3)*:1-7.

[227] Rivolta C, Sharon D, DeAngelis MM & Dryja TP. (2002). Retinitis pigmentosa and allied diseases: numerous diseases, genes, and inheritance patterns. *Hum Mol Genet., 11(10)*:1219-27.

[228] Musarella MA. (1990). Mapping of the X-linked recessive retinitis pigmentosa gene. A review. *Ophthalmic Paediatr Genet., 11(2)*:77-88.

[229] Wright AF. (1990). Towards the identification of genes in X-linked retinitis pigmentosa. In NN Osborne and GJ Chader (eds.), *Progress in Retinal Research*, Vol. 9, Pergamon Press, Oxford , pp. 197-227.

[230] Fishman GA, Farber MD & Derlacki DJ. (1988). X-linked retinitis pigmentosa. Profile of clinical findings. *Arch Ophthalmol., 106(3)*:369-75.

[231] Cideciyan AV & Jacobson SG. (1994). Image analysis of the tapetal-like reflex in carriers of X-linked retinitis pigmentosa. *Invest Ophthalmol Vis Sci., 35(11)*: 3812-24.

[232] Meindl A, Dry K, Herrmann K, Manson F, Ciccodicola A, Edgar A, Carvalho MR, Achatz H, Hellebrand H, Lennon A, Migliaccio C, Porter K, Zrenner E, Bird A, Jay M, Lorenz B, Wittwer B, D'Urso M, Meitinger T & Wright A. (1996). A gene (RPGR) with homology to the RCC1 guanine nucleotide exchange factor is mutated in X-linked retinitis pigmentosa (RP3). *Nat Genet., 13(1)*:35-42.

[233] Roepman R, van Duijnhoven G, Rosenberg T, Pinckers AJ, Bleeker-Wagemakers LM, Bergen AA, Post J, Beck A, Reinhardt R, Ropers HH, Cremers FP & Berger W. (1996). Positional cloning of the gene for X-linked retinitis pigmentosa 3: homology with the guanine-nucleotide-exchange factor RCC1. *Hum Mol Genet., 5(7)*: 1035-41.

[234] Schwahn U, Lenzner S, Dong J, Feil S, Hinzmann B, van Duijnhoven G, Kirschner R, Hemberger M, Bergen AA, Rosenberg T, Pinckers AJ, Fundele R, Rosenthal A, Cremers FP, Ropers HH & Berger W. (1998). Positional cloning of the gene for X-linked retinitis pigmentosa 2. *Nat Genet., 19(4)*:327-32.

[235] Melamud A, Shen GQ, Chung D, Xi Q, Simpson E, Li L, Peachey NS, Zegarra H, Hagstrom SA, Wang QK & Traboulsi EI. (2006). Mapping a new genetic locus for X linked retinitis pigmentosa to Xq28. *J Med Genet., 43(6)*:e27.

[236] Hardcastle AJ, Thiselton DL, Zito I, Ebenezer N, Mah TS, Gorin MB & Bhattacharya SS. (2000). Evidence for a new locus for X-linked retinitis pigmentosa (RP23). *Invest Ophthalmol Vis Sci., 41(8)*:2080-6.

[237] Gieser L, Fujita R, Göring HH, Ott J, Hoffman DR, Cideciyan AV, Birch DG, Jacobson SG & Swaroop A. (1998). A novel locus (RP24) for X-linked retinitis pigmentosa maps to Xq26-27. *Am J Hum Genet., 63(5)*:1439-47.

[238] Bhattacharya SS, Wright AF, Clayton JF, Price WH, Phillips CI, McKeown CM, Jay M, Bird AC, Pearson PL, Southern EM, et al. (1984). Close genetic linkage between X-linked retinitis pigmentosa and a restriction fragment length polymorphism identified by recombinant DNA probe L1.28. *Nature. May 17-23;309(5965)*:253-5.

[239] Musarella MA, Burghes A, Anson-Cartwright L, Mahtani MM, Argonza R, Tsui LC & Worton R. (1988). Localization of the gene for X-linked recessive type of retinitis pigmentosa (XLRP) to Xp21 by linkage analysis. *Am J Hum Genet., 43(4)*:484-94.

[240] Ott J, Bhattacharya S, Chen JD, Denton MJ, Donald J, Dubay C, Farrar GJ, Fishman GA, Frey D, Gal A, et al. (1990). Localizing multiple X chromosome-linked retinitis pigmentosa loci using multilocus homogeneity tests. *Proc Natl Acad Sci U. S A., 87(2)*:701-4.

[241] Inglehearn CF, Tarttelin EE, Keen TJ, Bhattacharya SS, Moore AT, Taylor R & Bird AC. (1998). A new dominant retinitis pigmentosa family mapping to the RP18 locus on chromosome 1q11-21. *J Med Genet., 35(9)*:788-9.

[242] Vervoort R, Lennon A, Bird AC, Tulloch B, Axton R, Miano MG, Meindl A, Meitinger T, Ciccodicola A & Wright AF. (2000). Mutational hot spot within a new RPGR exon in X-linked retinitis pigmentosa. *Nat Genet. Aug;25(4)*:462-6.

[243] Breuer DK, Yashar BM, Filippova E, Hiriyanna S, Lyons RH, Mears AJ, Asaye B, Acar C, Vervoort R, Wright AF, Musarella MA, Wheeler P, MacDonald I, Iannaccone A, Birch D, Hoffman DR, Fishman GA, Heckenlively JR, Jacobson SG, Sieving PA & Swaroop A. (2002). A comprehensive mutation analysis of RP2 and RPGR in a North American cohort of families with X-linked retinitis pigmentosa. *Am J Hum Genet., 70(6)*:1545-54.

[244] Sharon D, Sandberg MA, Rabe VW, Stillberger M, Dryja TP & Berson EL. (2003). RP2 and RPGR mutations and clinical correlations in patients with X-linked retinitis pigmentosa. *Am J Hum Genet., 73(5)*:1131-46.

[245] Neidhardt J, Glaus E, Lorenz B, Netzer C, Li Y, Schambeck M, Wittmer M, Feil S, Kirschner-Schwabe R, Rosenberg T, Cremers FP, Bergen AA, Barthelmes D, Baraki H, Schmid F, Tanner G, Fleischhauer J, Orth U, Becker C, Wegscheider E, Nürnberg G, Nürnberg P, Bolz HJ, Gal A & Berger W. (2008). Identification of novel mutations in X-linked retinitis pigmentosa families and implications for diagnostic testing. *Mol Vis., 14*:1081-93.

[246] Ferreira PA. (2005). Insights into X-linked retinitis pigmentosa type 3, allied diseases and underlying pathomechanisms. *Hum Mol Genet., 14 Spec No. 2*:R259-67.

[247] Francke U. Ochs HD, de Martinville B, Giacalone J, Lindgren V, Disteche C, Pagon RA, et al. (1985). Minor Xp21 chromosome deletion in a male associated with expression of Duchenne muscular dystrophy, chronic granulomatous disease, retinitis pigmentosa, and McLeod syndrome. *Am J Hum Genet 37*:250-267

[248] Musarella MA, Anson-Cartwright L, McDowell C, Burghes A, Coulson SE, Worton RG & Rommens JM. (1991). Physical mapping at a potential X-linked retinitis pigmentosa locus (RP3) by pulsed field gel electrophoresis, *Genomics, 11(2)*:263-272.

[249] Fujita R, Bingham E, Forsythe P, McHenry C, Aita V, Navia BA, Dry K, Segal M, Devoto M, Bruns G, Wright AF, Ott J, Sieving PA & Swaroop A. (1996). A recombination outside the BB deletion refines the location of the X linked retinitis pigmentosa locus RP3. *Am J Hum Genet. Jul;59(1)*:152-8.

[250] Hong DH, Pawlyk B, Sokolov M, Strissel KJ, Yang J, Tulloch B, Wright AF, Arshavsky VY & Li T. (2003). RPGR isoforms in photoreceptor connecting cilia and the transitional zone of motile cilia. *Invest Ophthalmol Vis Sci., Jun;44(6)*: 2413-21.

[251] Boylan JP & Wright AF. (2000). Identification of a novel protein interacting with RPGR. *Hum Mol Genet. Sep 1;9(14)*:2085-93.

[252] Zhao Y, Hong DH, Pawlyk B, Yue G, Adamian M, Grynberg M, Godzik A & Li T. (2003). The retinitis pigmentosa GTPase regulator (RPGR)- interacting protein: subserving RPGR function and participating in disk morphogenesis. *Proc Natl Acad Sci U S A. Apr 1;100(7)*:3965-70.

[253] Gerber S, Perrault I, Hanein S, Barbet F, Ducroq D, Ghazi I, Martin-Coignard D, Leowski C, Homfray T, Dufier JL, Munnich A, Kaplan J & Rozet JM. (2001). Complete exon-intron structure of the RPGR-interacting protein (RPGRIP1) gene allows the identification of mutations underlying Leber congenital amaurosis. *Eur J Hum Genet. Aug;9(8)*:561-71.

[254] Miano MG, Valverde D, Solans T, Grammatico B, Migliaccio C, Cirigliano V, DeBernardo C, Ventruto V, Meitinger T, Wright A, Del Porto G, Baiget M, D'Urso M & Ciccodicola A. (1998). Two novel mutations in the retinitis pigmentosa GTPase regulator (RPGR) gene in X-linked retinitis pigmentosa (RP3). Mutations in brief no. 172. Online. *Hum Mutat., 12(3)*:212-3.

[255] Miano MG, Testa F, Strazzullo M, Trujillo M, De Bernardo C, Grammatico B, Simonelli F, Mangino M, Torrente I, Ruberto G, Beneyto M, Antinolo G, Rinaldi E, Danesino C, Ventruto V, D'Urso M, Ayuso C, Baiget M & Ciccodicola A. (1999). Mutation analysis of the RPGR gene reveals novel mutations in south European patients with X-linked retinitis pigmentosa. *Eur J Hum Genet. Sep;7(6)*: 687-94.

[256] Neidhardt J, Glaus E, Barthelmes D, Zeitz C, Fleischhauer J & Berger W. (2007). Identification and characterization of a novel RPGR isoform in human retina. *Hum Mutat. Aug;28(8)*: 797-807.

[257] Hong DH & Li T. (2002). Complex expression pattern of RPGR reveals a role for purine-rich exonic splicing enhancers. *Invest Ophthalmol Vis Sci., Nov;43(11)*: 3373-82.

[258] Vervoort R & Wright. (2002). AF. Mutations of RPGR in X-linked retinitis pigmentosa (RP3). *Hum Mutat. May;19(5)*:486-500.

[259] Miano MG, Testa F, Filippini F, Trujillo M, Conte I, Lanzara C, Millán JM, De Bernardo C, Grammatico B, Mangino M, Torrente I, Carrozzo R, Simonelli F, Rinaldi E, Ventruto V, D'Urso M, Ayuso C & Ciccodicola A. (2001). Identification of novel RP2 mutations in a subset of X-linked retinitis pigmentosa families and prediction of new domains. *Hum Mutat. Aug;18(2)*:109-19.

[260] Tian G, Bhamidipati A, Cowan NJ & Lewis SA. (1999). Tubulin folding cofactors as GTPase-activating proteins. GTP hydrolysis and the assembly of the alpha/beta-tubulin heterodimer. *J Biol Chem. Aug 20;274(34)*:24054-8.

[261] Veltel S, Gasper R, Eisenacher E & Wittinghofer A. (2008). The retinitis pigmentosa 2 gene product is a GTPase-activating protein for Arf-like 3. *Nat Struct Mol Biol. Apr;15(4)*:373-80.

[262] Schwahn U, Paland N, Techritz S, Lenzner S & Berger W. (2001). Mutations in the X-linked RP2 gene cause intracellular misrouting and loss of the protein. *Hum Mol Genet. May 15;10(11)*:1177-83.

[263] De La Chappelle A, Alitalo T & Forsius H. (1994). X-linked juvenile retinoschisis. In: Wright AF, Jay B, eds. *Molecular Genet Inherited Eye Disorders*. Switzerland: Harwood Academic Publishers, 339-57.

[264] Yassur Y, Nissenkorn I, Ben-Sira I, Kaffe S & Goodman RM. (1982). Autosomal dominant inheritance of retinoschisis. *Am J Ophthalmol. Sep;94(3)*:338-43.

[265] Gieser and Falls. (1961). Hereditary retinoschisis. *Am J Ophthalmol. Jun;51*:1193-200.

[266] Arden GB, Gorin MB, Polkinghorne PJ, Jay M & Bird AC. (1988). Detection of the carrier state of X-linked retinoschisis. *Am J Ophthalmol. Jun 15;105(6)*:590-5.

[267] Haas J. (1898). Ueber das Zusammen vorkommen von Veranderungen der Retina und Choroidea. *Arch Augenheilkd, 37*:343-8.

[268] Pagenstecher H. (1913). Uebereine unterdem Bildeder Natzhauterblosung verlaufende, erbiche Erkankung der Retina. *Graefes Arch Clin Exp Ophthalmol, 86*:457-62.

[269] Thomson E. (1932). Memorandum regarding a family in which neuroretinal disease of an unusal kind occurred in only males. *Brit J Ophthalmol, 16*:681-6.

[270] Mann I & Macrae A. (1938). Congenital vascular veils in the vitreous. *Brit J Ophthalmol, 22*:1-10.

[271] Jager GM. (1953). A hereditary retinal disease. *Trans Ophthalmol Soc UK, 73*:617-619.

[272] Kaplan J, Pelet A, Hentati H, Jeanpierre M, Briard ML, Journel H, Munnich A & Dufier JL. (1991). Contribution to carrier detection and genetic counselling in X linked retinoschisis. *J Med Genet. Jun;28(6)*:383-8.

[273] Newton MS, Collyer S & Phillips CI (1991). Male proband with X linked retinoschisis apparently inherited from his father's family. *J Med Genet. Mar;28(3)*:213-4.

[274] Yanoff M, Kertesz Rahn E & Zimmerman LE. (1968). Histopathology of juvenile retinoschisis. *Arch Ophthalmol. Jan;79(1)*:49-53.

[275] Forsius H, Eriksson A & Vainio-Mattila B. (1963). sex-related hereditary Retinoschisis in 2 families in Finland. *Klin Monatsbl Augenheilkd., 143*:806-16.

[276] Sorsby A, Klein M, Gann JH & Siggins G. (1951). Unusual retinal detachment possibly sex-linked. *Br J Ophthalmol. Jan;35(1)*:1-10.

[277] Sauer CG, Gehrig A, Warneke-Wittstock R, Marquardt A, Ewing CC, Gibson A, Lorenz B, Jurklies B & Weber BH. (1997). Positional cloning of the gene associated with X-linked juvenile retinoschisis. *Nat Genet. Oct;17(2)*:164-70.

[278] The Retinoschisis Consortium (http://www.dmd.nl/rs/index.html).

[279] Wang T, Zhou A, Waters CT, O'Connor E, Read RJ & Trump D. (2006). Molecular pathology of X linked retinoschisis: mutations interfere with retinoschisin secretion and oligomerisation. *Br J Ophthalmol. Jan;90(1)*:81-6.

[280] Grayson C, Reid SN, Ellis JA, Rutherford A, Sowden JC, Yates JR, Farber DB & Trump D. (2000). Retinoschisin, the X-linked retinoschisis protein, is a secreted photoreceptor protein, and is expressed and released by Weri-Rb1 cells. *Hum Mol Genet, 9*:1873-9.

[281] Eksandh LC, Ponjavic V, Ayyagari R, Bingham EL, Hiriyanna KT, Andreasson S, Ehinger B & Sieving PA. (2000). Phenotypic expression of juvenile X-linked retinoschisis in Swedish families with different mutations in the XLRS1 gene. *Arch Ophthalmol, 118*:1098-104.

[282] Pimenides D, George ND, Yates JR, Bradshaw K, Roberts SA, Moore AT & Trump D. (2005). X-linked retinoschisis: clinical phenotype and RS1 genotype in 86 UK patients. *J Med Genet, 42*:e35.

[283] Simonelli F, Cennamo G, Ziviello C, Testa F, de Crecchio G, Nesti A, Manitto MP, Ciccodicola A, Banfi S, Brancato R & Rinaldi E. (2003). Clinical features of X linked juvenile retinoschisis associated with new mutations in the XLRS1 gene in Italian families. *Br J Ophthalmol, 87*:1130-4.

[284] Wu WW, Wong JP, Kast J & Molday RS. (2005). RS1, a discoidin domain-containing retinal cell adhesion protein associated with X-linked retinoschisis, exists as a novel disulfide-linked octamer. *J Biol Chem*, *280*:10721-30.
[285] Iannaccone A, Mura M, Dyka FM, Ciccarelli ML, Yashar BM, Ayyagari R, Jablonski MM & Molday RS. (2006). An unusual X-linked retinoschisis phenotype and biochemical characterization of the W112C RS1 mutation. *Vision Res*, *46*:3845-52.
[286] Jablonski MM, Dalke C, Wang X, Lu L, Manly KF, Pretsch W, Favor J, Pardue MT, Rinchik EM, Williams RW, Goldowitz D & Graw J. (2005). An ENU-induced mutation in Rs1h causes disruption of retinal structure and function. *Mol Vis*, *11*: 569-81.
[287] Zeng Y, Takada Y, Kjellstrom S, Hiriyanna K, Tanikawa A, Wawrousek E, Smaoui N, Caruso R, Bush RA & Sieving PA. (2004). RS-1 Gene delivery to an adult Rs1h knockout mouse model restores ERG b-wave with reversal of the electronegative waveform of X-linked retinoschisis. *Invest Ophthalmol Vis Sci*, *45(9)*:3279-85.
[288] Weber BH, Schrewe H, Molday LL, Gehrig A, White KL, SeeligerMW, Jaissle GB, Friedburg C, Tamm E & Molday RS. (2002). Inactivation of the murine X-linked juvenile retinoschisis gene, Rs1h, suggests a role of retinoschisin in retinal cell layer organization and synaptic structure. *Proc Natl Acad Sci USA*, *99*: 6222-7.
[289] Min SH, Molday LL, Seeliger MW, Dinculescu A, Timmers AM, Janssen A, Tonagel F, Tanimoto N, Weber BH, Molday RS & Hauswirth WW. (2005). Prolonged recovery of retinal structure/function after gene therapy in an Rs1h-deficient mouse model of x-linked juvenile retinoschisis. *Mol Ther*. *Oct;12(4)*:644-51.
[290] Goedbloed J. (1942). Mode of inheritance in choroideremia, *Ophthalmologica 104*, pp. 308-315.
[291] Waardenburg, PJ. (1942). Choroideremia als Erbmerkmal, *Acta Ophthalmol. 20*, pp. 235-274.
[292] Cremers FPM, van de Pol DJ, van Kerkhoff LP, Wieringa B & Ropers HH. (1990). Cloning of a gene that is rearranged in patients with choroideremia, *Nature 347*, pp. 674-677.
[293] Ohba N & Isashiki Y. (2000). Clinical and genetic features of choroideremia. *Jpn J Ophthalmol*. *May;44(3)*:317.
[294] Cremers FPM, van de Pol TJR, Diergaarde PJ, Wieringa B, Nussbaum RL, Schwartz M, Ropers H-H. (1989). Physical fine mapping of the choroideremia locus using Xq21 deletions associated with complex syndromes. *Genomics 4*: 41-46.
[295] Heckenlively JR & Bird AC. (1988). Choroideremia. In Heckenlively JR (ed): Retinitis Pigmentosa. *Philadelphia: JB Lippincott*, pp 176-187.
[296] Lesko JG, Lewis RA & Nussbaum RL. (1987). Multipoint linkage analysis of loci in the proximal long arm of the human X chromosome: Application to mapping the choroideremia locus. *Am J Hum Genet 40*:303-311.
[297] Fraser GRF & Friedmann AI. (1968). Choroideremia in a female. *BMJ 2*:732-734.
[298] Harris GS & Miller JR. (1968). Choroideremia. Visual defects in a heterozygote. *Arch Ophthalmol 80*:423-429.

[299] Kaplan J, Gilgenkrantz S, Dufier JL & Frézal J. (1989). Choroideremia and ovarian dysgenesis associated with an X;7 de novo, balanced translocation. Human Gene Mapping 10: Tenth International Workshop on Human Gene Mapping. *Cytogenet Cell Genet 51*:1022.

[300] Siu VM, Gonder JR, Jung JH, Sergovich FR & Flintoff WF. (1990). Choroideremia associated with an X-autosomal translocation. *Hum Genet 84*:459-464.

[301] Lewis RA, Nussbaum RL & Ferrell R. (1985). Mapping X-linked ophthalmic diseases: Provisional assignment of the locus for choroideremia to Xq13-q24. *Ophthalmology 92*:800-806.

[302] Nussbaum RL, Lewis RA, Lesko JG, Ferrell R. (1985). Choroideremia is linked to the restriction fragment length polymorphism DXYS1 at Xq13-21. *Am J Hum Genet 37*:473-481.

[303] Sankila E-M, Lehner T, Eriksson AW, Forsius H, Kärnä J, Page D, Ott J & de la Chapelle A. (1989). Haplotype and multipoint linkage analysis in Finnish choroideremia families. *Hum Genet 84*:66-70.

[304] Gal A, Brunsmann F, Hogenkamp D, Rüther K, Ahlert D, Wienker TF, Hammerstein W & Pawlowitzki IH. (1986). Choroideremia-locus maps between DXS3 and DXS11 on Xq. *Hum Genet. Jun;73(2)*:123-6.

[305] Rosenberg T, Niebuhr E, Yang H-M, Parving A & Schwartz M. (1987). Choroideremia, congenital deafness and mental retardation in a family with an X chromosomal deletion. *Ophthalm Paediatr Genet 8*:139-143.

[306] Schwartz M, Yang H-M, Niebuhr E, Rosenberg T & Page DC. (1988). Regional localization of polymorphic DNA loci on the proximal long arm of the X chromosome using deletions associated with choroideremia. *Hum Genet 78*: 156-160.

[307] Cremers FPM, Brunsmann F, van de Pol TJR, Pawlowitzki IH, Paulsen K, Wieringa B & Ropers H-H. (1987). Deletion of the DXS165 locus in patients with classical choroideremia. *Clin Genet 32*:421-423.

[308] Cremers FPM, van de Pol TJR, Wieringa B, Collins FS, Sankila EM, Siu VM, Flintoff WF, Brunsmann F, Blonden LAJ & Ropers H-H. (1989b). Chromosomal jumping from the DXS165 locus allows molecular characterization of four microdeletions and a de novo chromosome X/13 translocation associated with choroideremia. *Proc Natl Acad Sci USA 86*:7510-7514.

[309] Cremers FPM, Brunsmann F, Berger W, van Kerkhoff EPM, van de Pol TJR, Wieringa. B, Pawlowitzki IH & Ropers H-H. (1990a). Cloning of the breakpoints of a deletion associated with choroideremia. *Hum Genet 86*:61-64.

[310] Cremers FPM, Sankila E-M, Brunsmann F, Jay M, Jay B, Wright A, Pinckers AJLG, Schwartz M, van de Pol TJR, Wieringa. B, de la Chapelle A, Pawlowitzki IH & Ropers H-H. (1990b). Deletions in patients with classical choroideremia vary in size from 45 kb to several megabases. *Am J Hum Genet 47*:622-628.

[311] Merry DE, Jänne PA, Landers JE, Lewis RA & Nussbaum RL. (1992). Isolation of a candidate gene for choroideremia. *Proc Natl Acad Sci USA 89*:2135-2139.

[312] van Bokhoven H, van den Hurk JAJM, Bogerd L, Philippe C, Gilgenkrantz S, de Jong P, Ropers H-H & Cremers FPM. (1994). Cloning and characterizaton of the human choroideremia gene. *Hum Mot Genet 3*:1041-1046.

[313] Seabra MC, Goldstein JL, Südhof TC & Brown MS. (1992b). Rab geranylgeranyl transferase: A multisubunit enzyme that prenylates GTP-binding proteins terminating in Cys-X-Cys or CysCys. *J Biol Chem 267*:14497-14503.

[314] Seabra MC, Brown MS, Goldstein JL. (1993). Retinal degeneration in choroideremia: Deficiency of Rab geranylgeranyl transferase. *Science 259*: 377-381.

[315] Andres DA, Seabra MC, Brown MS, Armstrong SA, Smeland TE, Cremers FPM, Goldstein JL. (1993). cDNA cloning of component A of Rab geranylgeranyl transferase and demonstration of its role as a Rab escort protein. *Cell 73*: 1091-1099.

[316] Alexandrov K, Horiuchi H, Steele-Mortimer O, Seabra MC & Zerial M. (1994). Rab escort protein-1 is a multifunctional protein that accompanies newly prenylated rab proteins to their target membranes. *EMBO J 13*:5262-5273.

[317] Waldherr M, Ragnini A, Schweyen RJ & Boguski MS. (1993). MRS6–yeast homologue of the choroideraemia gene. *Nature Genet 3*:193-194.

[318] Jiang Y & Ferro-Novick S. (1994). Identification of yeast component A: Reconstitution of the geranylgeranyltransferase that modifies Ypt1p and Sec4p. *Proc Natl Acad Sci USA 91*:4377-4381.

[319] Jiang Y, Rossi G & Ferro-Novick S. (1993). Bet2p and Mad2p are components of a prenyltransferase that adds geranylgeranyl onto Ypt1p and Sec4p. *Nature 366*: 84-86.

[320] Benito-Moreno RM, Miaczynska M, Bauer BE, Schweyen RJ & Ragnini A. (1994). Mrs6p, the yeast homologue of the mammalian choroideraemia protein: Immunological evidence for its function as the Ypt1p Rab escort protein. *Curr Genet 27*:23-25.

[321] Fujimura K, Tanaka K, Nakano A & Toh-e A. (1994). The Saccharomyces cerevisiae MS14 gene encodes the yeast counterpart of component A of Rab geranylgeranyltransferase. *J Biol Chem 269*:9205-9212.

[322] Ragnini A, Teply R, Waldherr M, Voskova A & Schweyen RJ. (1994). The yeast protein Mrs6p, a homologue of the rabGDI and human choroideraemia proteins, affects cytoplasmic and mitochondrial functions. *Curr Genet 26*:308-314.

[323] Cremers FPM, Armstrong SA, Seabra MC, Brown MS & Goldstein JL. (1994). REP-2, a Rab escort protein encoded by the choroideremia-like gene. *J Biol Chem 269*:2111-2117.

[324] Teboul M, Mujica P, Chery M, Leotard B & Gilgenkrantz S. (1989). Translocations X-autosomes équilibrées et retard mental. Contribution à la cartographie des retards mentaux liés à L'X (à l'exclusion de L'X-FRA). *J Genet Hum 37*:179-195.

[325] van den Hurk JA, Schwartz M, van Bokhoven H, van de Pol TJ, Bogerd L, Pinckers AJ, Bleeker-Wagemakers EM, Pawlowitzki IH, Rüther K, Ropers HH & Cremers FP. (1997). Molecular basis of choroideremia (CHM): mutations involving the Rab escort protein-1 (REP-1) gene. *Hum Mutat., 9(2)*:110-7.

In: The Human Genome: Features, Variations...
Editor: Akio Matsumoto and Mai Nakano

ISBN: 978-1-60741-695-1

Chapter 3

Role of Extrachromosomal Elements in HL-60 Human Leukemia Cells

Tetsuo Hirano [a] ***and Kazunari K. Yokoyama*** [b, c]

[a] Life Science Group, Graduate School of Integrated Arts and Sciences, Hiroshima University, 1-7-1 Kagamiyama, Higashihiroshima, Hiroshima 739-8521, Japan

[b] Gene Engineering Division, RIKEN BioResource Center, 3-1-1 Koyadai, Tsukuba, Ibaraki 305-0042, Japan

[c] Center of Excellence for Environmental Medicine, Graduate Institute of Medicine, Kaohsiung Medical University, 100 Shih-Chuan 1st Road, San Ming District, Kaohsiung 80708, Taiwan

Abstract

Gene amplification is a cytogenetic abnormality frequently observed in cancer cells. It often occurs as a double minute chromosomes (dmin) consisting of a repeat of unit (amplicon) whose origin stems from a different chromosome. The number of dmins in each cell is unstable because they do not contain the centromeres. We found a large extrachromosomal element (LEE) that is a novel form of gene amplification in human myeloid leukemia HL-60 cells. In fact, HL-60 cells contained dmins, on which the *MYC* oncogene was amplified at early passages of cell culture; however, cells harboring LEEs emerged during long continuous culture, and finally replaced dmin-containing cells in late passages. Although the LEE resembled dmin in that it lacked alphoid sequence, it persisted in cells stably, unlike what is observed for dmin. The stabilization of the extrachromosomal element is believed to be involved in the progress of malignancy in many cancers. Thus, we think that LEEs provide a good experimental model to study the potential relationship between the stabilization of extrachromosomal elements and the progression of cancers. Here we summarize the molecular nature of LEE and its possible therapeutic use of LEEs for neoplastic cells.

Introduction

It is widely accepted that genomic instability is a major driving process that promotes initiation and progression of various cancers by generating cytological abnormalities that include not only chromosome number (aneuploidy) but also aberrant chromosomes and gene amplification [1]. Gene amplification is believed to promote malignancy in cancer cells by increasing the copy number of certain oncogenes and sometimes by producing unusual fusion genes during the progression of the neoplastic cancer cells [2, 3 and references therein]. Gene amplifications are often observed as visible structures (e.g., dmins and homogeneously stained regions (HSRs)), which consist of a repetitive DNA unit, or amplicon. A dmin is an independent structure, which looks like a tiny acentric chromosome, whereas an HSR is embedded in other chromosome [7, 8]. The number of dmins in each cell is unstable because of the lack of centromeres, which ensure even distribution to daughter cells during cell division. By contrast, HSRs are stably transmitted through mitosis, similarly to normal chromosomes [4]. A dmin is thought to be initiated as a fine and invisible structure produced by improper genomic replication or misrepair of a DNA lesion [5, 6]. The breakage-fusion-breakage cycle is a compelling model for the generation and expansion of gene amplification [7, 8]. This model provides a good explanation for the inverted repeats frequently observed in amplicons. In this case, double-strand breaks (DSB) are necessary for the initiation of gene amplification. Thus, a failure in the mechanism that checks DSBs may be a cause of gene amplification [9, 10]. DNA duplication generated by irrelevant intrachromosomal recombination can also initiate abnormal amplification. Once formed, the duplicated DNA sequence can expand as tandem repeats by sequential homologous recombination [3].

Deregulated expression of the *MYC* oncogene contributes to genomic instability [11, 12]. Forced expression of *MYC* reportedly produces temporal gene amplification [13]. The generation of extrachromosomal elements is induced by forced expression of c-*myc* in murine plasmacytomas [12, 14]. The stability of episomal extrachromosomal vectors based on the Epstein-Barr virus depends on the expression of c-*myc* [16]; however, these extrachromosomal elements regress when the expression of c-*myc* halts [16, 17]. Moreover, the long-term persistence of extrachromosomal elements requires a disorder in the mechanism of verification of the progression of mitosis, which includes p53 and its related protein
[15, 18, 19]. p53 undoubtedly participates in the genomic instability observed in leukemia and other tumors [20, 21]. Finally, the stabilization of extrachromosomal elements sometimes requires the scaffold/matrix attachment (S/MAR) as a cis-element [22]. It should be noted that the mechanism of generation and maintenance of gene amplification may vary according to type of cancer [3].

Large Extrachromosomal Elements (LEEs)

It is well known that HL-60 acute myeloid leukemia cells contain dmins, on which the *MYC* oncogene amplifies [5]. The number of dmins in a cell is not constant, as dmins do not have a centromere, which is necessary for the even segregation to daughter cells during cell division [23]. Because each dmin enters the daughter cell in a stochastic manner, theoretically the number of dmins should follow a binomial distribution. In fact, the distribution of the number of dmins in HL-60 cells was consistent with this theory [24, and our unpublished observations]. Up to 5% of dmin-containing HL-60 cells in culture displayed arrested growth and spontaneous differentiation. The number of copies of the *MYC* gene in these cells was significantly reduced. This observation suggests that the cells in which the number of dmins is below a certain threshold in the distribution, stop dividing [23-25].

Besides unequal distribution, hydroxyurea, which is an anticancer drug, also reduces the number of dmins by inducing micronuclei, which trap extrachromosomal elements and expel them from the cell [24]. In this situation, the population of terminally differentiated cells showing arrested division increased in the culture [24]. The organic solvent, dimethylsulfoxide (DMSO), is also able to induce the micronuclei-mediated exclusion of extrachromosomal elements [24]. In the presence of the drug, cells in which the number of dmins is reduced to a certain level are terminally differentiated and are maintained in quiescence thereafter, even when the drug is removed from the culture medium [24]. By contrast, cells containing many dmins that have not yet been excluded begin to divide without delay after removal of DMSO from the culture medium. These results suggest that a certain level of expression of the gene(s) encoded on extrachromosomal elements must be important for the maintenance of the growing state of HL-60 cells [26]. Interestingly, the population number of the spontaneously differentiating cells decreased after prolonged culture and fell to zero at approximately passage 100 [23]. By this late passage stage, all cells in culture had become resistant to the terminal differentiation induced by DMSO, and immediately resumed their growth after removal of the drug [23, 24]. Karyotype analysis revealed the absence of classical dmins and the existence of extrachromosomal structures larger than dmins in these cells. The *MYC* gene located on dmins in early-passage cells had relocated to the larger structures (Fig. 1) [24].

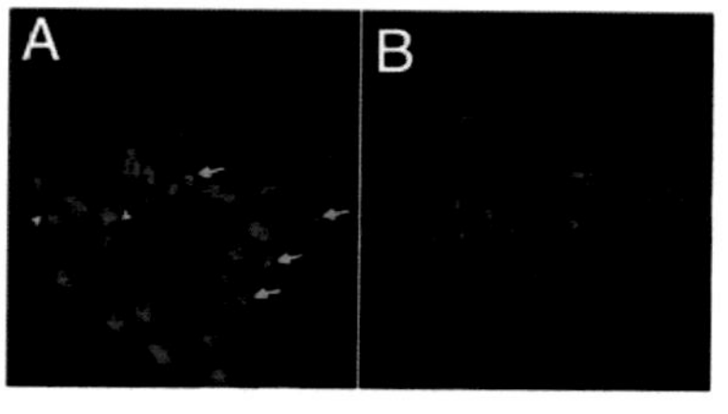

Figure 1. FISH images of high-passage (A) and low-passage (B) HL-60 cells using a *MYC* probe. Arrows indicate four LEEs and the arrowhead on chromosome 8 indicates the *MYC* locus. Many extrachromosomal dmins were seen in the low-passage cells.

We termed these structures "Large Extrachromosomal Elements" (LEEs). A minor population of cells harboring LEEs appeared at middle passages (30-40) but finally completely replaced dmin-containing cells in late passages (> 100). Two or four (typically four) LEEs were observed per cell in late-passage HL-60 cultures. The length of LEEs was microscopically estimated as 1/3 of that of the group G chromosomes (chromosomes 21 and 22). We did not detect an alphoid sequence in LEEs, which is necessary to construct centromeres in normal cells and in dmins, by using a specific probe [24]. This feature of LEEs definitely demonstrated that they differ from homogeneous stained regions (HSRs), which are embedded in chromosomes containing a centromere. LEEs were stably transmitted to the daughter cells after mitosis and were not excluded via the action of micronuclei. For this reason, late-passage HL-60 cells no longer spontaneously and terminally differentiate after DMSO treatment [24, 27].

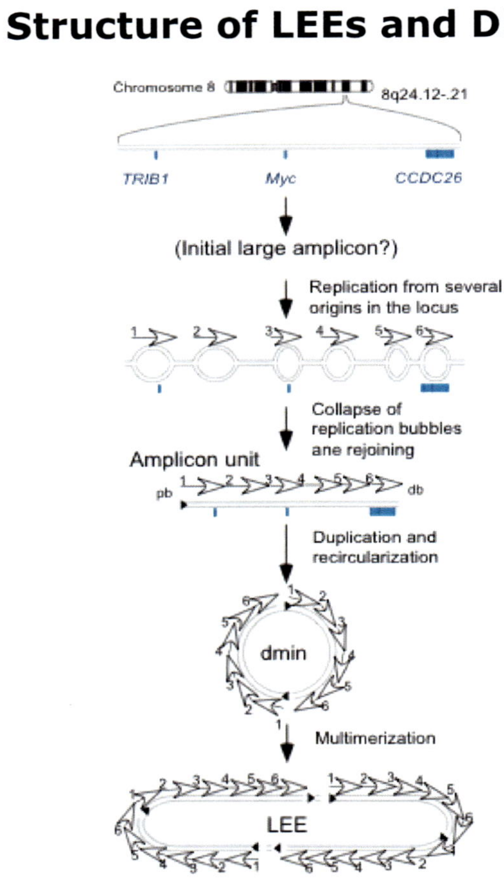

Figure 2. Model for the generation of the discontinuous amplicon unit of dmins and LEEs in HL-60 cells. Replication bubbles initiated from multiple origins on 8q24 locus rejoin to form an amplicon unit. Locations of the three genes (*TRIB1*, *MYC* and *CCDC26*) are indicated. Excision of an initial large amplicon may precede replication. The arrows numbered 1 through 6 indicate amplified regions separated with gaps. Note that the lengths of the various regions are not to scale. Once the amplicon unit has formed, its multimerization results in larger structures that include dmins and LEEs.

We isolated and analyzed LEEs by the method of chromosome sorting using a flow cytometer, to thoroughly clarify the nature of their amplicon. A shotgun library was obtained from the chromosomal fraction. Comparison with the human genome database showed that many clones from the library included sequences corresponding to the 4.4 Mb area of the human 8q24 locus [26]. The 4.4 Mb range coincided with the amplicon reported in acute myeloid leukemia (AML) and myelodysplastic syndrome (MDS) patients [28, 29]. Further fluorescence in situ hybridization (FISH) analysis using human BAC clones and quantitative PCR revealed that the amplicon of LEE and dmin was composed of gapped segments of the 8q24 genomic region (Fig. 2). This structure, which was not observed in other *MYC* amplified cells including Colo320 colon cancer cells, is suggested to be specific to leukemia cells [26]. We are currently analyzing whether this gapped amplicon exists in other AML/MDS samples.

So far, no differences between LEE and dmin have been found on their sequences; rather, the only difference between these elements seems to reside in the number of amplicons included in each of the structures, i.e., two or three for dmins and 10-15 for LEEs (Fig. 2) [26].

Genes Encoded on the Amplicon

Several genes were found in the amplicon of LEEs and dmins [26]. The observations that the *MYC* is also amplified in colon carcinoma, glioma, and other tumors suggest that enhanced expression of this gene plays a crucial role in neoplastic change of these cells [30]. In addition to *MYC*, we observed the amplification of other genes in HL-60 cells. *TRIB1*, another gene found at the chromosome 8q24 locus, is amplified in HL-60 cells. Amplification of this gene is also found in dmin-containing AML/MDS samples. This suggests a prominent role for this gene in leukemia especially in this disease.

Recently, various aspects of the physiological function of the *TRIB1* gene have been reported: *TRIB1* may be important for the migration of heart smooth muscle cells [31]; furthermore, abnormal expression of lipopolysaccharide (LPS)-induced genes related to the Toll-like receptors-mediated natural immune system is observed in *Trib1*-deficient mouse cells [32]. Although the molecular function of Trib1 protein remains unclear, this gene is thought to regulate the activity of mitogen-activated protein kinases (MAPKs) as a scaffold protein in the MAPK signal transduction pathway. Trib1 protein may regulate certain transcription factors that are targets of MAPK in normal and neoplastic cells [33].

The retinoic acid modulator (*RAM*) gene (or *CCDC26*) is also among the amplified and transcribed genes in HL-60 cells. *RAM* is frequently amplified and transcribed in dmin-positive AML/MDS patients [28, 34]. Mutation of this gene is reported to interrupt cell differentiation induced by retinoic acid; however, the molecular function of this gene remains unclear. Moreover, it is uncertain whether the open reading frame deduced from the sequence of this gene is actually translated. There is a possibility that its transcripts are functional by themselves and act as regulatory RNA molecules [34].

Further research is required to assess whether these genes actually contribute to AML/MDS and to the differentiation of myeloid cells [35]; however, the observation that a similar set of amplicon genes was transcribed in many dmin-positive AML/MDS cells suggests that these genes may play an important role in the development of some types of leukemia.

Inhibition of the expression of the genes encoded in the 8q24 locus via differentiation in low-passage HL-60 cells is probably a direct effect of the reduction of the copy number of the genes amplified on dmins. However, explanation by wide-range gene regulation affecting the expression of all genes on the whole span of the locus is still possible [36].

Stability of LEEs

Many lines of evidence support the contention that the stabilization of extrachromosomal elements is involved in the progress of malignancy in various cancers [37]; therefore, it is very important to clarify the mechanism for the stability of LEEs.

HL-60 cells carry null mutations of the *TP53* gene,which may be important for the generation and/or maintenance of LEEs and dmins [38]. Loss of function of p53, the *TP53*-encoded protein, and deregulated expression of *MYC* can cooperate in the formation and maintenance of extrachromosomal elements [6, 15, 36].

Dmins are unstable because they are not distributed in an equal manner to both daughter cells during cell division due to the absence of centromeres. By contrast, LEEs are stable in continuous culture, although almost all DNA sequences of LEEs seem to originate from dmins. This nature of LEEs is reminiscent of presence of neocentromere (NC). NC is found on a pathologically unusual chromosome and is considered to stabilize chromosomal fragments that have lost their original centromere, by acting as an epigenetic functional centromere [40]. A functional centromere like NC can account for the stability of LEEs. Epigenetic mechanisms, which include DNA methylation, the acetylation and phosphorylation of histones, and the poly(ADP-ribosyl)ation and ubiquitination of chromatin proteins, are keys to understand the nature of LEEs because they are considered to have a great influence on the stability of chromosomes and on gene expression [41, 42].

Application of LEEs as Vectors for Cancer Gene Therapy

It remains to be determined whether maintenance of LEEs requires the absence of p53. If LEEs are maintained stably only in cancer cells and are excluded from normal cells, this type of element can be used as a potential vector for the treatment of cancer. A gene therapy strategy for cancer involves the introduction of p53, which is a tumor suppressor gene, into cancer cells to either kill these cells or suppress their growth. Ideally, a vector for cancer treatment should be maintained only in cancer cells and should be eliminated from normal cells soon after introduction, to limit the manifestation of side effects [43]. An adenovirus-based vector has been put to practical use recently and is currently the most advanced vector for cancer treatment. One of the most successful vectors is based on an adenovirus deficient for the E1B 55 kDa early gene. This vector is maintained in cancer cells deficient for p53 function [43]. A functionally active p53 in normal cells suppresses the replication of viruses and limits side effects; however, the nature of adenovirus vector restricts the number and size of genes that can be introduced into cells. By contrast, LEEs would be suitable for massive gene transfer because of their size. If it is confirmed that a LEE is maintained specifically in cancer cells and is eliminated from normal cells, it will constitute an ideal vector for the treatment of cancer, which implies a great potential for its applications. The maintenance of LEEs is also likely dependent on the enhanced expression of the *MYC* gene; LEEs should be maintained preferentially in cancer cells in which *MYC* is strongly expressed; thus, LEEs can be highly valuable vectors, similarly to the vector deficient for p53 [44, 45]. To date, however, there are no reports of the development of a cancer therapy vector that targets *MYC*.

Conclusions

We think that LEE provides a good model system to study the potential relationship between some types of abnormal chromosomes and the progression of the malignancy of cancers. The method of sorting unusual extrachromosomal elements via the use of flow cytometry will contribute significantly to the cytogenetic analysis of cancer cells. This technique, which allows the determination of abnormal chromosome patterns, does not require prominent expertise and can be more suitable for handling bulk clinical samples through automated computer diagnosis in the near future. The isolation of a marker chromosome using this method, in combination with comparative genomic hybridization (CGH), may lead to the enhancement of cytogenetic research when compared with conventional approaches. The development of sorting equipment and fluorescent materials that bind specifically to chromosomes will allow the detection of abnormal karyotypes at higher resolution.

The accumulation of knowledge on the relevance of chromosome abnormalities in cancer will deepen our understanding of this disease. Advances in technologies that manipulate chromosomes will promote the development of a new nonviral vector for cancer treatment, among which LEE is a prominent candidate.

References

[1] Paulovich, A. G., Toczyski, D. P. & Hartwell, L. H. (1997). When checkpoints fail. *Cell*, *88*, 315-21.

[2] Myllykangas, S., Böhling T. & Knuutila, S. (2007). Specificity, selection and significance of gene amplifications in cancer. *Semin. Cancer Biol.*, *17*, 42-55.

[3] Albertson, D. G., (2006). Gene amplification in cancer. *Trends Genet.*, *22*, 447-55.

[4] Eckhardt, S. G., Dai, A, Davidson, K. K., Forseth, B. J., Wahl, G. M. & Von Hoff, D. D. (1994). Induction of differentiation in HL60 cells by the reduction of extrachromosomally amplified c-myc. *Proc. Natl. Acad. Sci. U. S. A.*, *91*, 6674-8.

[5] Von Hoff, D. D., Needham-VanDevanter, D. R., Yucel, J., Windle, B. E. & Wahl, G. M. (1988). Amplified human MYC oncogenes localized to replicating submicroscopic circular DNA molecules. *Proc. Natl. Acad. Sci. U. S. A.*, *85*, 4804-8.

[6] Elias, L. & Berry, C. O. (1991). Induction of differentiation by tumour necrosis factor in HL-60 cells is associated with the formation of large DNA fragments. *Leukemia*, *5* , 879-85.

[7] Myllykangas, S. & Knuutila S., (2006). Manifestation, mechanisms and mysteries of gene amplifications. *Cancer Lett.*, *232*, 79-89.

[8] Reddy, K. S. (2007). Double minutes (dmin) and homogeneously staining regions (hsr) in myeloid disorders: a new case suggesting that dmin form hsr in vivo. *Cytogenet. Genome Res.*, *119*, 53-9.

[9] Hahn, P. J. (1993). Molecular biology of double-minute chromosomes. *Bioessays*, *15*, 477-84.

[10] Blow, J. J. & Gillespie, P. J. (2008). Replication licensing and cancer—a fatal entanglement? *Nat. Rev. Cancer*, *8*, 799-806.

[11] To, K. K., Koshiji, M., Hammer, S. & Huang, L. E. (2005). Genetic instability: the dark side of the hypoxic response. *Cell Cycle*, *4*, 881-2.

[12] Taylor, C., Jalava, A. & Mai, S. (1997). c-Myc dependent initiation of genomic instability during neoplastic transformation. *Curr. Top. Microbiol. Immunol.*, *224*, 201-7.

[13] Wade, M. & Wahl, G. M. (2006). c-Myc, genome instability, and tumorigenesis: the devil is in the details. *Curr. Top. Microbiol. Immunol.*, *302*, 169-203.

[14] Taylor, C. & Mai, S. (1998). c-Myc-associated genomic instability of the dihydrofolate reductase locus in vivo., *Cancer Detect Prev.*, *22*, 350-6.

[15] Prochownik, E. V. & Li, Y. (2007). The ever expanding role for c-Myc in promoting genomic instability. *Cell Cycle*, *6*, 1024-9.

[16] Marcu, K. B., Bossone, S. A. & Patel, A. J. (1992) myc function and regulation. *Annu Rev Biochem.*, 61, 809-60.

[17] Pelengaris, S. & Khan, M. (2003). The many faces of c-MYC. *Arch. Biochem. Biophys.*, *416*, 129-36.

[18] Efeyan, A. & Serrano, M. (2007). p53: guardian of the genome and policeman of the oncogenes. *Cell Cycle*, *6*, 1006-10.

[19] Yin, X. Y., Grove, L., Datta, N. S., Long, M. W. & Prochownik, E. V. (1999). C-myc overexpression and p53 loss cooperate to promote genomic instability. *Oncogene*, *18*, 1177-84.

[20] Lunghi, P., Costanzo, A., Salvatore, L., Noguera, N., Mazzera, L., Tabilio, A., Lo-Coco, F., Levrero, M. & Bonati, A. (2006). MEK1 inhibition sensitizes primary acute myelogenous leukemia to arsenic trioxide-induced apoptosis. *Blood*, *107*, 4549-53.

[21] Tomasini, R., Mak, T. W. & Melino, G. (2008). The impact of p53 and p73 on aneuploidy and cancer. *Trends Cell Biol.*, *18*, 244-52.

[22] Lufino, M. M., Manservigi, R. & Wade-Martins, R. (2007). An S/MAR-based infectious episomal genomic DNA expression vector provides long-term regulated functional complementation of LDLR deficiency. *Nucleic Acids Res.*, *35*, e98.

[23] Haque, M. M., Hirano, T., Itoh, N. & Utiyama, H. (2001). Evolution of large extrachromosomal elements in HL-60 cells during culture and the associated phenotype alterations. *Biochem. Biophys. Res. Commun.*, *288*, 592-6.

[24] Haque, M. M., Hirano, T., Nakamura, H. & Utiyama, H. (2001). Granulocytic differentiation of HL-60 cells, both spontaneous and drug-induced, might require loss of extrachromosomal DNA encoding a gene(s) not c-MYC. *Biochem. Biophys. Res. Commun.*, *288*, 586-91.

[25] Shimizu, N., Nakamura, H., Kadota, T., Kitajima, K., Oda, T., Hirano, T. & Utiyama, H. (1994). Loss of amplified c-myc genes in the spontaneously differentiated HL-60 cells. *Cancer Res.*, *54*, 3561-7.

[26] Hirano, T., Ike, F., Murata, T., Obata, Y., Utiyama, H. & Yokoyama, K. K. (2008). Genes encoded within 8q24 on the amplicon of a large extrachromosomal element are selectively repressed during the terminal differentiation of HL-60 cells. *Mutat. Res.*, *640*, 97-106.

[27] Kitajima, K., Haque, M., Nakamura, H., Hirano, T. & Utiyama, H. (2001). Loss of irreversibility of granulocytic differentiation induced by dimethyl sulfoxide in HL-60 sublines with a homogeneously staining region. *Biochem. Biophys. Res. Commun.*, *288*, 1182-7.

[28] Storlazzi, C. T., Fioretos, T., Surace, C., Lonoce, A., Mastrorilli, A., Strömbeck, B., D'Addabbo, P., Iacovelli, F., Minervini, C., Aventin, A., Dastugue., N. Fonatsch, C., Hagemeijer, A., Jotterand, M., Mühlematter, D., Lafage-Pochitaloff, M., Nguyen-Khac, F., Schoch, C., Slovak, M. L., Smith, A., Solè, F., Van Roy, N., Johansson, B. & Rocchi, M. (2006). MYC-containing double minutes in hematologic malignancies: evidence in favor of the episome model and exclusion of MYC as the target gene. *Hum. Mol. Genet.*, *15*, 933-42.

[29] Storlazzi, C. T., Fioretos, T., Paulsson, K., Strömbeck, B., Lassen, C., Ahlgren, T., Juliusson, G., Mitelman, F., Rocchi, M. & Johansson, B. (2004). Identification of a commonly amplified 4.3 Mb region with overexpression of C8FW, but not MYC in MYC-containing double minutes in myeloid malignancies. *Hum. Mol. Genet.*, *13*, 1479-85.

[30] Brison, O. (1993). Gene amplification and tumor progression. *Biochim. Biophys. Acta*, *1155*, 25-41.

[31] Sung, H. Y., Guan, H., Czibula, A., King, A. R., Eder, K., Heath, E., Suvarna, S. K., Dower, S. K., Wilson, A. G., Francis, S. E., Crossman, D. C. & Kiss-Toth, E. (2007). Human tribbles-1 controls proliferation and chemotaxis of smooth muscle cells via MAPK signaling pathways. *J. Biol. Chem.*, *282*, 18379-87.

[32] Yamamoto, M., Uematsu, S., Okamoto, T., Matsuura, Y., Sato, S., Kumar, H., Satoh, T., Saitoh, T., Takeda, K., Ishii, K. J., Takeuchi, O., Kawai, T. & Akira, S. (2007). Enhanced TLR-mediated NF-IL6 dependent gene expression by Trib1 deficiency. *J. Exp. Med.*, *204*, 2233-9.

[33] Kiss-Toth, E., Bagstaff, S. M., Sung, H. Y., Jozsa, V., Dempsey, C., Caunt, J. C., Oxley, K. M., Wyllie, D. H., Polgar, T., Harte, M., O'Neill, L. A., Qwarnstrom, E. E. & Dower, S. K. (2004). Human tribbles, a protein family controlling mitogen-activated protein kinase cascades. *J. Biol. Chem.*, *279*, 42703-8.

[34] Yin, W., Rossin, A., Clifford, J. L. & Gronemeyer, H. (2006). Co-resistance to retinoic acid and TRAIL by insertion mutagenesis into RAM. *Oncogene*, *25*, 3735-44.

[35] Christacos, N. C., Sherman, L., Roy, A., DeAngelo, D. J. & Dal Cin, P. (2005). Is the cryptic interstitial deletion of 8q24 surrounding MYC a common mechanism in the formation of double minute chromosome? *Cancer Genet. Cytogenet.*, *161*, 90-2.

[36] Miele, A. & Dekker, J. (2008). Long-range chromosomal interactions and gene regulation. *Mol. Biosyst.*, *4*, 1046-57.

[37] Knuutila, S., Björkqvist, A. M., Autio, K., Tarkkanen, M., Wolf, M., Monni, O., Szymanska, J., Larramendy, M. L., Tapper, J., Pere, H., El-Rifai, W., Hemmer, S., Wasenius, V. M., Vidgren, V. & Zhu, Y. (1998). DNA copy number amplifications in human neoplasms: review of comparative genomic hybridization studies. *Am. J. Pathol.*, *152*, 1107-23.

[38] Wolf, D. & Rotter, V. (1985). Major deletions in the gene encoding the p53 tumor antigen cause lack of p53 expression in HL-60 cells. *Proc. Natl. Acad. Sci. U. S. A.*, *82*, 790-4.

[39] Mai, S. & Mushinski, J. F. (2003). c-Myc-induced genomic instability. *J. Environ. Pathol. Toxicol. Oncol.*, *22*, 179-99.

[40] Capozzi, O., Purgato, S., Verdun di Cantogno, L, Grosso, E., Ciccone, R., Zuffardi, O., Della Valle, G. & Rocchi, M. (2008). Evolutionary and clinical neocentromeres: two faces of the same coin? *Chromosoma*, *117*, 339-44.

[41] Choo, K. H. (2000). Centromerization. *Trends Cell Biol.*, *10*, 182-8.

[42] Brown, J. D., Golden, D. & O'Neill, R. J. (2008). Methylation perturbations in retroelements within the genome of a Mus interspecific hybrid correlate with double minute chromosome formation. *Genomics*, *91*, 267-73.

[43] Kirn, D. (2001). Clinical research results with dl1520 (Onyx-015), a replication-selective adenovirus for the treatment of cancer: what have we learned? *Gene Ther.*, *8*, 89-98.

[44] Jackson, D. A., Juranek, S. & Lipps, H. J. (2006). Designing nonviral vectors for efficient gene transfer and long-term gene expression. *Mol. Ther.*, *14*, 613-26.

[45] Lufino, M. M., Edser, P. A, & Wade-Martins R. (2008). Advances in high-capacity extrachromosomal vector technology: episomal maintenance, vector delivery, and transgene expression. *Mol. Ther. 16*, 1525-38.

In: The Human Genome: Features, Variations... ISBN: 978-1-60741-695-1
Editor: Akio Matsumoto and Mai Nakano

Chapter 4

Molecular Basis of Human Coagulopathies

Isis S.R. Carter*[*]*, Ann Y.K. Wong, Mark R. Bleackley, Ganna Vashchenko, Heather D.E. Fox and Ross T.A. MacGillivray

Centre for Blood Research and Department of Biochemistry & Molecular Biology, University of British Columbia, Vancouver, BC, V6T 1Z3, Canada

Abstract

Recent advances in technology have greatly improved our understanding of the molecular basis of inherited disorders. Prior to the development of recombinant DNA technology, only inherited disorders where there was an abundance of gene product were conducive to study. With the advent of recombinant DNA technology, specific hybridization probes became available that allowed the detection and diagnosis of rarer disorders using Southern blot analysis and manual DNA sequencing of cloned DNA fragments. The development of the polymerase chain reaction and automated DNA sequence analysis has since extended predictive DNA testing to most inherited disorders where the non-functional gene has been identified. Further technological advances led to the completion of the Human Genome Project (HGP) in 2003. Since that time, next-generation DNA sequencers hold the potential of the fast and cheap determination of an individual's genomic sequence leading to the promise of personalized medicine. In this review, we discuss the application of these advancing technologies on the molecular basis of human coagulopathies including (1) the relatively frequent X-linked hemophilias including the gene rearrangement that is the cause of up to 50% of the cases of hemophilia A, (2) the rarer autosomal bleeding disorders, and (3) the complex coagulopathies caused by indirect effects of mutations in non-clotting factor genes. These studies have revealed a multitude of genetic mutations that give rise to coagulopathies.

[*] These authors contributed equally to this manuscript. This work was supported in part by a grant from the CBS-CIHR Program in Blood Utilization and Conservation Research awarded to RTAM. ISRC and MRB were supported by graduate studentships from the Strategic Training Program in Transfusion Science (funded by the Canadian Institutes of Health Research and the Heart & Stroke Foundation of Canada). AYKW was supported by a graduate studentship from Canadian Blood Services.

When individual genomic sequencing becomes routine, the coagulopathies will be amenable to diagnostic testing with subsequent genetic counseling on the consequences of the findings.

Introduction

Historical aspects of blood coagulation The history of blood coagulation dates back to 1835 with the introduction of the terms *hematology* by Hewson and *anemia* by Andral [1]. From the 19th century onwards, bleeding disorders were recognized in males whose mothers and sisters remained unaffected. The work of Buchanan and later Schmidt on hydrocoele coagulation by the addition of serum gave rise to the *classical theory of blood coagulation* by Schmidt in 1892, in which he suggested that fibrin ferment (thrombin) could be formed from a precursor in the plasma when exposed to tissue extracts. The *fibrin ferment* then caused purified fibrinogen to coagulate. Hopff conceived the term *hemophilia* in 1828 [1]; however, it was not until 1910 that Addis showed the correction of hemophiliac blood coagulation by the addition of normal plasma. Based on this experiment, hemophiliac blood was shown to contain both thrombin and fibrinogen with the absence of an unknown plasma component being responsible for the bleeding disorder. [1].

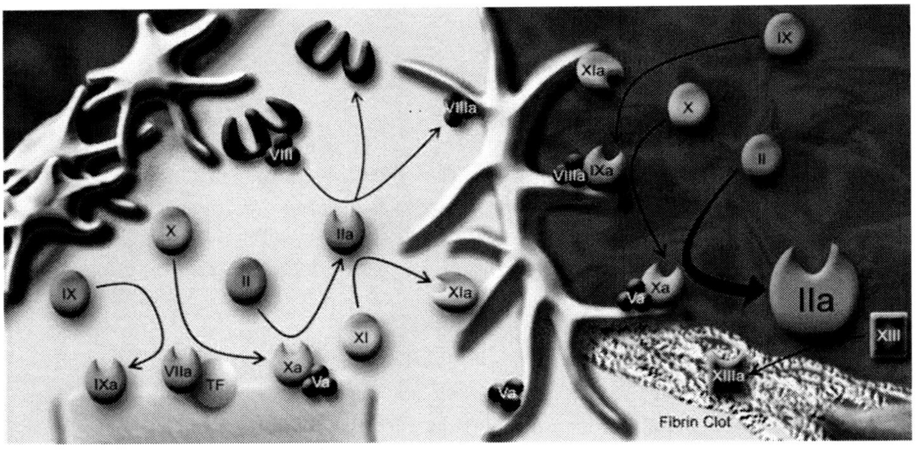

Figure 1. Cell-based model of blood coagulation (reviewed in [5]). This model defines coagulation into 3 stages: initiation, amplification and propagation. During initiation, a small amount of FIX and FX are activated by FVIIa-TF complex on TF-bearing cells. Upon adhesion at the site of injury, platelets are partially activated, promoting the release of partially activated FV (FVa*). FVa* complexes with FXa on TF-bearing cells and is completely activated by FXa or other proteases. The FVa-FXa prothrombinase complex converts a small amount of prothrombin (II) to thrombin (IIa), which plays a major role in amplification. Thrombin completes platelet activation, releases activated FVIIIa from vWF to platelet surfaces, and activates FV and FXI on platelet surface. Clot formation is propagated through a series of events on activated platelet surface: FIX activated by FVIIa-TF complex or FXIa complexes with FVIIIa to form the Xase complex; Xase activates FX to FXa, which joins FVa to form the prothrombinase complex and activates II to IIa; IIa converts fibrinogen to fibrin and activates FXIII to FXIIIa to cross-link the fibrin clot.

In 1947, Owren [2] introduced the Roman numeral system for the coagulation factors to clarify the names and uniformly identify the coagulation factors; however, it was not until the 1960s that this system of nomenclature was adopted. Owren also demonstrated that there could be more than one type of hemophilia through the identification of a patient whose bleeding disorder was unresponsive to tissue factor, leading to the discovery of factor V. The following fifteen years saw the discovery of factors VII, IX, and X through the analysis of patient blood sample clotting times, which could be normalized by challenging with blood samples from other patients, indicating deficiencies in different factors [2]. Finally, in the 1960s, two similar models for coagulation were proposed: the *waterfall* sequence for intrinsic blood clotting [3] and the *coagulation cascade* [4]. In the following decade, factors XI, XII, XIII were discovered, and the clotting factors were purified to homogeneity from bovine and human blood plasma. Biochemical studies of purified proteins in many different laboratories revealed the biochemical mechanisms responsible for clotting factor activation, inactivation, fibrinogenesis and fibrinolysis [2]. Coagulation is now understood to occur in three main stages: (1) vascular damage causes initiation of platelet aggregation and formation of a platelet plug, (2) the coagulation cascade is initiated on the activated platelet membrane surface, and (3) an insoluble network of fibrin is generated and secured to the platelet plug. Following cessation of bleeding, wound healing and clot fibinolysis occur. It has also become clear that coagulation does not occur in the liquid phase as originally proposed, but rather occurs on exposed endothelial and activated platelet membrane surfaces. Recently, a cell-based model of coagulation has been proposed by Monroe and Hoffman [5] that accounts for the spatial control of the coagulation response through the generation of active factors on membranous surfaces. The initiation, amplification and propagation phases of coagulation are summarized in Figure 1.

Recent technological advances in biotechnology Prior to the development of recombinant DNA technology in the mid-1970s, molecular genomic studies were limited to sequences that occurred in repetitive DNA such as the ribosomal RNA genes. Although crude studies could be completed on these multi-copy genes, several advances allowed single copy genes to be characterized. The development of bacterial hosts and plasmid vectors, the discovery and purification of type II restriction endonucleases and the purification of DNA ligases provided the basic tools to cut and isolate DNA fragments from different sources, and then to join them together in a plasmid vector. Standard microbiological cloning methods then allowed the isolation of cloned bacterial colonies that all contained the same plasmid DNA molecule containing foreign (or recombinant) DNA. For the first time, single copy genes could be isolated in sufficient quantities to allow physical-chemical characterization. Other developments included rapid DNA sequencing methods by either chemical modification of bases or enzymatic synthesis in the presence of chain terminators. In this way, the DNA sequences of cloned DNA fragments could be determined. This led to an explosion of scientific knowledge in many areas of biological medicine including the coagulopathies. Recombinant DNA methods were used to isolate complementary DNA (cDNA) molecules coding for the coagulation factors. cDNA sequence analysis and the genetic code then allowed the prediction of the complete amino acid sequences of even rare proteins such as human factor VIII.

The availability of cloned cDNAs allowed their use as hybridization probes to characterize and isolate the corresponding genes by using Southern blot analysis (the identification of specific DNA restriction fragments in a complex mixture), genomic library screening (using plasmid and phage libraries that contained DNA fragments corresponding to the complete genome) and heteroduplex analysis in the electron microscope. These studies revealed the presence of intron sequences that interrupted the coding regions present in the cDNA sequences. Regulatory sequences associated with gene expression could also be characterized. By the late 1980s, the human genes for most of the coagulation factors had been cloned and characterized. With these DNA sequences as references, the dysfunctional genes from individuals with bleeding problems could also be isolated from libraries made specifically from the affected individual's DNA. Although this analysis was laborious and time-consuming, it soon became obvious that inherited coagulopathies were due to altered DNA sequences caused by missense and nonsense mutations, splicing mutations, deletions, insertions and gene rearrangements.

Once the molecular basis of a coagulopathy had been determined, DNA diagnosis of other family members became possible. Initially, this was often achieved by Southern blot analysis of restriction fragment length polymorphisms (RFLPs). These RFLPs were the result of the causal mutation altering the short DNA sequence that was recognized by a particular type II restriction endonuclease. The analysis could be used both if the mutation destroyed a restriction site or created a new site. However, due to the strict sequence requirements of most type II restriction endonucleases, not all mutations created or destroyed a site. To overcome this restriction, linked RFLPs were used with Southern blot analysis. In this case, a RFLP close to the affected gene (*a linked marker*) was characterized within a particular family pedigree, and specific RFLPs were assigned to specific alleles within that family. Once the disease allele had been identified, its inheritance within the pedigree could usually be followed using Mendelian inheritance. This analysis was complicated when the linked marker was far enough away from the gene that the probability of cross-over during meiosis became significant. This possibility caused a degree of uncertainty in the DNA diagnosis that could be quantified by knowing the distance from the linked marker to the disease gene locus.

DNA genomic studies and DNA diagnosis were revolutionized by the development of the polymerase chain reaction (PCR) in the mid-1980s. PCR is a method whereby a specific DNA sequence can be rapidly amplified as long as the DNA sequence of the region is known. PCR utilizes short oligonucleotides to prime DNA synthesis at two specific sequences that flank the region of interest. By alternating DNA thermal denaturation with new enzymatic DNA synthesis (primed by the specific oligonucleotide primers), a DNA sequence could be amplified over a million-fold in a few hours. The PCR had the advantage of being very sensitive, very rapid and suitable for automation. The combination of PCR and automated DNA sequence analysis in the 1990s enabled mutated genes to be readily characterized leading to an explosion of knowledge at this time (see Figure 2).

Subsequent improvements in PCR, oligonucleotide synthesis and DNA sequence analysis led to the completion of the HGP in 2003 [6]. This in turn led to the establishment of the new discipline of bioinformatics that is involved in interpreting the huge amounts of data that can be generated with modern technology. Although the HGP was completed at great expense, recent advances have allowed the determination of individual genome sequences for relatively modest prices in the $20,000 range [7]. As DNA sequencing technology advances even more, the goal of determining an individual genome sequence for $1,000 is now approaching reality.

Personalized accounts of many of these technology developments and genomic breakthroughs can be found in the Nobel Prize acceptance speeches (http://nobelprize.org/nobel_prizes/) of the key contributors including Arber, Nathans and Smith (restriction endonucleases, 1978), Gilbert and Sanger (DNA sequence analysis, 1980), Roberts and Sharp (split genes, 1993) and Mullis (PCR, 1993).

Molecular basis of bleeding disorders Bleeding disorders caused by mutations of the coagulation factor genes are considered single gene disorders, and may be classified as autosomal, sex-linked, dominant or recessive. Autosomal dominant mutations require only one of the two autosomal copies of a gene to be mutated for the development of the condition; thus, from an affected parent, there a 50% chance that a child will inherit the mutated gene. Examples of autosomal dominant coagulopathies include von Willebrand's disease and VKORC1 deficiency. In contrast, autosomal recessive disorders require both autosomal copies of the gene to be mutated for the bleeding phenotype to be expressed. In this case, an affected individual has a 25% chance of inheriting two mutations from parents who are carriers for one mutation each. Examples of autosomal recessive coagulopathies include mutations in prothrombin, factor V, factor VII and others listed in Table 1. X-linked recessive disorders are caused by mutations in genes on the X chromosome. Due to the absence of a compensatory X chromosome, males are more frequently affected than females for recessive disorders, and females are usually carriers for the disorder passed on to their sons. Examples of this type of disorder are hemophilias A and B caused by factor VIII and IX deficiency, respectively (see Table 1).

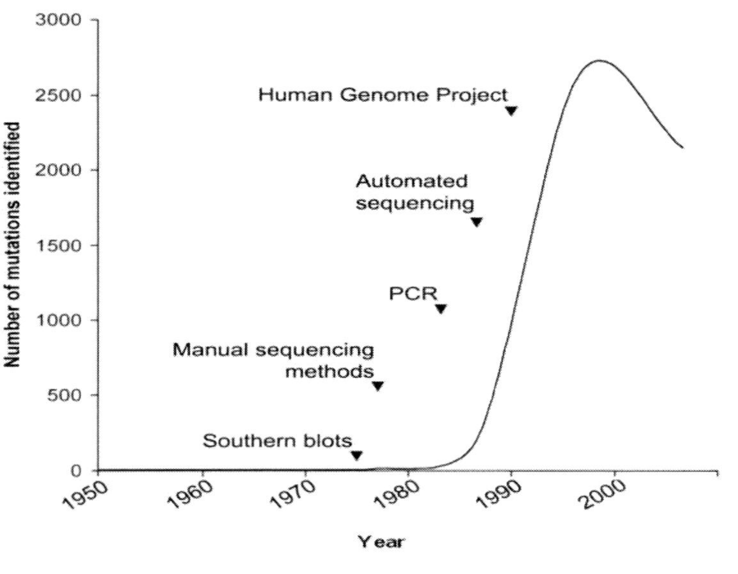

Figure 2. Number of mutations identified as technology advances since 1950. The sum of all mutations identified annually in genes involved in coagulation based on database available on ISTH indicates that the development of sequencing methods and PCR has facilitated identification of genetic defects. The number peaks with the completion of the Human Genome Project.

Table 1. Proteins involved in procoagulation. For the mode of inheritance, AD, AR and XR denote autosomal dominant, autosomal recessive and X-linked recessive, respectively.

Protein deficiency	Gene	NCBI mRNA reference	Chromosomal location	Gene size (kb)	# exons	Protein type	Mode of inheritance
Prothrombin	*F2*	NM_000506	11p11-q12	20.3	14	Zymogen for a Serine protease	AR
FV	*F5*	NM_000130	1q24.2	72.3	25	Procofactor	AR
FVII	*F7*	NM_000131	13q34	14.2	9	Zymogen for a Serine protease	AR
FVIII	*F8*	NM_019863	Xq28	186	26	Procofactor	XR
FIX	*F9*	NM_000133	Xq27.1-q27.2	33.5	8	Zymogen for a Serine protease	XR
FX	*F10*	NM_000504	13q34	26.7	8	Zymogen for a Serine protease	AR
FXI	*F11*	NM_000128	4q35.2	22.6	15	Zymogen for a Serine protease	AR
FXIIIA	*F13A*	NM_000129	6p25.3-p24.3	176.6	15	Zymogen for a gransglutaminase	AR
FXIIIB	*F13B*	NM_001994	1q31-q32.1	28	12	Carrier subunit for FXIIIA	AR
MCFD2	*MCFD2*	NM_139279	2p21	13.9	4	Cofactor to ERGIC-53	AR
LMAN1/ ERGIC-53	*LMAN1*	NM_005570	18q21.3-18q22	29.4	13	ER/Golgi transport protein	AR
vWF	*vWF*	NM_000552	12p13.3	178	52	Glycoprotein	AD or AR
VKORC1	*VKORC1*	NM_024006	16p12-q21	5.1	3	Vitamin K epoxide reductase complex, subunit 1	AD
GGCX	*GGCX*	NM_000821	2p11.2	12.4	15	Gamma-glutamyl carboxylase	AR
Fibrinogen Aα	*FGA*	NM_000508	4q31.3	7.6	6	Plasma glycoprotein	AR
Fibrinogen Bβ	*FGB*	NM_005141	4q31.3	8.1	8	Plasma glycoprotein	AR
Fibrinogen γ	*FGG*	NM_000509	4q32.1	8.5	10	Plasma glycoprotein	AR

Scope of review In this review, we have limited the discussion to those coagulation factors that are directly involved in fibrin polymer formation at the injury site. As such, historical coagulation factors associated with the initiation of the intrinsic pathway such as factor XII, high molecular weight kininogen and prekallikrein will not be discussed here as these proteins are no longer considered to be involved in blood coagulation (see [5] for a discussion). Although they can result in bleeding problems, non-coagulopathies such as platelet disorders and mutations in factors causing hypercoagulability are beyond the scope of this review. Similarly, proteins involved in other aspects of hemostasis (such as anticoagulation and fibrinolysis) are also not included in this review. Table 2 summarizes the types of mutations that are currently believed to cause human coagulopathies.

Table 2. Number of unique coagulopathy mutations described to date on the ISTH mutation registries.

Deficiency	Gene	Number of unique mutations			
		Total	Point mutation	Deletion	Insertion
FVIII	*F8*	1233[*1]	797	345	911
FIX	*F9*	925[2,3]	756[2]	134	353
vWF	*vWF*	358[4]	302[4]	36	20
PT	*F2*	44	38	5	1
FV	*F5*	55	30	19	6
FVII	*F7*	147	129	16	2
FX	*F10*	87	77	9	1
FXI	*F11*	120[5]	107	8[5]	5
FXIII	*F13A*	77[6]	56	15[6]	6
	F13B	4	1	2	1
Fibrinogen	*FGA*	88	49	29	107
	FGB	51	45	5	1
	FGG	72	65	6	1
Combined FV-FVIII	*LMAN1*	28	14	10	4
	MCFD2	9	6	3	0

[*] The numbers of mutations were calculated from the whole mutation database; inversions that account for 50% of all severe cases are not included.
[1]Twelve of the insertion entries also involve deletions.
[2]For four of the point mutation entries, two mutations were found on the same allele; one involves three mutations that were found on the same allele.
[3]One of the insertion entries also involves a deletion.
[4]For one of the point mutation entries, two mutations were found on the same allele
[5]One of the deletion entries also involves another deletion plus a point mutation
[6]One of the deletion entries also involves an insertion
[7]Two insertions also involve a deletion

X-Linked Coagulopathies – The Common Hemorrhagic Disorders

Bleeding disorders are caused by deficiencies or defects in specific coagulation factors such as factor VIII (FVIII) in hemophilia A, factor IX (FIX) in hemophilia B and von Willebrand factor (vWF) in von Willebrand's disease (vWD). Being the most commonly inherited coagulation factor disorders, hemophilia and vWD together account for 95% to 97% of all cases [8]. While vWD affects approximately 1% of World's population with 1 in 1,000 patients presenting clinically significant disease [9], hemophilia is estimated to have over half a million individuals affected globally with hemophilia A displaying a five-fold higher frequency than hemophilia B [10]. Neither type of hemophilia exhibits ethnic nor geographic predisposition [8].

Hemophilia A Hemophilia A is also known as the Royal disease because of its appearance in the royal families of England, Spain, Germany and Russia resulting from a mutation originating in Queen Victoria. Being an X-linked recessive trait, hemophilia A is found at a frequency of 1 in every 5,000 males [11]. The disease is caused by defects in the *F8* gene which maps to the long arm of the X chromosome and spans a region of 186kb at position Xq28. When activated by thrombin, FVIIIa is involved in the amplification of FX activation for coagulation to proceed to completion [10] (see Figure 1). Up to half of the severe hemophilia A cases (equating to about 30% of all cases) are attributed to an inversion of intron 22 [12]. The less frequent inversion of intron 1 and large or whole gene deletions account for 5% of the severe cases. These inversion events occur through intra-chromosomal homologous recombination between a 9.5kb region in intron 22 (int22h-1) and one of the two copies of homologous repeat (int22h-2 or int22h3) located downstream of the *F8* gene or between 1kb region in intron 1 (int1h-1) and its repeat upstream of exon 1 (int1h-2) [11, 13, 14]. The remaining 45% of the severe cases and the moderate and mild hemophilia A cases are caused mainly by point mutations (85% missense; 15% nonsense) plus incidences of small deletions or insertions [10, 14]. As shown in Table 2, a total of 1233 unique mutations of all types have been identified with the majority being point mutations. Details of *F8* gene mutations can be found at the HAMSTeRS database: http://europium.csc.mrc.ac.uk/WebPages/Main/main.html.

Hemophilia B Less common than hemophilia A, hemophilia B is sometimes called Christmas disease after the first patient described with this bleeding disorder [15]. Hemophilia B is an X-linked recessive bleeding disorder that occurs in 1 in 25,000 males [8]. The disease is caused by the presence of dysfunctional FIX protein or the absence of FIX which plays an important role in activating FX to ensure sufficient FXa is present to sustain coagulation [10] (see Figure 1). The *F9* gene spans a region of 33.5kb at Xq27 on the long arm of the X chromosome. As shown in Table 2, 925 unique mutations are reported on the hemophilia B mutation database: http://www.kcl.ac.uk/ip/petergreen/haemBdatabase.html. The mutations are predominantly single nucleotide substitutions, among which 60% are missense mutations [10, 11]. Founder mutations contribute to 20% to 30% of the milder hemophilia B cases. Short deletions or insertions in the *F9* gene comprise approximately 7% of all mutations on the database, while around 3% of the reported *F9* genetic defects are caused by gross deletions or complex rearrangements [10].

von Willebrand Disease von Willebrand Disease (vWD) is another common bleeding disorder resulting from defects in the *VWF* gene. Located on chromosome 12 at 12p13.3, the *VWF* gene is approximately 178kb in length and contains 52 exons [9]. The *VWF* gene encodes von Willebrand factor (vWF) protein that plays an important role in the mediation of platelet adhesion to reactive surfaces, platelet aggregation and thrombus growth [16] (see Figure 1). Additionally, 1 to 2% of vWF multimers in plasma are bound via their D' domain to the pro-coagulant FVIII, thereby providing protection from proteolytic degradation and enabling transport of FVIII to the site of vascular injury [9]. vWD was first described in 1926, and is now recognized as the most common inherited bleeding disorder in humans [17] affecting up to 1% of the population in certain areas [9]. Characterized by low levels of functional vWF in plasma, the condition is classified into three types that differ by their severity and mode of inheritance. Type 1 vWD, the least severe form of the disease, is autosomal dominant, whereas Type 3, the most severe form, shows autosomal recessive inheritance. A moderate form of the disorder, designated as Type 2 vWD, presents as autosomal dominant or recessive trait depending on the nature of the mutation. The majority of cases of Type 2 vWD are due to missense mutations found in the A1, A2, and D'-D3 domains [18]. Partial and total gene deletions have been reported to cause Type 3 vWD along with missense mutations, and tend to be distributed throughout the whole vWF gene [9]. For a comprehensive review on the types of vWF deficiency and their functional consequences see [19]. A public database of vWF gene mutations can be found at http://www.hgmd.cf.ac.uk/ac/gene.php?gene=VWF. As shown in Table 2, a total of 358 mutations have been described.

Autosomal Recessive Coagulopathies – The Rare Coagulation Disorders

Rare autosomal coagulation deficiencies (RACDs) are usually the result of defects in the genes encoding the coagulation proteins, causing bleeding manifestations in the general population ranging from approximately 1 in 1 - 2,000,000 for FII and FXIII deficiency to 1 in 500,000 for FVII deficiency [20]. However, in populations where consanguineous marriages are more common (such as in some Muslim countries and Southern India), the frequencies of rare bleeding disorders are higher due to rare autosomal recessive traits occurring in homozygosity. For example, recessive coagulation disorders in Iran occur approximately 10 times more frequently than in Western countries [21]. Additionally, FXI deficiency is more prevalent in countries with large Ashkenazi Jewish populations [22]. Due to the heterogeneity of mutations in these genes causing RACDs, full mutational screening is required to ascertain a molecular diagnosis and in most cases remains a research endeavor rather than a standard clinical laboratory test.

Prothrombin Also known as coagulation factor II, prothrombin is the zymogen for thrombin, the terminal serine protease in the coagulation cascade. FXa in the prothrombinase complex (FXa and FVa in the presence of phospholipids and calcium) activates prothrombin via limited proteolysis to form thrombin (see Figure 1). Thrombin cleaves soluble fibrinogen to form insoluble fibrin monomers that aggregate to form a clot. The prothrombin (*F2*) gene is located on chromosome 11 (11p11) and is comprised of 14 exons spanning 21kb of genomic DNA. The prevalence of prothrombin deficiency is very rare, reported to be 1 per 1 – 2,000,000 in the general population [23]. A quantitative deficiency, known as hypoprothrombinemia is characterized by reduced activity and antigen levels in plasma [24]. A qualitative deficiency, dysprothrombinemia is characterized by normal antigen and reduced activity levels. As shown in Table 2, to date there have been 44 mutations published, including 35 missense, 2 nonsense, 2 splicing and 6 insertions/ deletions. An updated mutation database is found at www.med.unc.edu/isth/mutations-databases/ mutations_rare_bleeding_disorders.html. The most frequent mutations in the prothrombin gene are missense mutations in patients with hypoprothrombinemia [25]. Several cases of dysprothrombinemia have been reported; however, there have been no cases of true aprothrombinemia (zero prothrombin levels) reported. A murine model of aprothrombinemia has confirmed that absence of prothrombin is lethal in mice [26, 27]. However, interestingly Acharya *et al.* [28] found 10 homozygous prothrombin deficient subjects with a median FII activity of 0.03U/ml^{-1}, of which only 40% had provoked bleeding manifestations.

Tissue Factor Also known as coagulation factor III or thromboplastin, tissue factor (TF) is a cell surface glycoprotein present on subendothelial cells, platelets and leukocytes that acts as a high affinity receptor for plasma factor VII. Upon exposure of TF to the vasculature during injury, factor VII and TF form a complex allowing the activation of factor VII to factor VIIa, and the initiation of the coagulation cascade [29] (see Figure 1). TF is encoded by the *F3* gene, consisting of six exons spanning 12kb located at 1p22-p21. There have been two mutations reported in the *F3* gene: one insertion in a non-coding region associated with a thrombotic phenotype, and the other a regulatory mutation affecting gene expression. Both mutations are listed at the Human Gene Mutation Database: http://www.hgmd.cf.ac.uk/ac/gene.php?gene=F3. The paucity of mutations for TF probably reflects the central importance of this protein to coagulation initiation. Congruently, TF null mouse knockout studies have demonstrated that lack of TF results in embryonic lethality [30].

Factor V Activated factor V (FVa) is a non-enzymatic cofactor for FXa in the prothrombinase complex that acts to enhance FXa activation of prothrombin to thrombin (see Figure 1). FV is found in plasma in an inactive form that is activated by thrombin. The factor V (*F5*) gene is composed of 25 exons spanning approximately 80kb on chromosome 1 (1q23). Inherited FV deficiency has a prevalence of approximately 1 in 1,000,000 in the general population. Approximately 75% of cases are characterized as quantitative deficiencies, measured as a concomitant reduction of antigen and activity levels. However, approximately 25% of patients have a qualitative deficiency indicating production of a dysfunctional protein [31]. FV mutations have been found to cause both bleeding and clotting disorders and FV Leiden is a very common pro-thrombotic mutation that is carried by up to 5% of the general Caucasian population [32]. As shown in Table 2, 55 mutations have been reported for FV to cause bleeding. A database of published mutations for FV can be found at http://www.med.unc.edu/isth/mutations-databases/FactorV_2007numbers.html. The majority of the mutations reported are unique; however, the Italian population has a higher incidence of the Tyr1702Cys mutation [33]. The clinical phenotype for FV deficiency is variable. Severe FV deficiency is known as parahemophilia and is characterized by moderate to severe bleeding. This is most often expressed as type 1 deficiency, with low or unmeasurable FV antigen and activity levels [25]. Homozygous deficiency is often recognized during childhood and is associated with severe bleeding requiring intervention management [31]. However, mouse FV knockout studies indicate FV is essential as FV-deficient mice die either during mid-embryogenesis or of massive hemorrhage within two hours of birth [34]. Heterozygous patients with FV activity levels approximately 50% of normal are often asymptomatic.

Factor VII FVII is a zymogen that circulates in plasma and is activated by limited proteolysis. In a complex with TF on the surface of exposed endothelial cell membranes, FVIIa is a serine protease that serves to initiate the coagulation cascade by activating FX to FXa and FIX to IXa (see Figure 1). The nine exons of *F7* gene span 12kb on chromosome 13 (13q34), closely located to the FX gene. FVII deficiency is the most common autosomal recessive coagulation disorder with an incidence of 1 in 500,000 in the general population [35]. Two databases of known mutations can be found at http://www.med.unc.edu/isth/mutations-databases/FactorVII_2007.html and http://europium.csc.mrc.ac.uk [36]. As shown in Table 2, there have been 147 novel mutations described in the literature. An additional 22 other mutations have been published but have not been described in detail [37]. The majority of patients have a quantitative deficiency, characterized by low antigen and activity levels. However, qualitative deficiencies have also been described. Animal studies have shown that FVII deficiency is compatible with normal embryonic growth, although fatal perinatal bleeding has been described [38]. Patients with undetectable FVII activity are usually homozygous or compound heterozygous for mutations that affect expression of the *F7* gene while missense mutations generally cause mild to moderate deficiency [23].

Factor X FXa is a serine protease responsible for prothrombin activation to thrombin. FXa activity is enhanced by forming a complex with its cofactor FVa in the prothrombinase complex (see Figure 1). The *F10* gene is encoded by eight exons, spanning over 2.2kb on chromosome 13q34; the *F10* gene is located 2.8kb downstream of the *F7* gene. FX was discovered in the 1950's and named Stuart-Prower factor after the analysis of patients with prolonged prothrombin times [39]. The incidence of FX deficiency is about 1 in 1,000,000 in the general population; this rarity is thought to be due to the central role of FX in the coagulation pathway [39]. Patients with a severe FX deficiency are among the most seriously affected of patients with RACDs, and develop severe bleeding symptoms early in life. Similarly, animal studies have shown that targeted inactivation of the *F10* gene producing FX knockout mice usually results in embryonic lethality, with any mice surviving birth suffering fatal neonatal bleeding [40]. To date, 87 mutations have been described in the literature, of which 70 are missense mutations (mainly in exon 8 encoding the catalytic domain). The remainder are splice mutations and insertions or deletions. There have been no nonsense mutations reported. A database of mutations can be found on the ISTH website at http://www.med.unc.edu/isth/mutations-databases/FactorX_2007.html [41]. None of the missense mutations are predicted to lead to the production of a truncated protein or loss of protein production. This is supported by the observation that patients usually have low activity but normal antigen levels, indicating a qualitative deficiency [23].

Factor XI FXIa is a serine protease that was originally thought to be involved in the contact initiation stages of the intrinsic coagulation cascade, but is now recognized to be involved in maintenance of the coagulation cascade once the extrinsic pathway has been inhibited [5]. FXI circulates as a zymogen that is activated by small amounts of thrombin generated at a site of injury. The physiological function of FXIa is to activate FIX and maintain the coagulation response [42] (see Figure 1). The *F11* gene is located on chromosome 4 (4q35), with 15 exons located over 23kb [23]. As shown in Table 2, there have been 120 mutations published. A database of FXI mutations can be found on the ISTH website: http://www.med.unc.edu/isth/mutations-databases/FactorXI_2007.html [41]. FXI deficiency differs from the other RACDs due to the presence of two common mutations in Ashkenazi and Iraqi Jews, that together are responsible for the compound heterozygosity commonly observed in Jewish populations [25]: (1) a nonsense Glu117Stop mutation in exon 5 causes low to undetectable FXI levels in homozygotes, and (2) a missense Phe283Leu mutation in exon 9 leads to defective secretion, resulting in FXI levels of approximately 10% [25]. The majority of the cases in the literature are patients of Ashkenazi Jewish origin, as the frequency for heterozygosity in this population can be as high as 8% [23]. In non-Jewish populations, unique FXI mutations occur throughout the gene [23]. Additionally, a founder missense mutation has been found through haplotypes analysis of non-Jewish populations from France and the United Kingdom (reviewed in [25]). The genotype-phenotype relationship for FXI deficiency is not as apparent as other RACDs as patients with severe FXI deficiency may present with only mild bleeding [23]. Consistent with patient observations, FXI knockout mice can achieve adulthood without bleeding tendencies [43].

Factor XIII Participating in the final reaction of the coagulation pathway, factor XIIIa (FXIIIa) is a transglutaminase that catalyzes peptide bond formation between γ chains of fibrin polymers. Plasma FXIII is activated to FXIIIa by thrombin. Thrombin also converts fibrinogen to fibrin monomers which spontaneously polymerize to form fibrin polymers that comprise the clot. The fibrin polymers are further strengthened by the introduction of *iso* peptide bonds between adjacent γ-glutamyl and ε-lysine residues in the γ chains resulting in a tough, insoluble, cross-linked fibrin clot [22]. FXIIIa also cross-links fibrin α chains to proteins on the exposed extracellular matrix thereby anchoring the growing clot to the wound site (see Figure 1). Intracellular FXIII is present in monocytes and platelets as a homodimer composed of 2 A subunits ($FXIIIA_2$) while plasma FXIII is a non-covalently linked heterotetramer of two catalytic A subunits and two carrier B subunits ($FXIIIA_2B_2$). The FXIII A subunit is encoded by the *F13A1* gene on chromosome 6 (6p25.3-p24.3) with 15 exons spanning over 200kb while the *F13B* gene encoding the FXIII B subunit is found over 12 exons spanning 28kb on chromosome 1 (1q31-q32.1) [23]. Inherited FXIII deficiency is as rare as prothrombin deficiency, with an incidence of approximately 1 in 2,000,000 in the general population [8]. Over 200 cases of inherited FXIII deficiency have been reported [22]. To date, 81 unique mutations have been described and are listed in two online databases: http://www.f13-database.de [44] and http://www.med.unc.edu/isth/mutations-databases/mutations_rare_bleeding_disorders.html [41]. As shown in Table 2, mutations have been found throughout the two FXIII A and B subunits although FXIII deficiency is most commonly attributed to mutations in the catalytic FXIII A subunit. Four mutations have been described for FXIII B subunit. Patients with FXIII deficiency present with an increased tendency to ooze from fibrin clots and delayed wound healing [22]. FXIII A knockout mice demonstrate spontaneous miscarriage in addition to prolonged bleeding and wound healing [45-47]. Patients deficient in FXIII A have been described with quantitative deficiency only, diagnosed by undetectable levels of plasma and platelet FXIII A antigen or activity. FXIII B deficient patients have reduced but detectable FXIII B levels, usually 5-40% in plasma levels [22]. To our knowledge, no qualitative FXIII deficiency (characterized by normal antigen levels but low activity levels) has been described.

Fibrinogen Fibrinogen is a 340kD soluble hexameric glycoprotein composed of two copies each of three subunits (Aα, Bβ, and γ) linked by disulfide bonds. Fibrinogen is cleaved by thrombin to form insoluble fibrin monomers that spontaneously polymerize (see Figure 1). FXIIIa catalyses the covalent cross-linking of polymerized fibrin monomers to form the strong fibrin clot. The fibrinogen α, β and γ chains are encoded by three genes (*FGA*, *FGB*, *FGG*) located as a cluster on the long arm of chromosome 4 (4q28) [31]. The *FGA* gene consists of 6 exons located over 5.4kb, and encodes two isoforms (α and αE) produced by alternate splicing at the 3' end. The α isoform is found in 99% of transcripts, encoded by the first 5 exons. The αE isoform is encoded by all 6 exons, resulting in an additional 236 amino acids. The *FGB* gene is 8.2kb in length, with 8 exons coding for the Bβ chain. The *FGG* gene consists of 10 exons located over 8.4kb and alternate splicing of transcripts results in two isoforms (γ and γ'). The γ' isoform is found in approximately 15% of transcripts, and enhances FXIII binding to thrombin [31]. Inherited fibrinogen deficiency due to mutations in any of the three chains presents as dysfibrinogenemia, hypofibrinogenemia or afibrinogenemia. Severe hypofibrinogenemia and afibrinogenemia (resulting in low to no detectable circulating fibrinogen) has been attributed to deletions, nonsense mutations, splicing mutations and also missense mutations [25]. Dysfibrinogenemia is transmitted as an autosomal dominant trait while hypofibrinogenemia and afibrinogememia are transmitted as autosomal recessive traits. As shown in Table 2, there have been 211 unique mutations published of which 49 mutations are found in *FGA*, 45 in *FGB* and 65 in *FGG*. A database of mutations can be found in the Fibrinogen Database at http://www.geht.org/databaseang/fibrinogen [48].

Multiple Deficiencies

Combined FV/VIII deficiency Patients with combined FV/VIII deficiency have concomitantly low but detectable levels of both coagulation factors [23]. In approximately two thirds of cases, this combined deficiency is caused by mutations in the *ERGIC-53* gene encoding an Endoplasmic Reticulum/Golgi Intermediate Compartment based chaperone protein involved in the intracellular trafficking of both FV and FVIII [25]. The *ERGIC-53* gene, also known as *LMAN1*, is located on chromosome 18 (18q21.3-q22) and encoded by 13 exons. As shown in Table 2, of the 28 mutations reported, 14 are insertions/ deletions, 7 splice, and 7 nonsense mutations. An updated database of mutations can be found at http://www.med.unc.edu/isth/mutations-databases/mutations_rare_bleeding_disorders.html.
Other combined FV/VIII deficiencies have been attributed to mutations in a gene encoded on chromosome 2 (2p21) called Multiple Coagulation Factor Deficiency 2 gene (*MCFD2*). The *MCFD2* gene encodes a protein that acts as a cofactor to ERGIC-53 for intracellular transport of FV and FVIII [25]. There have been nine reported mutations in *MCFD2* of which there are equal numbers each for splicing, frameshift and missense mutations. A database of mutations can be found at the ISTH website: http://www.med.unc.edu/isth/mutations-databases/FVandVIII_2007.html.

Vitamin K dependant coagulation factor deficiencies The vitamin K dependent coagulation factors include FII, FVII, FIX, FX and the anticoagulants protein C, protein S and protein Z. Other vitamin K dependent proteins include bone proteins such as matrix GLA protein and osteocalcin. Multiple deficiencies of vitamin K dependent proteins have been attributed to defects in enzymes involved in the post-translational modification (γ-carboxylation of Glu residues to Gla residues) that is essential for Ca^{2+} binding and subsequent membrane binding by these proteins [25]. To date, there have been 13 mutations reported for the gene encoding vitamin K epoxide reductase complex subunit 1 (VKORC1); these mutations can be found at the human gene mutation database: http://www.hgmd.cf.ac.uk/ac/gene.php?gene=VKORC1. There are 11 known mutations in the γ-glutamyl carboxylase gene (*GGCX*) listed on the database at: http://www.hgmd.cf.ac.uk/ac/gene.php?gene=GGCX. The symptoms of these *inherited* deficiencies of the vitamin K dependent proteins are mimicked by *acquired* vitamin K deficiency that can be induced by dietary vitamin K deficiency or by drugs such as warfarin.

Advances in DNA Sequence Analysis – Towards Personalized Medicine

For the past thirty years the majority of DNA sequencing has been performed using the chain termination method developed by Sanger [49]. In this method, both deoxyribonucleotides and limiting amounts of dideoxyribonucleotides are incorporated into the newly synthesized DNA strand leading to defined chain termination due to the absence of a 3' hydroxyl group in the dideoxyribonucleotides. Since the original method was published in 1977, several advances have been published including the use of fluorescent nucleotides, improved DNA polymerases, automation, improved detection technologies and bioinformatics. These advances led to increased through-put and DNA sequence read length along with decreased cost and time requirements. The HGP was the paragon of Sanger sequencing technology that also drove many of the advances. Even with the recent development of new DNA sequencing technologies, most laboratories still use the Sanger method to determine the nucleotide sequences of DNA fragments. Most DNA sequencing facilities use automated capillary electrophoresis machines in simultaneous arrays of 96 or 384 columns. Read lengths now reach over 1,000 bp with the average cost of a sequence ranging from 50 cents to $20. Other than the inversion of intron 22 in the *F8* gene (which is detected by PCR), determination of the mutations leading to coagulopathies currently utilizes this technology [12, 50]. Combined with recombinant DNA technology and developments in protein expression systems, the Sanger DNA sequencing technology has played a vital role in the dissection of the molecular biology of the coagulation cascade. The contributions of various technology advances to the discovery of gene mutations causing coagulopathies are shown in Figure 2.

An alternate method to the Sanger sequencing technologies was published around the same time by Maxam and Gilbert [51]. This method differed from the Sanger method in that the sequences of purified DNA fragments were determined directly by radiolabelling one end of the DNA and then using chemical modification reactions to cleave the DNA strand in base-specific ways. Initially, the Maxam and Gilbert method was more popular than the Sanger method and played an important role in the characterization of a number of coagulation factor genes [52, 53]. However, the aforementioned advances in Sanger sequencing technology led to its emergence as the preferred method.

Chain termination sequence analysis has now reached a point close to the maximum potential at the lowest cost. As a result, the scientific community is developing new technologies based on entirely different methods of determining DNA sequences. Many of these technologies have recently become accessible to the public at competitive cost effectiveness and are changing the way we determine DNA sequences.

Microelectrophoresis is an effective miniaturization of the traditional chain termination method. By exploiting microfabrication techniques, a single microfluidic circuit has been assembled that brings together clone isolation, amplification, chain termination extension and electrophoresis on a nanoliter scale [54]. This *Lab on a Chip* [55] technology may become a cost effective method for point of care molecular diagnosis.

Sequencing by hybridization (SBH) is a technology that is proving to be a useful tool for re-sequencing genomes to detect differences between an individual and the reference genome. An array of defined oligonucleotides is printed on a chip and labeled sample DNA is hybridized to the chip. The oligonucleotides on the chip differ at each base to be sequenced, thus allowing for detection of which base is at a particular position with respect to the reference genome sequence [56]. This allows detection of mutations in an individual genome and has led to the discovery of a number of novel single nucleotide polymorphisms (SNPs), [57]. Alternative methods to SBH use the ligation of short DNA probes to adapter oligonucleotides to detect specific bases [58]. Shotgun sequencing, involving rolling circle amplification and hybridization to a universal array of 582 pentamer probes, gives a hybridization spectrum that can be compared to the reference genome [59]. Simple chemistry, the lack of enzymes and the use of only dilute solutions of probes keep the cost of shotgun SBH down while maintaining high throughput. Complete Genomics (Mountain View, CA) employs a three-tiered DNA fragment library to identify structural elements of the genome such as Alu repeats, LINEs and SINEs along with independent sequencing of diploid chromosomes. DNA nanoball technology combined with the self assembling array has allowed Complete Genomics to claim the first $1,000 price for completing the sequence of an individual human genome.

MALDI-TOF mass spectrometry (MS), a technology usually associated with proteomics, also has some applications in DNA sequence analysis. At first, MS was used for SNP genotyping [60] but has shown to have the capacity for DNA sequence analysis [61]. Read lengths for MS DNA sequencing have generally been limited to under 100bp [56] but there are specific applications in DNA sequencing that MS is proving to be useful for. These include quantitative DNA analysis, expression analysis, insertions/ deletions detection, methylation analysis and direct RNA sequence analysis [62].

Cyclic array sequencing is at the forefront of next generation sequencing technologies. Key advances that have contributed to the advancement of cyclic array sequencing include emersion amplification, pyrosequencing and the capacity to resolve 2D data outputs from sequencing by synthesis reactions. In emulsion PCR, fragmented sample DNA is ligated to an adapter and then amplified in an oil emulsion where each droplet of water contains the PCR reagents. Each droplet, or polony, contains a single template DNA molecule thereby eliminating competition between templates for a limited number of polymerase molecules and resulting in bias-free amplification [63]. Two studies published in 2005 displayed the potential for next generation sequencing. The first used emulsion PCR and pyrosequencing - a *sequencing by synthesis* method that exploits the release of pyrophosphate from added nucleotides and a luciferase reporter protein to detect sequence data [64]. This method allowed assembly of the *Mycoplasma genitalium* genome at 96% coverage with 99.96%

accuracy in a single 4 hour run [65]. The second study demonstrated the re-sequencing of the genome of an evolved *Esherishia coli* strain using emulsion PCR where one primer was attached to a 1μm paramagnetic bead followed by DNA sequencing by ligation, where a labeled oligonucleotide was ligated to a sequencing primer that had been hybridized to the adapter. This approach yielded approximately 30,100,000 bp of DNA sequence data at 99.7% media raw accuracy [66]. Commercial cyclic array platforms are currently available. These include: (1) 454/Roche which uses emulsion PCR with beads followed by pyrosequencing, (2) Solexa/Illumina which uses solid phase amplification and sequencing by reversible terminators and (3) SOLiD/ Applied Biosystems which uses emulsion PCR with beads followed by sequencing by ligation [67].

First proposed in the 1980's [68], nanopore sequencing involves the passing of ssDNA through a membrane protein such as α-hemolysin [69] which acts as a nanopore embedded in a membrane. Electrical current is applied to the system which forces the DNA to pass through the nanopore resulting in base-specific fluctuations in the conductance of the membrane [56, 70]. At the present time this technology is still in its infancy; however, a vast amount of research is underway to eliminate the problems as there is the potential for nanopore sequencing to emerge as the most cost-effective DNA sequencing method currently proposed. For a more in depth discussion of nanopore technology see [71].

Transmission electron microscopy has also been shown to have applications in DNA sequence analysis. By synthesizing a DNA strand that contains three or four heavy atoms and which is complementary to the target DNA, the DNA hybrid can be observed using a specialized transmission electron microscope. In this way, it is possible to discriminate between the labeled bases due to differences in size. Development of this technology is being carried out by ZS Genetics who claim read lengths of 10,000 to 20,000 bp at a rate of 1.7 billion bp per day [70].

Diagnosis and Treatment of Coagulopathies in the Post-Genomic Era

In 2003, the completion of the HGP opened up many new approaches to studying inherited disorders for the benefit of both scientists and patients. One important outcome of the HGP was the recognition of the importance of SNPs [72] as illustrated by the abundance of SNP databases (for example www.ncbi.nlm.nih.gov/SNP). A *polymorphism* is defined as the least common allele occurring in 1% or more of the population while a *mutation* is a rare difference that occur in less than 1% of the population and usually results in disease [73]. Mutations are responsible for Mendelian diseases – single gene disorders - while SNPs are thought to be one of the major causes of inter-individual variability potentially affecting both disease severity and drug response. This variability can also be affected by other factors such as dietary status and lifestyle, age, and concurrent disease. These undefined complexities sometimes make diagnosis and drug prescription more art than science.

Even in the case of simple monogenic disease (caused by the dysfunction of a single gene), there can be a number of modifier genes that may affect disease progression [74]. The situation is far more complicated in the case of complex diseases where the diagnostic features– called quantitative trait loci (QTL) – result from the activity of two or more genes [75].

Genome-bound variability in drug response is the new discipline of *pharmacogenomics* [76]. One well-studied example of drug reaction variability is the response to warfarin which is used clinically for the treatment of hypercoagulation [77]. The way warfarin affects an individual's coagulation system depends greatly on the allele composition in both the *CYP2C9* and *VKORC1* genes. The *CYP2C9* gene encodes a warfarin metabolizing enzyme whereas the product of the *VKORC1* gene is the enzymatic target of warfarin inhibition. As discussed earlier, the synthesis of the vitamin K-dependent clotting factors requires the γ-carboxylation of specific Gly residues. In this reaction, the reduced form of vitamin K (vitamin K hydroquinone) is converted into vitamin K 2,3-epoxide. The vitamin K epoxide reductase complex (VKORC) regenerates vitamin K hydroquinone thus providing the cofactor for γ-carboxylation. Two common allele variants of *CYP2C9* are associated with low-dose warfarin requirements and increased risk of bleeding [78] while different SNPs in *VKORC1* may result in either warfarin resistance or excessive bleeding upon warfarin treatment [79].

Historically, a *map-based* approach (positional cloning) was widely used for the investigation of gene polymorphism; however, entering the postgenomic era has allowed a step towards a *DNA sequence-based* approach [80]. This includes the sequencing of candidate gene(s) to reveal SNPs associated with peculiarities of the disease in the symptoms or therapeutic response of certain patient(s). While positional cloning may be an easier way to reveal SNPs when the candidate gene is known, utilization of a sequence based approach is also suited to the discovery of new susceptibility genes and even new pathways. With methods of SNPs detection becoming cheaper and more sensitive, a sequence-based approach is rapidly replacing map-based methods [81].

DNA sequence data collected in statistically representative groups can be further used to determine linkage disequilibrium, the non-random association of alleles at different polymorphic sites in a gene. In turn, the association between representative SNPs for a region of the genome with high linkage disequilibrium and a certain phenotype can reveal a set of phenotype-specific biomarkers; these are known as tag SNPs [82]. The identification of tag SNPs allows the characterization of short regions of interest in the genome thereby avoiding the time-consuming and expensive whole-genome sequencing. Analysis of tag SNPs can also lead to the identification of associated or causal genes thereby giving an insight into the molecular mechanisms of common diseases and drug response. The International HapMap Project aims to catalog the SNPs common to the human genome so that others may use the information to discover genes responsible for certain disorders (http://www.hapmap.org/).

Conclusion

During the past 30 years, advances in biotechnology have revolutionized molecular medicine culminating with the completion of the HGP in 2003. Since then, new research advances promise to make an affordable, personal genome sequence a reality. In combination with advances in metabolomics [83], proteomics, transcriptomics [84] and bioinformatics, a new era of high quality, individually-tailored diagnosis and treatment of inherited diseases is attainable in the near future. These developments have the potential to improve human health but must be accompanied by protection of this confidential data and appropriate genetic counseling on the implications of personalized medicine to society.

References

[1] Malone, PC; Agutter, PS. Chapter 2: The Coagulation Cascade and the Consensus Model of DVT. *The Aetiology of Deep Venous Thrombosis: A Critical, Historical and Epistemological Surve.*: Edition 1, Springer; 2008. p. 11-27.

[2] Malone, PC; Agutter, PS. Chapter 5: Coagulation and its Disorders: A History of Haematological Research. *The Aetiology of Deep Venous Thrombosis: A Critical, Historical and Epistemological Survey*1. Edition 1, Springer; 2008. p. 57-67.

[3] Davie, EW; Ratnoff, OD. Waterfall Sequence for Intrinsic Blood Clotting. *Science*, 1964, 145:1310-2.

[4] MacFarlane, RG. An Enzyme Cascade in the Blood Clotting Mechanism, and Its Function as a Biochemical Amplifier. *Nature*, 1964, 202, 498-9.

[5] Monroe, DM; Hoffman, M. What does it take to make the perfect clot? *Arteriosclerosis, Thrombos. Vasc. Biol.*, 2006, 26:41-8.

[6] Collins, FS; Morgan, M; Patrinos, A. The Human Genome Project: Lessons from Large-Scale Biology. *Science*, 2003, 300:286 - 90.

[7] Church, GM. Genomes for ALL. *Scientific American*, 2006, 294:46-54.

[8] Peyvandi, F; Jayandharan, G; Chandy, M; Srivastava, A; Nakaya, SM; Johnson, MJ; et al. Genetic diagnosis of haemophilia and other inherited bleeding disorders. *Haemophilia*, 2006, Suppl 3:82-9.

[9] Federici, AB. Diagnosis of inherited von Willebrand disease: a clinical perspective. *Semin. Thrombos. Hemostas.*, 2006, 32:555-65.

[10] Bolton-Maggs, PH; Pasi, KJ. Haemophilias A and B. *Lancet*, 2003, 361:1801-9.

[11] Graw, J; Brackmann, HH; Oldenburg, J; Schneppenheim, R; Spannagl, M; Schwaab, R. Haemophilia A: from mutation analysis to new therapies. *Nature Rev. Genet.*, 2005, 6:488-501.

[12] Lakich, D; Kazazian, HH; Jr. Antonarakis, SE; Gitschier, J. Inversions disrupting the factor VIII gene are a common cause of severe haemophilia A. *Nature Genet.*, 1993, 5: 236-41.

[13] Oldenburg, J; El-Maarri, O. New insight into the molecular basis of hemophilia A. *Int. J. Hematol.*, 2006, 83:96-102.

[14] Castaldo, G; D'Argenio, V; Nardiello, P; Zarrilli, F; Sanna V; Rocino A; et al. Haemophilia A: molecular insights. *Clin. Chem. Lab. Med.: CCLM / FESCC*, 2007, 45: 450-61.

[15] Taylor, SA; Duffin, J; Cameron, C; Teitel, J; Garvey, B; Lillicrap, DP. Characterization of the original Christmas disease mutation (cysteine 206----serine): from clinical recognition to molecular pathogenesis. *J. Thrombos. Haemostas.*, 1992, 67:63-5.

[16] Ruggeri, ZM. The role of von Willebrand factor in thrombus formation. *Thrombos. Res.*, 2007, 120 Suppl 1:S5-9.

[17] Peake, I; Goodeve, A. Type 1 von Willebrand disease. *J. Thrombos. Haemostas.,* 2007, Jul, 5 Suppl 1:7-11.

[18] Michiels, JJ; Berneman, Z; Gadisseur, A; van der Planken, M; Schroyens, W; van de Velde, A, et al. Classification and characterization of hereditary types 2A, 2B, 2C, 2D, 2E, 2M, 2N, and 2U (unclassifiable) von Willebrand disease. *Clin. Appl. Thrombos./Hemostas.*, 2006, 12:397-420.

[19] Lillicrap, DP. Von Willebrand Disease: An Introduction For The Primary Care Physician. *Treatment of Hemophilia,* 2009, 47 (Downloaded from: http://www.wfh.org).

[20] Peyvandi, F; Kaufman, RJ; Seligsohn, U; Salomon, O; Bolton-Maggs, PH; Spreafico, M; et al. Rare bleeding disorders. *Haemophilia*, 2006, Jul, 12 Suppl 3:137-42.

[21] Lak, M; Sharifian, R; Peyvandi, F; Mannucci, PM. Symptoms of inherited factor V deficiency in 35 Iranian patients. *Br. J. Haematol.*, 1998 103:1067-9.

[22] Peyvandi, F; Cattaneo, M; Inbal, A; De Moerloose, P; Spreafico, M. Rare bleeding disorders. *Haemophilia*, 2008, 14 Suppl 3:202-10.

[23] Peyvandi, F; Duga, S; Akhavan, S; Mannucci, PM. Rare coagulation deficiencies. *Haemophilia*, 2002 8:308-21.

[24] Akhavan, S; Mannucci, PM; Lak, M; Mancuso, G; Mazzucconi, MG; Rocino, A; et al. Identification and three-dimensional structural analysis of nine novel mutations in patients with prothrombin deficiency. *Thrombos. Haemostas.*, 2000 84: 989-97.

[25] Mannucci, PM; Duga, S; Peyvandi, F. Recessively inherited coagulation disorders. *Blood*, 2004 104:1243-52.

[26] Sun, WY; Witte, DP; Degen, JL; Colbert, MC; Burkart, MC; Holmback, K; et al. Prothrombin deficiency results in embryonic and neonatal lethality in mice. *Proc. Natl. Acad. Sci. U.S.A.,* 1998 95:7597-602.

[27] Mullins, ES; Kombrinck, KW; Talmage, KE; Shaw, MA; Witte, DP; Ullman, JM, et al. Genetic elimination of prothrombin in adult mice is not compatible with survival and results in spontaneous hemorrhagic events in both heart and brain. *Blood*, 2008 113:696-704.

[28] Acharya, SS; Coughlin, A; Dimichele, DM, North American Rare Bleeding Disorder Study G. Rare Bleeding Disorder Registry: deficiencies of factors II, V, VII, X, XIII, fibrinogen and dysfibrinogenemias. *J. Thromobos. Haemostas.,* 2004 2:248-56.

[29] Tilley, R,; Mackman N. Tissue factor in hemostasis and thrombosis. *Semin. Thrombos. Hemostas.*, 2006, 32:5-10.

[30] Bugge, TH; Xiao, Q; Kombrinck, KW; Flick, MJ; Holmbäck, K; Danton, MJ; et al. Fatal embryonic bleeding events in mice lacking tissue factor, the cell-associated initiator of blood coagulation. *Proc. Natl. Acad. Sci. U.S.A.*, 1996 , 93:6258-63.

[31] Bolton-Maggs, PH; Perry, DJ; Chalmers, EA; Parapia, LA; Wilde, JT; Williams, MD; et al. The rare coagulation disorders-review with guidelines for management from the United Kingdom Haemophilia Centre Doctors' Organisation. *Haemophilia,* 2004, 10: 593-628.

[32] Horne, MKI; McCloskey, DJ. Factor V Leiden as a common genetic risk factor for venous thromboembolism. *J. Nurs. Scholar.*, 2006 38:19-25.

[33] Asselta, R; Tenchini, ML; Duga, S. Inherited defects of coagulation factor V: the hemorrhagic side. *J. Thromobos. Haemostas.*, 2006, 4:26-34.

[34] Cui, J; O'Shea, KS; Purkayastha, A; Saunders, TL; Ginsburg, D. Fatal haemorrhage and incomplete block to embryogenesis in mice lacking coagulation factor V. *Nature*, 1996 384:66-8.

[35] Peyvandi, F; Mannucci, PM; Lak, M; Abdoullahi, M; Zeinali, S; Sharifian, R; et al. Congenital factor X deficiency: spectrum of bleeding symptoms in 32 Iranian patients. *Br. J. Haematol.*, 1998 102:626-8.

[36] McVey, JH; Boswell, E; Mumford, AD; Kemball-Cook, G; Tuddenham, EG. Factor VII deficiency and the FVII mutation database. *Human Mut.*, 2001 17:3-17.

[37] Mariani, G; Herrmann, FH; Dolce, A; Batorova, A; Etro, D; Peyvandi, F; et al. Clinical phenotypes and factor VII genotype in congenital factor VII deficiency. *Thromobos. Haemostas.*, 2005 93:481-7.

[38] Rosen, ED; Chan, JC; Idusogie, E; Clotman, F; Vlasuk, G; Luther, T; et al. Mice lacking factor VII develop normally but suffer fatal perinatal bleeding. *Nature*, 1997 390:290-4.

[39] Uprichard, J; Perry, DJ. Factor X deficiency. *Blood Rev.*, 2002 16:97-110.

[40] Dewerchin, M; Liang, Z; Moons, L; Carmeliet, P; Castellino, FJ; Collen, D; et al. Blood coagulation factor X deficiency causes partial embryonic lethality and fatal neonatal bleeding in mice. *Thromobos. Haemostas.*, 2000 83:185-90.

[41] Seligsohn, U, Zivelin, A; Inbal, A. Inherited deficiences of coagulation factors II, V, VII, X, XI and XIII and combined deficiencies of factors V and VII and of the vitamin K-dependent coagulation factors. *In:* Lichtman MA, Beutler E, Kipps TJ, Seligsohn U, Kaushansky K, Prchal JT, *Eds., Williams Hematology, Chapter 116, McGraw-Hill, New York,* 2006.

[42] Franchini, M; Veneri, D; Lippi, G. *Inherited factor XI deficiency: a concise review.* Hematol., 2006 11:307-9.

[43] Gailani, D; Lasky, NM; Broze, GJ, Jr. A murine model of factor XI deficiency. *Blood Coag. Fibrinolys.*, 1997 8:134-44.

[44] Ivaskevicius, V; Seitz, R; Kohler, HP; Schroeder, V; Muszbek, L; Ariens, RA; et al. International registry on factor XIII deficiency: a basis formed mostly on European data. *Thromobos. Haemostas.*, 2007 97:914-21.

[45] Lauer, P; Metzner, HJ; Zettlmeissl, G; Li, M; Smith, AG; Lathe, R; et al. Targeted inactivation of the mouse locus encoding coagulation factor XIII-A: hemostatic abnormalities in mutant mice and characterization of the coagulation deficit. *Thromobos. Haemostas.*, 2002 88:967-74.

[46] Inbal, A; Lubetsky, A; Krapp, T; Castel, D; Shaish, A; Dickneitte, G, et al. Impaired wound healing in factor XIII deficient mice. *Thromobos. Haemostas.*, 2005 94:432-7.

[47] Koseki-Kuno, S; Yamakawa, M; Dickneite, G; Ichinose A. Factor XIII A subunit-deficient mice developed severe uterine bleeding events and subsequent spontaneous miscarriages. *Blood*, 2003 102:4410-2.

[48] Hanss, M; Biot, F. A Database For Human Fibrinogen Variants. *Ann. N.Y. Acad. Sci.*, 2001 936:89-90.

[49] Sanger, F; Air, GM; Barrell, BG; Brown, NL; Coulson, AR; Fiddes, CA; et al. Nucleotide sequence of bacteriophage phi X174 DNA. *Nature*, 1977 265:687-95.

[50] Naylor, J; Brinke, A; Hassock, S; Green, PM; Giannelli F. Characteristic mRNA abnormality found in half the patients with severe haemophilia A is due to large DNA inversions. *Hum. Molec. Genet.*, 1993 2:1773-8.

[51] Maxam, AM; Gilbert, W. A new method for sequencing DNA. *Proc. Natl. Acad. Sci. U.S.A.*, 1977 74:560-4.

[52] MacGillivray, RTA; Degen, SJ; Chandra, T; Woo, SL; Davie, EW. Cloning and analysis of a cDNA coding for bovine prothrombin. *Proc. Natl. Acad. Sci. U.S.A.*, 1980 77:5153-7.

[53] Chung, DW; Rixon, MW; MacGillivray, RTA; Davie, EW. Characterization of a cDNA clone coding for the beta chain of bovine fibrinogen. *Proc. Natl. Acad. Sci. U.S.A.*, 1981 78:1466-70.

[54] Paegel, BM; Blazej, RG; Mathies RA. Microfluidic devices for DNA sequencing: sample preparation and electrophoretic analysis. *Curr. Opin. Biotechnol.*, 2003 14:42-50.

[55] Blazej, RG; Kumaresan, P; Mathies, RA. Microfabricated bioprocessor for integrated nanoliter-scale Sanger DNA sequencing. *Proc. Natl. Acad. Sci. U.S.A.*, 2006 103:7240-5.

[56] Shendure, JA; Porreca, GJ; Church, GM. Overview of DNA sequencing strategies. *Curr. Protocol. Mol. Biol.*, 2008 Chapter 7:Unit 7 1.

[57] Patil, N; Berno, AJ; Hinds, DA; Barrett, WA; Doshi, JM; Hacker, CR, et al. Blocks of limited haplotype diversity revealed by high-resolution scanning of human chromosome 21. *Science*, 2001 294:1719-23.

[58] Drmanac, R; Drmanac, S; Chui, G; Diaz, R; Hou, A; Jin, H; et al. Sequencing by hybridization (SBH): advantages, achievements, and opportunities. *Adv. Biochem. Eng./Biotechnol.*, 2002 77:75-101.

[59] Pihlak, A; Bauren, G; Hersoug, E; Lonnerberg, P; Metsis, A; Linnarsson, S. Rapid genome sequencing with short universal tiling probes. *Nature Biotechnol.*, 2008 26:676-84.

[60] Gut, IG. DNA analysis by MALDI-TOF mass spectrometry. *Human Mut.*, 2004 23:437-41.

[61] Edwards, JR; Ruparel, H; Ju, J. Mass-spectrometry DNA sequencing. *Mut. Res.*, 2005 573:3-12.

[62] Ragoussis, J; Elvidge, GP; Kaur, K; Colella, S. Matrix-assisted laser desorption/ionisation, time-of-flight mass spectrometry in genomics research. *PLoS Genetics*, 2006 2:e100.

[63] Schuster, SC. Next-generation sequencing transforms today's biology. *Nature Meth.*, 2008 5:16-8.

[64] Ronaghi, M. Pyrosequencing sheds light on DNA sequencing. *Genome Res.*, 2001 11:3-11.

[65] Margulies, M; Egholm, M; Altman, WE; Attiya, S; Bader, JS; Bemben, LA, et al. Genome sequencing in microfabricated high-density picolitre reactors. *Nature*, 2005 437:376-80.

[66] Shendure, J; Porreca, GJ; Reppas, NB; Lin, X; McCutcheon, JP; Rosenbaum, AM; et al. Accurate multiplex polony sequencing of an evolved bacterial genome. *Science*, 2005 309:1728-32.

[67] Rusk, N; Kiermer, V. Primer: Sequencing--the next generation. *Nature Meth.*, 2008 5:15.

[68] Deamer, DW; Akeson M. Nanopores and nucleic acids: prospects for ultrarapid sequencing. *Trends Biotechnol.*, 2000 18:147-51.

[69] Kasianowicz, JJ; Brandin, E; Branton, D; Deamer, DW. Characterization of individual polynucleotide molecules using a membrane channel. *Proc. Natl. Acad. Sci. U.S.A.*, 1996 93:13770-3.

[70] Gupta, PK. Single-molecule DNA sequencing technologies for future genomics research. *Trends Biotechnol.*, 2008 26:602-11.

[71] Branton, D; Deamer, DW; Marziali, A; Bayley, H; Benner, SA; Butler, T, et al. The potential and challenges of nanopore sequencing. *Nature Biotechnol.*, 2008 26:1146-53.

[72] Brookes, AJ. The essence of SNPs. *Gene*, 1999 234:177-86.

[73] Marez, D; Legrand, M; Sabbagh, N; Guidice, JM; Spire, C; Lafitte, JJ, et al. Polymorphism of the cytochrome P450 CYP2D6 gene in a European population: characterization of 48 mutations and 53 alleles, their frequencies and evolution. *Pharmacogenetics*, 1997 7:193-202.

[74] Ginsburg, D. Identifying novel genetic determinants of hemostatic balance. *J. Thromobos. Haemostas.*, 2005 3:1561-8.

[75] Lemmerhirt, HL; Broman, KW; Shavit, JA; Ginsburg, D. Genetic regulation of plasma von Willebrand factor levels: quantitative trait loci analysis in a mouse model. *J. Thromobos. Haemostas.*, 2007 5:329-35.

[76] Goldstein, DB; Tate, SK; Sisodiya, SM. Pharmacogenetics goes genomic. *Nature Rev. Genet.*, 2003 4:937-47.

[77] Kamali, F. Genetic influences on the response to warfarin. *Curr. Opin. Hematol.*, 2006 13:357-61.

[78] Aithal, GP; Day, CP; Kesteven, PJ; Daly, AK. Association of polymorphisms in the cytochrome P450 CYP2C9 with warfarin dose requirement and risk of bleeding complications. *Lancet*, 1999 353:717-9.

[79] Rieder, MJ; Reiner, AP; Gage, BF; Nickerson, DA; Eby, CS; McLeod, HL, et al. Effect of VKORC1 haplotypes on transcriptional regulation and warfarin dose. *New Eng. J. Med.*, 2005 352:2285-93.

[80] Peltonen, L. Positional cloning of disease genes: advantages of genetic isolates. *Hum. Hered.*, 2000 50:66-75.

[81] Hacia, JG. Resequencing and mutational analysis using oligonucleotide microarrays. *Nature Genet.*, 1999 21(1 Suppl):42-7.

[82] Ahmadi, KR; Weale, ME; Xue, ZY; Soranzo, N; Yarnall, DP; Briley, JD, et al. A single-nucleotide polymorphism tagging set for human drug metabolism and transport. *Nature Genet.*, 2005 37:84-9.

[83] Keun, HC; Athersuch, TJ. Application of metabonomics in drug development. *Pharmacogenomics*, 2007 8:731-41.

[84] Collins, CD; Purohit, S; Podolsky, RH; Zhao, HS; Schatz, D; Eckenrode, SE, et al. The application of genomic and proteomic technologies in predictive, preventive and personalized medicine. *Vasc.Pharmacol.*, 2006 45:258-67.

In: The Human Genome: Features, Variations... ISBN: 978-1-60741-695-1
Editor: Akio Matsumoto and Mai Nakano

Chapter 5

Itinerant Genome

Branko Borštnik, Borut Oblak and Danilo Pumpernik
National Institute of Chemistry
Ljubljana, Slovenia

Abstract

An estimate of the information content of human genome is presented. The evolutionary processes are pictured as an information channel and the loss/gain of the information content is discussed. The evolutionary mechanisms such as nucleotide replacements, insertions, deletions and replication slippages that are shaping the human genome are discussed. The nucleotide replacement processes are modeled in terms of minimal number of parameters such as the transition/transversion ratio and some context dependent parameters. The variation of DNA sequences is scrutinized for the period since mammalian origin and for the time since the last common ancestor to human and chimpanzee. A special emphasis is given to the distinction between four sequence categories that emerge when the genomic sequences are partitioned into i) CpG rich islands, ii) Alu - type short interspersed repeats, iii) the cross section of these two categories and iv) the class of all the remaining genomic sequences. It is shown that although the Alu sequences seem to be a typical representative of "junk DNA" they participate in CpG islands, which are known to embed the regulatory regions and are thus the sequences of supreme functional importance.

1. Introduction

The diversity of human race has its source in the diversity of genetic material packed into the chromosomes 1 to 22, X and Y. All but identical twins are carriers of different genetic material. The genetic diversity is maintained by two mechanisms in the life cycle of germ cells: various types of mutations in the course of DNA replication and recombination during meiosis. The recombination process, where maternal and paternal chromosomes are recombined in a way that the nascent chromosome is formed as a mosaic of maternal and paternal segments is a rather deterministic process and represents a result of miraculous potency of generation of complexity. The stochasticity enters only through the process of determination of the crossing over sites. The mutational processes, on the other hand, such as the replacements, insertions, deletions and segmental duplications represent a kind of imperfection of the functioning of the cellular machinery in the form of inability to reproduce the DNA sequences with absolute fidelity. This, latter mechanism, represents thus the basic source of genomic variability of any species, the meiotic recombination just serves as a means of regrouping the genetic variants and to search for optimal combination of variants. In addition, the evolutionary process should be visualized as a composite process running on two levels. On the level of genotype takes place the submission component of the biological evolution. The submitted variants are later evaluated at the level of phenotype. The evaluation is performed on the ground of the Darwinian principle of the survival of the fittest. This means that the new variants of the genetic material are accepted or rejected – depending upon their influence on the selection coefficient that determines the number of progeny in the population dynamics of a species.

The landmarks that are of crucial importance for the studies of the last part of the evolutionary history that led to the existence of human species are the accomplishments in the field of the whole genome sequencing projects. Human genome was the first among the mammalian genomes to pass the initial phase of sequencing and partial annotation (Lander et al. 2001). Two years later the mouse genome was available (Waterston et al. 2002) and the rat genome was running in parallel (Gibbs et al. 2004). The comparison between the genomes belonging to the organisms with comparable complexity represents a basic approach to study the evolutionary phenomena. The most valuable data that enable the insight into the fine evolutionary details can be derived from the comparison of two or several closely related species. Since the subject of our interest is human species, the closest relatives are hominids or primates. The biological family Hominidae consists of the "great apes", what means the extinct and extant humans, chimpanzees, gorillas and orangutans. The chimpanzee genome and its comparison with human genome appeared in 2005 by Mikkelsen et al. Two years later also a great part of rhesus genome was sequenced (Gibbs et al. 2007) . The gorilla genome is in progress.

The fundamental question that has nearly no answer yet refers to the evolution of cognitive capabilities. During the primate evolution only humans developed unprecedented information processing capabilities, but nobody knows which part of DNA sequence is responsible for these abilities. It is known that on the level of phenotype the information processing capabilities are based upon more advanced architecture of neural wiring (Hill et al. 2005). The answer to the question regarding the genetic basis of consciousness and other advanced brain properties is usually searched in the context of positive selection (Nielsen et al. 2005). What one would hope to obtain would be a list of genes being expressed predominantly in the brain tissues. The gene coding regions are proclaimed to be positively selected when the relative number of nonsynonymous replacements exceeds the relative number of synonymous replacements. The analyses of approximately of one half of existing human genes and their comparison with the chimpanzee orthologs show (Nielsen et al. 2005) that one can indeed find several hundreds of genes that exhibit positive selection but with only marginal statistical significance. Only for a dozen of genes the statistical significance has a firm basis. The positive selection issue did not support the idea that the genes being expressed in brain tissues evolve fast. It seems more feasible that human cognitive capabilities go on the account of the differences in the regulatory regions either in the transcription factor binding sites (TFBS), promotor regions, miRNA or other gene regulatory elements (Griffith et al. 2008). One can thus understand that to capture the differences between human and other primates it is necessary to perform a lot of small steps in a careful and patient way. In this chapter we shall present some results that refer to some details regarding the evolution of genomic sequences within the mammalian region of the vertebrate phylogeny. We shall first make a glance of mammalian genomes, comment their information content and present the molecular evolution as a transmission of information through an information channel. Then we shall present the results of the analyses of mutational events that emerge from the comparison of the genomic DNA sequences.

2. Mammalian Genomes

The living forms of mammals exhibit well diversified organization of genetic material within their genomes. In every species the total amount of DNA is close to 3 billions of base pairs (3 Gbp) while the number of chromosomes exhibits a wide range of values as seen in Figure 1. Rearranging the DNA material does not represent a major issue in evolutionary process. In human history the last event of this type took place within the last few million years after the human - chimpanzee split. Chimpanzees and gorillas possess each 23 autosomes. In human evolutionary history two of chromosomes were fused and the result of fusion is known as the chromosome no. 2 and it is the second largest among the human chromosomes. The fusion site is well apparent in the DNA sequence (Ijdo 1991). The difference in the chromosome number is the only major change within the hominoids. The main differences come on the account of minor differences – point mutations, for instance, that are 30 million in number when comparing human and chimpanzee. The number of differences is approximately ten times greater than the number of differences between human individuals (Zhao et al. 2002, 2006). When studying the mammalian genomes one can also put under scrutiny mouse and rat. When a study is aimed towards the analyses of triple comparison, which enables to determine the directionality of mutational changes (Jiang 2006) one ususally processes the triple alignments of human, chimpanzee and macaque genomic sequences. The following features are put under scrutiny when related genomes are compared: i) genomic size; ii) the extent of differences in the way how DNA is distributed among the chromosomes; iii) the gene number and their organization; iv) the repeat content of the genome - the autonomous and nonautonomous transposable elements in particular; v) the number and the distribution of CpG islands.

As far as the genome size is concerned there are no drastic differences between mammals. Human and macaque genome sizes are both close to 2.9 Gbp - not to count the exact centromeric and telomeric repeat configurations. The rodent genomes are to some extent smaller comprising 2.75 Gbp in rat and 2.6 Mbp in mouse. In both pairs of organisms there are several chromosomes with highly similar content. 19 of 22 macaque autosomes can be mapped 1:1 with the corresponding human chromosomes. The remaining three chromosomes exhibit a mosaic composition. A similar situation is revealed also in the case of mouse - rat comparison. 13 mouse chromosomes have one to one correspondence with the analogous rat chromosomes and the remaining 7 chromosomes have the mosaic structure with two to three segments coming from the chromosomes that contribute to the DNA pool with mixed content. It is worth to note that human - mouse comparison shows that the two organisms pack the genetic material into the chromosomes in rather different manner. This is understandable because the common ancestor to primates and murids lived 70 to 80 Myrs ago.

The investigation of the gene content of different mammalian genomes lead to interesting findings. Early estimates of gene number in human genome was of the order of magnitude of 100,000. This was considered to be a fair overshoot with regard to the gene number of low complexity organisma such as worms that may possess between 10,000 and 20,000 genes. However, the results of the initial sequencing of human genome reduced the number of protein coding genes to 35,000 and more exact studies led to a further reduction to 23,000. This numbers is also a good estimate for the number of genes of other primates and rodents. There are of course major differences in gene content between various mammalian species. Mice and rats (Goodfrey 2004), for instance, are heavily dependent upon the recognition of odors and they possess close to 1000 functional olfactory receptor (OR) genes. Humans and chimpanzees (Malnic 2004, Go 2008), on the other hand, depend more on recognitions based upon the visual perceptions and possess less than 400 active OR genes.

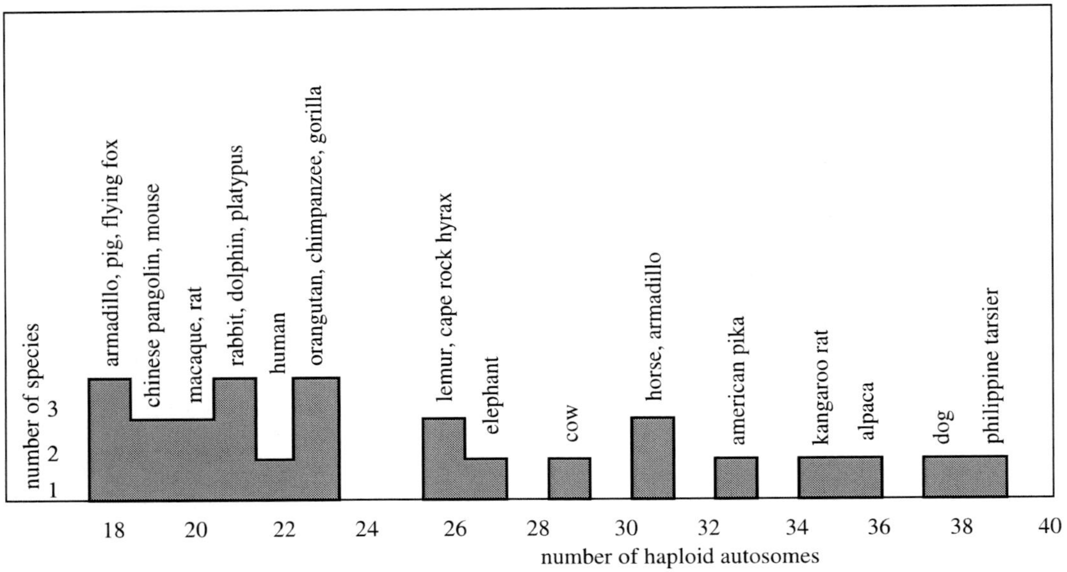

Figure 1. The distribution of number of haploid autosomes in 25 mammalian organisms whose genomes are sequenced or they are in the process of being sequenced.

The specificity of mammalian genomes is very high content of interspersed repeats. Long interspersed nuclear elements (LINEs) and short interspersed nuclear elements (SINEs) are two most abundant representatives of interspersed repeats. They are the product of transposon activities. LINEs are autonomous transposable elements - few of them are still active in the human genome. They also mediate the proliferation of nonautonomous elements such as SINEs. All the repeat elements contribute to nearly one half of the content of human genomes. The most numerous SINEs are the Alu sequences. In human genome there are more than a million of Alu elements. Full length Alu element is 280 bp long. Its internal structure has a form of an imperfect dimer with several inserts. One among the inserts contains two polyadenine stretches. Polyadenines are nearly always also present in the 3' flank of Alu sequences. Since the Alu elements themselves represent a significant portion of mammalian genomes they play an important role in genome evolution as we shall see in section 6. Comparing the human and chimpanzee genome it turns out that both species have a minor fraction of their own Alu lineages that were introduced in recent evolutionary history. The comparison between murids and rodents shows that in spite of the fact that their common ancestor possessed a significant number of Alu sequences the situation is different today. In rodents the SINEs mutate much faster than in primates so that the old sequences are hardly recognizable and consequently it is hard to perform the comparisons.

A phenomenon that gives most severe impact on the evolutionary processes in mammalian ascent is the increased mutability of CpG dinucleotides (Fryxell et al. 2000, 2005). The transitions C=>T and G=>A at the complementary strand comprise a quarter of single nucleotide replacements in the course of time since human and chimpanzee separated. The increased mutability goes on the account of cytosine methylation that makes the CpG dinucleotides vulnerable to mutations. The consequence of such an increased mutability of CpG dinucleotides is their reduced content. They are ten times less frequent than expected. There are, however, certain regions where CpG dinucleotides exhibit nearly normal density. Such regions are called CpG islands (CGI) (Ponger et al. 2001) that have important role since they are positioned in the promoter region of three quarters of human genes.

3. Classification of Mutational Changes

The mutational changes that affect the longest DNA segment are chromosome fusion or chromosome breakage processes and the pericentric inversion events. The processes just mentioned do not change the amount of DNA sequences. The process that increases the amount of DNA sequences are segmental duplications. The human genome contains approximately 5% of DNA sequences that are the result of segmental duplication. The segments of length 10 kbp and 50 kbp can be copied to another site on the same chromosome or to other chromosome. The duplications are running all the time. The sequential divergence of duplicated segments is proportional to the time that elapsed since the divergence occurred. The evolutionary history of human as well as rat genome unravels the fact that the segmental duplications are coming in bursts. In some cases the accelerated pace of the segmental duplications is connected with the speciation events. Human genome contains approximately 5% of duplicated segments while macaque genome exhibits only half of this amount. Within the murid lineage the segmental duplications were running with lower pace as in the case of primate lineage. The genome of rat contains close to 3% DNA in duplicated regions and the genome of mouse between 1% and 2%. Large scale (on the level between kilobase and megabase) deletions are also important events that are giving a significant quantitative contribution to the genome identity. By comparing the genomes one can find out what was deleted in a certain lineage. Human genome, for instance, lost in the period since the last common ancestor to human and chimpanzee roughly 8 Mbp of sequences in 150 events. Also the chimpanzee genome went through a similar sequence of events.

Most frequently the mutational changes only modify short nucleotide segments. In such a case one speaks about point mutations. One nucleotide replacements, deletions and insertions are their most numerous form. Point mutations are definitively the consequence of the susceptibility of molecular processes to the stochastic effects what cause the imperfections in the replication processes on the molecular scale. More complex organisms can afford better control in the form of proof reading mechanisms what enables them to achieve lower error thresholds. In mammals the fidelity of replicating the DNA sequences is of the order of magnitude of one per billion. However, the fidelity of DNA replication also varies along the chromosomes. There are so called mutational hot spots (Ollila 1996) where the mutations are accumulated faster than in average. An example of mutational hot spots are the sites where the repeat enrichment mechanism is operative (Borštnik et al. 2002, 2004, 2005). Human genome comprises approximately 2% of its content in the form of short tandem repeats (STR). This category of sequences, also termed short sequence repeats (SSR), microsatellites, variable number tandem repeats (VNTR) have the form of mononucleotide up to hexanucleotide tandem repeats. They are subjected to a special mutational mechanisms due to which the repeats change their length and is therefore called the repeat enrichment or repeat amplification mechanism. The repeat enrichment mechanism has its root in the fact that the translational symmetry of the template strand produces degenerate energetic states in the landscape of the template DNA strand enzyme and nascent DNA strand enzyme interactions. Such a degeneracy makes the mechanism of DNA synthesis vulnerable to errors in the form of so called slippage events that occur when the DNA ribbon slips out of order to a new position relative to the polymerase enzyme, forward or backward for one or several monomeric repeat units. The repeat enrichment mechanism can be explained in simple words as a process of subsequent copying a text through a mechanism which is susceptible to spontaneous random modification, which can generate repeats. When such repeats are copied, errors are generated as in the case when a typist, copying a string of characters that is several times repeated, would likely lengthen or shorten the number of repeats since the focus of his sight would accidentally slip to the left or to the right repeat monomer. If during the functioning of the cellular machinery the template strand slips backward, for instance, the resulting nascent strand will be prolonged; in the opposite case shortening will result. A similar mechanism is operative also in the case of another important molecular cellular process: DNA recombination at meiosis. Also in this case the repetitive DNA sequence represents a drawback since a vital phase of recombination is the pairing of complementary strands of highly homologous regions of maternal and paternal regions of DNA sequences. In repetitive regions the process of homologous pairing has multiple realizations and with a certain probability the recombination process results in the prolongation of one and the shortening of the other newly recombined strand. The above mentioned mechanisms become operative only above a certain threshold repeat length that is approximately ten base pairs.

The value of the threshold repeat length is approximately equal to the extent of the contact between the DNA helix and polymerase enzyme.

4. Information Content of the Human Genome

The information content of DNA sequences can be evaluated in terms of the fundamental expression for the Shannon entropy (Shannon 1948)

$$H = -\Sigma_i \, p_i \log_2 p_i$$

where H is the information content per unit of sequence and p_i is the probability of appearance of i-th DNA constituent. The smallest possible constituent is a single nucleotide and in such a case one would obtain H_{nt}=2 bits per nucleotide if all four nucleotides are equiprobable. The non-uniform composition and sequential correlations of DNA constituents are diminishing the information content. The effect of the correlations can be taken into account by dividing the sequence into the k-tuplets and taking the limit

$$H_{nt} = \lim\left(k \to \infty\right)\frac{1}{k}\sum_i p_{k,i} \log_2 p_{k,i} \tag{1}$$

where k is the word (unit of sequence) length. The full implementation of eq. (1) to genomic sequences is impossible. One can determine with confidence the $p_{k,i}$ values only for k<10. At higher k values the low accuracy of $p_{k,i}$ of rarely occurring words is hindering the calculations. In order to perform the limiting procedure required by eq. (1) properly, one should proceed with k values beyond the DNA correlation length. It was shown by Peng et al. long time ago (Peng et al. 1992) that the correlation length of DNA sequences has no upper limit for non-selective genomic sequences and we have shown (Borštnik et al. 1993) that for the exonic sequences the correlation length is several hundred base pairs. The results for human genome for k=1 to 6 show that the information content per nucleotide decays from 1.98 at k=1 to 1.93 at k=6 for non-selective genomic regions and from 1.93 at k=1 to 1.91 at k=6 for exons. The apparently lower H_{nt} value of exonic sequences is an artifact due to our inability to evaluate eq. (1) at large enough k values.

Another factor that also determines the amount of information carried by the genomic sequences is the total sequence length N_{nt}

$$H_{tot} = N_{nt}\, H_{nt} \tag{2}$$

Taking the time derivative of this equation (and considering that $\Sigma_i dp_i / dt = 0$ since $\Sigma_i p_i = 1$) one obtains

$$dH/dt = -\, dN_{nt}/dt\, \Sigma_i\, p_i \log_2 p_i - N_{nt}\, \Sigma_i\, dp_i/dt \log_2 p_i. \tag{3}$$

The first factor on the right hand side describes the rate of lengthening/shortening through insertions and deletions (indels) events. It turns out that the stochastic component of indel process is so strong that the results of the alignment of related sequences does not provide us sufficient data even to infer the sign of $dN_{nt}/\mathrm{d}t$. In what follows we shall not be concerned with indel processes but will focus upon the nucleotide replacements that may change the nucleotide populations and can thus be quantified by the $\mathrm{d}p_i/\mathrm{d}t$ terms in eq. (3).

5. Molecular Evolution as the Information Channel

The information theory defines the channel capacity as the maximum information that can be transmitted over a communication channel. The substantial component of the perpetuation of life represents the transmission of information. It is transmitted from one generation of living organism to the next one in the form of DNA sequences but also through the epigenetic mechanisms such as chromatine remodeling, small RNAs and cytosine methylation mechanisms. We shall be only interested in the in the transmission of information in the framework of genetic mechanisms. Each DNA replication represents the propagation of the genetic information from the existing to the next cell generation – be it somatic or germ cell line. Due to the error prone replication mechanism the next generation of DNA sequence can be modified. In the information theory formalism the identity of each DNA sequential element can be expressed in terms of the mutational replacement probability matrix $\boldsymbol{W}$. The matrix elements w_{ij} represent the probability that the sequential element of type j is copied in next generation in a way that it obtains the identity of the element of i-th type. Computationally the sequential elements are usually taken in the form of mononucleotides , dinucleotides (Michel 2007) or trinucleotides (Yampolsky et al. 2005).

If the elements of DNA sequence are taken in the form of mononucleotids the $\boldsymbol{W}$ matrix is a rang 4 matrix. Due to the strand symmetry requirements A,T as well as C,G have to appear in a symmetric way and the $\boldsymbol{W}$ matrix can be expressed in terms of four different matrix elements representing the probabilities that the nucleotides retain their identities, or that one class is replaced in the replication process by the nucleotides of another class. The nucleotides are classified as purines (R) or pyrimidines (Y) and with regard to the strength in hydrogen bonding between the base pairs that is strong (S) between C and G and weak (W) between A and T. The population of four nucleotides is defined with one parameter since the strand symmetry requires that $p_C=p_G=p$ and $p_A=p_T=P$. The nomalization condition is $p_C+p_G+p_A+p_T=1$. There are three constraints on four components of the composition vector. If one accepts the supposition that the evolutionary events are retaining the nucleotide composition in a steady state the mutation probability matrix and the composition vector are connected with the following equation

$$W p = p \tag{4}$$

The explicit form of this equation is the following

$$\begin{bmatrix} 1-w_{WE}-2w_{WU} & w_{SE} & w_{WU} & w_{SU} \\ w_{WE} & 1-w_{SE}-2w_{SU} & w_{WU} & w_{SU} \\ w_{WU} & w_{SU} & 1-w_{WE}-2w_{WU} & w_{SE} \\ w_{WU} & w_{SU} & w_{WE} & 1-w_{SE}-2w_{SU} \end{bmatrix} \begin{bmatrix} P \\ p \\ P \\ p \end{bmatrix} = \begin{bmatrix} P \\ p \\ P \\ p \end{bmatrix}. \quad (5)$$

The indices 1 through 4 run over T,C,A,G nucleotides, respectively. The E and U subscripts to the W matrix elements refer to the pairs of Equal and Unequal nucleotides with respect to purine/pyrimidine character, respectively. The S and W subscripts refer to strong/weak bonding classes. Equation (5) represents the constraints upon the parameters by means of which ***W*** matrix and the ***p*** vector is defined. The most straightforward case appears when the ***W*** matrix is known and the composition vector is unknown. A standard algebraic procedure enables one to determine the composition vector. It is more suitable to take an alternative approach in which we try to solve the following problem: Find the mutation probability matrix, given the overall mutability (μ), the transition/transversion ratio (α) and the (C+G) genomic content (2p). Solving the equation (5) one gets the following solution:

$$W_{SE} = \mu\alpha/(4p(2+\alpha))\ ;\ W_{WE} = \mu\alpha/(4P(2+\alpha))\ ;\ W_{SU} = \mu/(4p(2+\alpha));$$

$$W_{WU} = \mu/(4P(2+\alpha))$$

The alignment of human and chimpanzee genome leads us to the values of the parameters. Humans and chimpanzees exhibit approximately 30 million single nucleotide replacements. This leads to the value μ=10-8 per base pair per year. Transitions are approximately four times more frequent than transversions (α=4) and the C+G content of human or chimpanzee genome is approximately 41% ($2p$=0.41).

In the process of cell division the information content is diminishing because of replication errors. The amount of information that is lost is proportional to the deviation of the replication mechanism from the ideal fidelity. In the case of human species the replication fidelity is on the level of a few misplaced nucleotide per billion replication steps. A more quantitative approach proceeds over the information theoretical formalism. The amount of information transmitted over a channel – in this case from t-th to (t+1)-th generation of cell line is equal to

$$I(p_t, p_{t+1}) = H(p_t) + H(p_{t+1}) - H(p_t, p_{t+1}) \quad (6)$$

where $H(p_t)$ and $H(p_{t+1})$ are the information content per unit of sequence of the genomes of t-th and (t+1)-th generations, respectively and $H(p_t,p_{t+1})$ is their joint entropy defined in the following way

$$H(p_t, p_{t+1}) = -\Sigma_{ij} Q_{ij} \log_2 Q_{ij} \quad (7)$$

where the elements of the ***Q*** matrix are the probabilities that the element of type i that enters into the information channel is transformed into an element of type j. This probability can be expressed as the product of the probability that i-th sequential element is subjected to the replication process and the probability that it will appear in the next generation as an element of type j

$$Q_{ij} = p_i W_{ji}. \tag{8}$$

Let us illustrate the situation on the case of the information transfer between human generations. The rate of decay of the genetic information can be quantified on the basis of equation (6). Taking the above mentioned parameters for μ and p the I value results in I =1.868 bits per nucleotide for the period since the human - chimpanzee divergence. Let us recall that in section 4 the information content per nucleotide is reported to exceed the value 1.9 bits per nucleotide. The information theory can thus only predict a substantial decrease of information that is caused by the mutations. We know that the evolutionary process can lead to more complex organisms whose genomes possess higher and higher information content, but simple theory of information can not cope with this phenomenon.

6. Analyses of the Genomic changes through the Analyses of the Dinucleotide Replacements

Aligning the two related sequences one can count the differences and increment the counters ordered in the form of dinucleotide replacement matrix **A**. If one finds the dinucleotide of i-th type in one sequence and the dinucleotide j at the same site along the other sequence the A_{ij} element is incremented. The indices i and j run from 1 to 16 (AA, AC... through TT). Let $N=\Sigma_i\Sigma_j A_{ij}$ denote the total number of dinucleotides. The composition vector of dinucleotides can be expressed in terms of A_{ij} in the following way

$$p_j = \Sigma_i A_{ij}/N. \tag{9}$$

The replacement count matrix elements A_{ij} are connected with the mutation probability matrix elements (see equation (4) of the preceding section)

$$W_{ij} = A_{ij}/(f_j N). \tag{10}$$

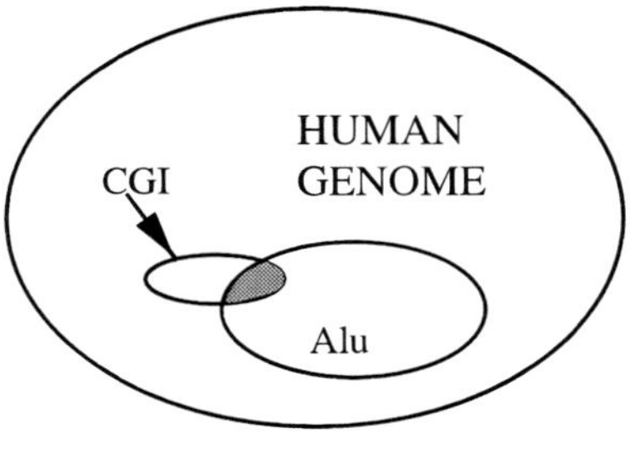

Figure 2. The big ellipse represents the entire human genome. The two small ellipses represent the amount of small interspersed repeats (SINE) of Alu type and the amount of CpG islands (CGI) and their intersection that is shaded.

Equation (4) can be exploited to construct the model of the mutational process - this means to define a scheme how the point mutations are running. After one constructs the mutation probability matrix elements, the equation (4) can be treated as a system of linear equations for unknown values of the components of the composition vector. Since (4) has an apparent form of the eigenvalue equation, the determinant should be equal to zero for the nontrivial solution, what means that the rank of the matrix is n-1 and the equations are not independent. One can thus remove an arbitrary line from (4) and replace it by the equation representing the normalization condition of the dinucleotide population
$\Sigma_i p_i = 1$. After solving the system of equations and getting the population vector one can calculate the model **A** matrix and subsequently one can compare the model values with the natural values. The model can be optimized by choosing some criterion to quantify the goodness of the model such as the evaluation of the correlation coefficients.

Our analyses of the replacements in the period of human - chimpanzee divergence were focused on the differences between the pace of the dinucleotide replacemets within different sequence categories such as CpG islands and interspersed repeats of Alu type. Alu sequences constitute roughly 10 % of human genome. Their genomic coordinates are available at http://www.ncbi.nlm.nih.gov. The CGI content can not be determined exactly because, by definition, a CpG rich DNA region possesses the status of CGI if the cytosines within the CpG dinucleotides escape methylation. This is not easy to find out. Saxonov et al. (2006) invented a special procedure that enables them to locate undoubtedly the CGIs. The standard computational procedures for CGI determination contain a number of free parameters usually the minimal length of the DNA segment, lower threshold of C+G content and the ratio of observed versus expected frequency of CpG dinucleotides. We ran the CPGIF algorithm of Ye et al. (2008) with two values of the CpG observed/expected parameter. In this way we generated the information about two sets of CGIs for the entire human genome. Each CGI was defined with 5' and 3' chromosome coordinates. Having thus the information about the locations of Alu repeats and CGIs we were able to scrutinize the mutational behavior of four categories of DNA sequences (see Figure 2): i) CpG islands, ii) Alu repeats, iii) the overlap of both two sequence classes and iv) the class of all the remaining sequences. In human genome the CPGIF algorithm leads to the CGI content in the interval between two and three percent, depending upon parameterization. The content of Alu sequences is between 10 and 11 percent. The amount of DNA sequences that Alus and CGIs have in common lies in the interval 0.8% to 1.2% of genomic sequences. The analyses of the results of the alignment of human and chimpanzee sequences partitioned in the above mentioned classes give us important information about the nature of mutational processes.

Let us first focus our attention to the results that emerge from the analyses of human – chimpanzee alignment of Alu repeats. By studying the repeats one can obtain two types of information. By comparing the Alu sequences within one species one obtains the information about the mutational events that took place after an element was transposed. Since the majority of transpositions occurred in early history of mammalian radiation before several tens of millions of years the comparison of Alu repeats located at various locations along the chromosome unravel the mutations that took place during this early period. On the other hand one can also obtain the information about the mutations taking place during the recent history of primate evolution by comparing the orthologous human and chimpanzee Alu elements –

see Figure 3. In Figure 4 the sequence dependence is plotted for a 35 bp long stretch of AluSx sequence. The consensus AluSx sequence is also displayed in the figure. The long term mutation probability is presented by full circles and the short term mutability by empty circles. The two mutabilities are normalized in a way to display comparable variations. One can see that the mutability exhibits a peak value at the location of CpG dinucleotides. In Figure 5 the short term and long term mutabilities of the AluSx fragment displayed in Fig. 4 are plotted on x and y axis, respectively. Each point represents one base pair. We can see that the points are grouped along the straight line that is plotted in order to emphasize correlation between long term and short term mutability. There are two groups of points. The upper right one refers to the increased mutability of CpG dinucleotides. The lower left group of points represent the non-CpG sites and exhibited approximately an order of magnitude lower mutability in both time intervals.

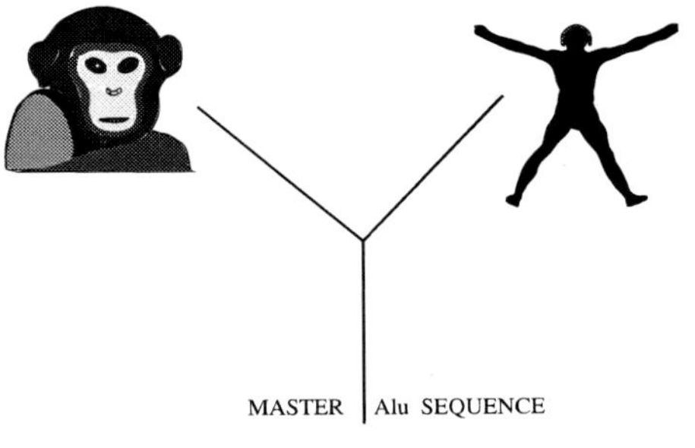

Figure 3. A symbolic evolutionary tree of Alu sequences. A master sequence that was inserted in a genome of our ancestors prior to the mammalian radiation was multiplied by the transposon activities and the end result is a multitude of Alu sequences in chimpanzee as well as in humans. By comparing the human – chimpanzee orthologous Alu repeats one obtains the information about the recent evolutionary history. The mutual comparison of all the human or all the chimpanzee Alus stemming from a common master sequence reveals the information about the entire evolutionary period (see Figs. 4 and 5).

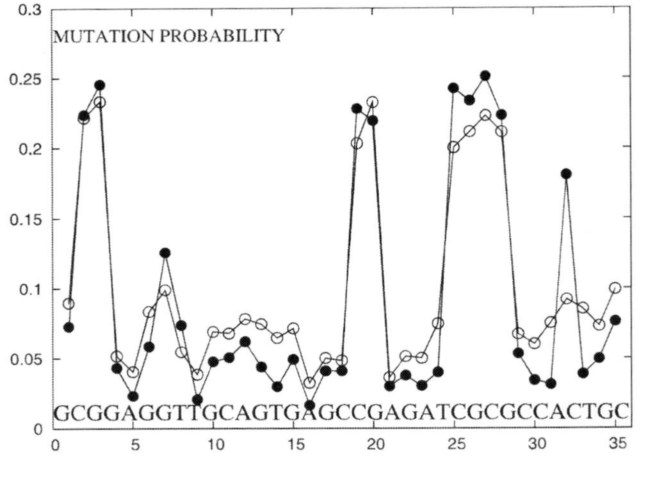

Figure 4. The mutabilities of the sites along the fragment of AluSx sequence. The long term mutation probability is presented by full circles and the short term mutability by empty circles. The two mutabilities are normalized in a way to display comparable variations.

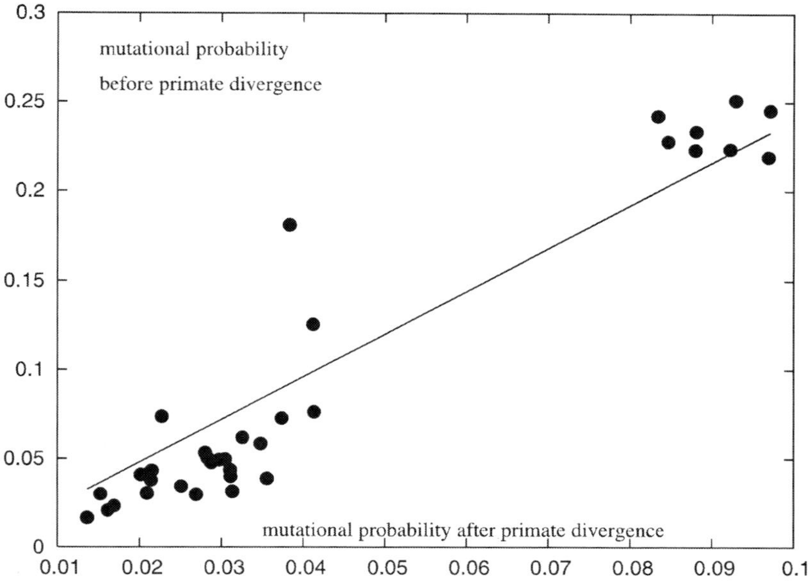

Figure 5. The coordinates of ach point represent the short term average mutability (horizontal axis) and long term average mutability (vertical axis) of the sites along the AluSx fragment.

The enhanced CpG mutability is only present in the DNA regions outside the CGIs, while inside the CGIs the cytosines are not methylated and their mutability should be comparable to the mutability of all the other dinucleotides. We can see in Figure 2 that a portion of Alu sequences is located within CGIs and the majority of Alus is located outside the islands. We analysed both subclasses of Alu sequences and also the remaining two classes (nonAlu/CGI and nonAlu/nonCGI) within the late evolutionary period after the human - chimpanzee divergence. The dinucleotide replacement frequency **A** matrices were determined and the procedure described at the beginning of this section was carried out. The modelling of the most abundant class of sequences, the nonAlu/nonCGI DNA regions that comprise close to 90% of the human genome revealed that the best fit is obtained by the transition over transversion ratio equal to 3.45 and that the CpG to TpG transitions are 9 times more frequent than the other replacements of the transition type. The high mutability of CpG dinucleotides has for the consequence low CpG content. The influx of CpG dinucleotides caused by the decay of other dinucleotides having CpG as a result can hold the CpG abundance in a steady state. The Alu/nonCGI class of sequences is close to nonAlu/nonCGI class as far as the mutational properties are concerned. The main difference comes on the account of the composition. Since the Alu sequences exhibit above average G+C content there are some minor differences in the parameterization of the **W** matrix, but the above mentioned two parameters referring to the transition over transversion ratio and the CpG decomposition enhancement remains approximately the same. The sequences within the CGIs exhibit reduced CpG mutability in comparison with nonCGI sequences, but still above the level of all the other dinucleotides. Here we shall try to unravel some details regarding the mutabilities of two subclasses of Alu sequences - those residing inside and those residing outside the CGIs. In Table 1 four characteristics of the four classes of genomic sequences are presented. In the first column one can see that the total content of Alu sequences is 10%, the content of CGIs is

3% and the overlap between the two classes is 1% of genomic material. The second column represents the CpG content. The content is close to the average dinucleotide content (1/16=6.25%) in CGI classes (the last two rows) and exhibits nearly tenfold CpG depletion outside the islands. The depletion is not so strong in Alu/nonCGI class of sequences. The CpG depletion is correlated with G+C content, which means that in G+C poor regions the depletion seems stronger. A more realistic measure of CpG depletion is obtained when the GpC dinucleotides are taken as a reference and one evaluates the CpG/GpC ratio that is presented in the third column of Table 1, where one can see that the numbers are more coherent. Within the CGIs the CpG dinucleotides are confined somewhere between 60 and 70 percent of the level of GpC dinucleotides. Outside the CGIs the depletion is stronger and the GpC dinucleotides only attain 20% of the GpC share in the dinucleotide composition. The rightmost column of Table 1 contains the probability that a CpG dinucleotide experiences the C=>T mutation and transforms to TpG or CpA, depending upon which strand the mutation occurs. One would expect that the probabilities of CpG decomposition would be inversely proportional to the CpG content, since high decomposition rate leads to low components of the composition vector. We see in the rightmost column of Table 1 that this is not the case. It is true that both, the NonAlu/CGI and Alu/CGI mutabilities exhibit lower values than those of nonCGI sequences, but the differences are not significant. The value of nonAlu/CGI is acceptable - it is nearly five times lower than the nonCGI average, but the Alu/CGI class of sequences possesses high mutability - comparable to the nonCGI values.

Table 1. Four characteristics of four classes of genomic sequences.

	Genome content	CpG content	CpG/GpC	W(CpG=> TpG, CpA)
NonAlu/NonCGI	88%	0.56%	0.15	0.15
Alu/NonCGI	9%	1.6%	0.23	0.125
NonAlu/CGI	2%	5.8%	0.67	0.04
Alu/CGI	1%	4.5%	0.57	0.1

Acknowledgement

This work was financed by the Slovenian Research Agency.

References

Borštnik, B. & Pumpernik D. (2002). Tandem repeats in protein coding regions of primate genes. *Genome Research, 12*, 909-915

Borštnik, B. & Pumpernik, D. (2005). Evidence on DNA slippage step-length distribution. *Physical Review E - Statistical, Nonlinear, and Soft Matter Physics, 71*, 031913/1-031913/7

Borštnik, B., Pumpernik, D. & Lukman, D. (1993), Analysis of apparent 1/f spectrum in DNA sequences. *Europhysics Letters, 23*, 389-394

Borštnik, B. & Pumpernik, D. (2004). Mutational dynamics of short tandem repeats in human genome. *Europhysics Letters, 65*, 290-296.

Fryxell, K. J. & Moon, W-J. (2005). CpG mutation rates in the human genome are highly dependent on local GC content. *Molecular Biology and Evolution, 22*, 650-658

Fryxell, K. J. & Zuckerkandl, E. (2000). Cytosine deamination plays a primary role in the evolution of mammalian isochores. *Molecular Biology and Evolution, 17*, 1371-1383

Gibbs, R. A., Rogers, J., et al. (2007). Evolutionary and biomedical insights from the rhesus macaque genome. *Science, 316*, 222-234

Gibbs, R. A., Weinstock, G. M., et al. (2004). Genome sequence of the brown norway rat yields insights into mammalian evolution. *Nature, 428*, 493-521

Go, Y. & Niimura, Y. (2008), Similar Numbers but Different Repertoires of Olfactory Receptor Genes in Humans and Chimpanzees. *Molecular Biology and Evolution, 25*, 1897–1907

Goodfrey, P. A., Malnic, B., & Buck, L. B. (2004). The mouse olfactory receptor gene family. *Proceedings of the National Academy of Sciences of the United States of America, 101*, 2186-2161

Griffith, O. L., Montgomery, S. B., et al. (2008). ORegAnno: an open-access community-driven resource for regulatory annotation. *Nucleid Acids Research, 36*, D107-D113

Hill, R. S. & Walsh, C. A. (2005). Molecular insights in human brain evolution. *Nature, 437*, 64-66

Ijdo, J. W., Baldini, A., Ward, D. C., Reeders, S. T. & Wells, R. A. (1991). Origin of human chromosome 2: an ancestral telomere-telomere fusion. *Proceedings of the National Academy of Sciences of the United States of America, 88,* 9051-9055

Jiang, C. & Zhao, Z. (2006). Directionality of point mutation and 5-methylcytosine deamination rates in the chimpanzee genome. *BMC Genomics*, *7*, 316 doi:10.1186/1471-2164-7-316

Lander, E. S., Linton, L. M., et al. (2001). Initial sequencing and analysis of the human genome. *Nature, 409*, 860-921

Malnic, B., Goodfrey, P. A. & Buck, L. B. (2004). The human olfactory receptor gene family. *Proceedings of the National Academy of Sciences of the United States of America, 101*, 2584-2589

Michel, C. J. (2007). Evolution probabilities and phylogenetic distance of dinucleotides. *Journal of Theoretical Biology, 249*, 271-277

Mikkelsen, T. S., Hillier, L. W., et al. (2005). Initial sequence of the chimpanzee genome and comparison with the human genome. *Nature, 437*, 69-87

Nielsen, R., Bustamante, C., Clark, A. G., Glanowski, S., Sackton, T. B., Hubisz, M. J., Fidel-Alon, A., Tanenbaum, D. M., Civello, D., White, T. J., Sninsky, J. J., Adams, M. D. & Cargill, M. (2005). A scan of positively selected genes in the genomes of human and Chimpanzee. *PLOS Biology, 3*, 0976-0985

Ollila, J., Lappalainen, I. & Vihinen, B. (1996). Sequence specificity in CpG mutation hotspots. *FEBS Letters, 396*, 119-122

Peng, C. K., Buldyrev, S. V., Goldberger, A. L., Havlin, S., Sciortino, F., Simons, M. & Stanley, H. E. (1992). Long-range correlations in nucleotide sequences. *Nature, 356*, 168-170

Ponger, L., Duret, L. & Mouchiroud, D. (2001). Determinants of CpG islands: Expression in early embrio and isochore structure. *Genome Research, 11*, 1854-1860

Saxonov, S., Berg, P. & Brutlag, D. L. (2006). A genome-wide analysis of CpG dinucleotides in the human genome distinguishes two distinct classes of promoters. *Proceedings of the National Academy of Sciences of the United States of America, 103*, 1412-1417

Shannon, C. E. (1948). A Mathematical Theory of Communication. *Bell System Technical Journal, 27*, 379-423, 623-656

Waterston, R. H., Lindbald-Toh, K. et al. (2002). Initial sequencing and comparative analysis of the mouse genome. *Nature, 420*, 520-562

Yampolsky, L. Y., Kondrashov, F. A. & Kondrashov, A. S. (2005). Distribution of the strength of selection against amino acid replacements in human proteins. *Human Molecular Genetics, 14*, 3191-3201

Ye, S., Asaithambi, A. & Liu, Y. (2008). CpGIF: an algorithm for the identification of CpG islands. *Bioinformation, 2*, 335-338

Zhao, Z. & Boerwinkle, E. (2002), Neighboring-nucleotide effects on single nucleotide polymorphisms: A study of 2.6 million polymorphisms across the human genome. *Genome Research, 12*, 1679-1686

Zhao, Z. & Zhang, F. (2006). Sequence context analysis of 8.2 million single nucleotide polymorphisms in the human genome. *Gene, 366*, 316-324

In: The Human Genome: Features, Variations...
Editor: Akio Matsumoto and Mai Nakano

ISBN: 978-1-60741-695-1

Chapter 6

Wobble Splicing: Subtle Alternative Splicing at Tandem Splice Sites in Human Genome

***Kuo-wang Tsai*[1] *and Wen-chang Lin*[1,2,*]**

[1]Institute of Biomedical Sciences, Academia Sinica, Taipei, Taiwan, Republic of China;
[2]Institute of Biomedical Informatics, School of Medicine, National Yang-Ming University, Taipei, Taiwan, Republic of China

Abstract

Alternative splicing is an important mechanism mediating the function and complexity of genes in multicellular organisms. Recently, a new splice-junction wobbling mechanism is discovered that generates subtle alterations in mRNA by indiscriminately selecting tandem donor sites (GTNGT) or acceptor sites (NAGNAG). It results in tri-nucleotides insertion/deletion in the transcripts that can escape from the nonsense-mediated decay surveillance due to the in-frame InDel event occurred in these mRNA without generating new and premature stop codon. The reading frame is not altered by the insertion or deletion of tri-nucleotides in the transcripts, therefore the resulting protein isoforms would be highly similar in sequences. Nonetheless, most of subtly changes in protein generated by wobble splicing could increase functional diversity of protein and some of theses wobble splicing isoforms might have functional impacts and disease implications in terms of cellular functions and regulations. Therefore, the wobble splicing phenomenon occurs mostly in a tissue and developmental stage–independent manner. Only a few wobble splicing genes are proven to be differentially spliced in tissues or developmental stages. Remarkably, most of this wobble-splicing process is likely due to stochastic splice site selection at tandem motif sequence. Here, we review recent progress in understanding functional aspects as well as the mechanism of wobble splicing at tandem motifs.

---

* Corresponding author Dr. Wen-chang Lin, Institute of Biomedical Sciences, Academia Sinica, Taipei 115, Taiwan. Phone: +(886)-2-2652-3967; Fax: +(886)-2-2782-7654; e-mail: wenlin@ibms.sinica.edu.tw.

Overview

Splicing and Traditional Alternative Splicing

Splicing is a mechanism to remove introns of precursor mRNA and to form mature mRNA, which can subsequently be translated to produce proteins. During the splicing process, the essential *cis* elements of an intron include the 5' donor site (GT), branch point sequence (BPS) [1], polypyrimidine tract (PPT) [2] and 3' acceptor site (AG) (Figure 1A). In higher eukaryoties, splicing is done through a series of reactions which are catalyzed by the spliceosome complex containing five small nuclear RNAs (U1, U2, U4/U6 and U5) and more than 150 protein factors [3-5]. First, the branchpoint attacks the 5' donor site to generate the splicing intermediates; then the released 5' exon attacks at the 3' acceptor site to generate the ligated exons and lariat intron.

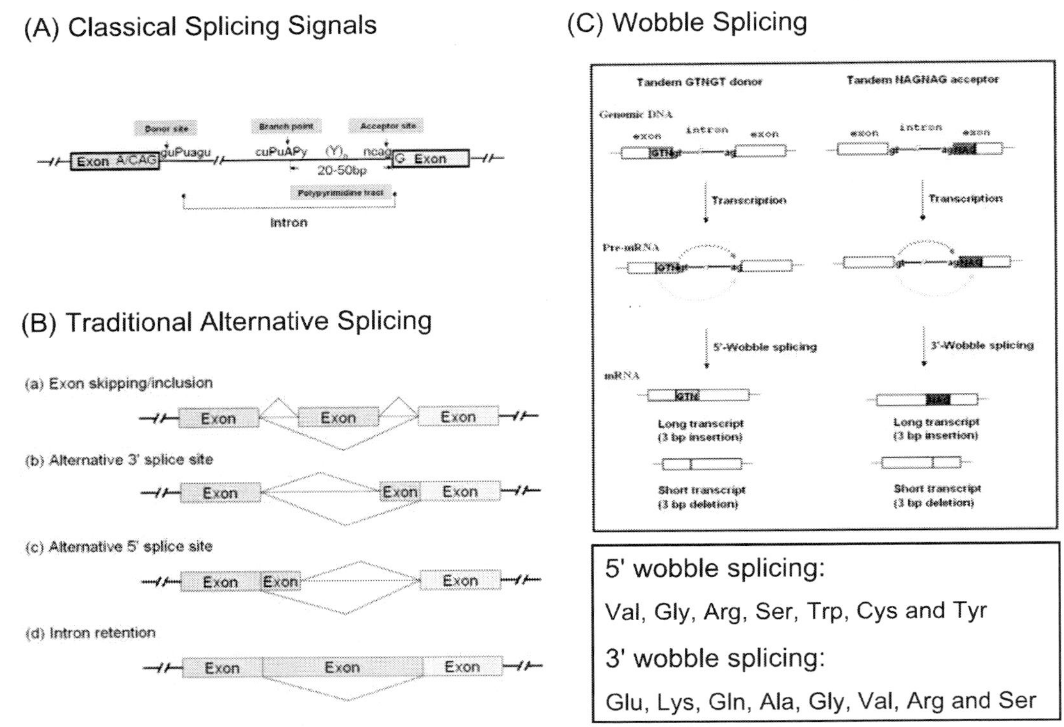

Figure 1. Schematic illustration of classical consensus splicing signals, traditional alternative splicing and subtle variant wobble splicing.

(A) The nearly invariant donor site (GT) and acceptor site (AG) dinucleotides at the intron ends, the polypyrimidine tract (Y)*n* preceding the acceptor site, and the A residue that serves as a branchpoint is shown in upper case letters. (Pu = A or G; Py = C or U) (B) Four common modes of alternative splicing. In each case, the alternative splicing path is indicated by a dotted line. a. exon skipping or inclusion, b. alternative 5' splice site, c. alternative 3' splice site and d. intron retention. (C) Schematic illustration of wobble splicing at tandem donor or acceptor sites The relative amino acids which are insertion/deletion or change by wobble splicing are shown in below panel.

Alternative splicing has recently emerged as a major mechanism of gene regulation in the human genome, occurring in perhaps 40-60% of human genes[6]. It serves as a main source of transcriptome and proteome diversity [7]. In many genes, the splicing process can create variant protein isoforms by creating different exon compositions of the same precursor mRNA. This phenomenon is then called alternative splicing. There are four modes of alternative splicing, including (A) exon skipping/inclusion, (B) alternative splice at 5'- splice site, (C) alternative splice at 3' splice site and (D) intron retention (Figure 1B). Exon skipping/inclusion is spliced out particular exon to alter the sequence of amino acids in the expressed protein ,which is one of the most frequent types of all alternative splicing events. In the intron retention model, instead of splicing out an intron, the intron is retained in the mRNA transcript, which may be created new animo sequence or distroyed reading frame to produce the non-functional protein. However, alternative splice at 5' and 3' splice sites contributing together to 25% of all alternative splicing events are used for a given exon resulting in frame shift or insertion/deletion of amino acids in the expressed protein[8].

Recently, Zavolan *et al.* [9] report that some alternative 5'/3'-splice sites are observed frequently only a few nucleotides apart, indicating that some of the variation may be caused by the random use of 5'- and 3'-splice sites within a short region around the exon boundaries. Alternative mRNA splicing can lead to small changes in protein structure through the insertion or deletion of small peptides [10]. In another study, Wen *et al.* [11] also reported that very short alternative splicing (VSAS) in the human genome might alter protein structures and thus influence protein function. Interesting, the shortest distance alternative at 5'- and 3'- has be reported recently, and such phenomenon is distribution genome-wide, which could result in protein isoforms with one amino acid insertion or deletion [12-20].

Wobble Splicing

Over the past few years, the analysis of alternative splicing using bioinformatics has emerged as an important new research topic [1, 9, 11, 13, 21]. Recently, Hiller *et al.* [13, 15] reported results showing that splicing sites with the genomic NAGNAG and GTNGT motif can cause trinucleotide insertion-deletion in transcripts and may play surprisingly complex roles in switching protein conformation and function. At the same time, we also discover similar phenomenon by searching and analyzing the human EST database [22]. If splicing junction containing GTNGT at 5' splice site or NAGNAG at 3' splice site, it probably generates long transcripts (3bp included) and short transcripts (3bp excluded) during splicing process (Figure 1C, upper panel). Such phenomenon is called wobble splicing. It seems that only few studies have thoroughly investigated such subtle variant alternative splicing event, since the reading frame is preserved and both transcripts and proteins are highly similar. However, we believe such wobble splicing phenomenon exists within many genes and that the mechanism of such splicing event is not clearly. In order to elucidate more about distribution of wobble splicing within human genome, we have performed a computational analysis using the Alternative Splicing Database (ASD)[23]. ASD is derived from EST entries and reports the use of >15,000 alternative splicing sites within the human transcriptome. 3' wobble splicing event at the NAGNAG tandem motif occurs with a higher frequency than 5' GTNGT splicing one, since the 3' end of introns has a more intricate set of regulatory elements. In human, 635 cases showed 3'-NAGNAG-based wobble splicing events, which has a widespread occurrence in 30% of human genes and is active in at least 5% of genes according to an EST database search [13]. On the other hand, only 46 cases is used GTNGT as splice donor to generate such InDel variants (Figure 2).

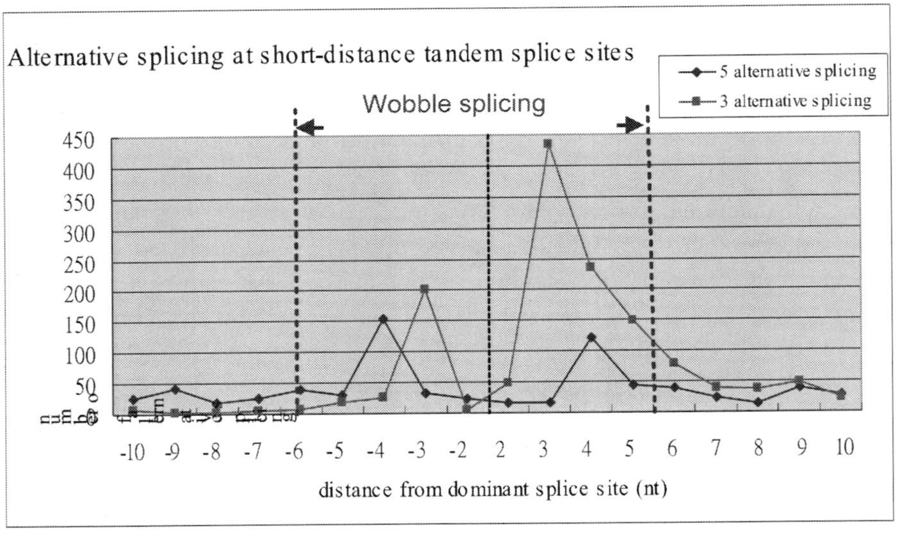

Figure 2. Distribution of human EST-confirmed alternative splicing at short-distance tandem splice sites. Wobble splicing is frequently generated by 5'- (blue line) and 3'- (red line) alternative splice at positions ranging from – 6 to + 6 from the dominant splice site. The number of occurrences of 5'- and 3'- alternative splicing events is obtained from a computational analysis using the Alternative Splicing Database. The blue dashed line indicates the range of wobble splicing and black dashed line indicates the position of dominant splice site. Alternative splice at upstream splice site from the exon / intron boundary are denoted as negative, those downstream are positive. The Y-axis indicates the numbers of alternative splice sites. The X-axis indicates the nucleotide distance between alternative and dominant splice sites.

Wobble Splicing Increase Protein Diversity

Alternative splicing can generate different isoforms which can differ in various functional aspects including protein-protein interaction, protein-DNA binding affinity, protein localization, protein half-life and enzyme activity. Although wobble splicing only subtly changes the protein sequence, it also slightly increase protein diversity.

According to reading frame as basis, there are three possible types to alter protein sequence by wobble splicing event, I. Single animo acid insertion/deletion, II. Replace dipeptide with a unrelated single amino acid and III. Stop codon insertion/deletion. By codon selection, seven amino acids, Val, Gly, Arg, Ser, Trp, Cys and Tyr, could involve in 5 GTNGT-based wobble splicing, and eight amino acids involve in 3 NAGNAG-based wobble splicing including Glu, Lys, Gln, Ala, Gly, Val, Arg and Ser (Figure 1C and 3). These amino acids may nonetheless be of functional relevance by changing hydrophobic/hydrophilic and charge, or Indel of important sequence for post-translation modifications, such as phosphorylation, ubiquitination and sumoylation. Although wobble splicing only subtly changes the protein sequence with one or two amino acid residues, it might increase functional diversity of protein: examples include NR3C1, DRPLA, PAX3, PAX7, IGF1R and ING4 [16, 24-27]. Interesting, some genes harbor more than one tandem splice sites in its splicing junction. Therefore, such genes may produce many subtly variant isoforms through additional wobble splicing events. Two independence of wobble splicing occur at adjacent introns in human MMLT4 gene (containing 3'- tandem acceptors at intron 2-3 and intron 3-4) (Figure 4A) or containing both tandem donor and acceptor within the same intron 4-5 of ING4 (Figure 4B). Serial acceptor sites exist at the splice boundary within FOXM1 intron 6-7 generating four wobble splicing by wobble splice at TAGCAGCAGCAG motif (Figure 4C). Such particular genes may tend to reinforce the diversity of protein through wobble splicing events.

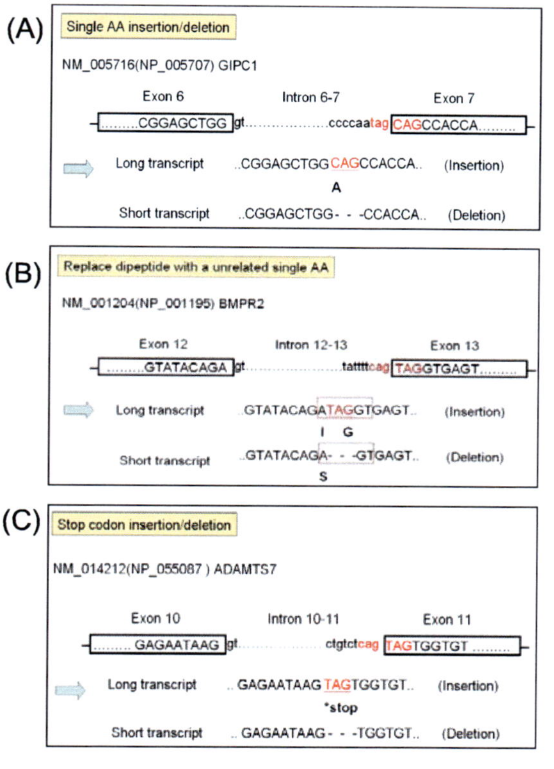

Figure 3. Illustrative examples of effects of 3' NAGNAG-based wobble splicing on proteins.

Indel of NAG trinucleotides probably has only a subtle change in the protein since the reading frame is not changed except involving indel of a stop codon. The effects of protein generated by NAGNAG-based wobble splicing are (A) indel of a single amino acid {example: GIPC PDZ domain containing family, member 1 (GIPC1), intron 6-7} (B) the exchange of a dipeptide and unrelated single amino acid {example: bone morphogenetic protein receptor type II (BMPR2), intron 12-13}, or (C) involving stop codon{example: ADAM metallopeptidase with thrombospondin type 1 motif 7 (ADAMTS7), intron 10-11} }. The partial exon is shown in boxes and dashed lines indicate intron. Exonic nucleotides are shown in upper case letters, intronic nucleotides in lower case letters and the red letters indicate the position of tandem splice sites. Missing residues in the transcript sequences are indicated by dashes.

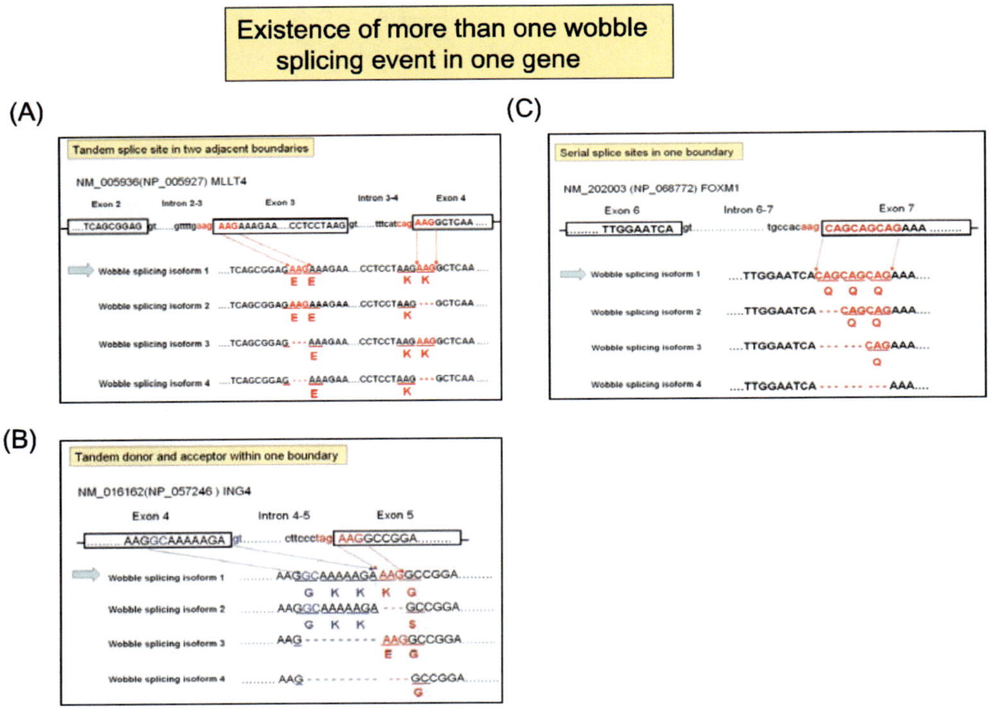

Figure 4. Existence of more than one tandem splice sites in one gene diversifies its protein. (A) Occurring two wobble splicing events at two tandem acceptor site (AAGAAG and CAGAAG) in the adjacent introns of the MMLT4 (myeloid/lymphoid or mixed-lineage leukemia) gene generates four types of wobble splicing transcripts. (B) Intron 4-5 in human ING4 (inhibitor of growth 4) gene containing 5'-GC(N)$_7$GT and 3'-TAGAAG produces four different wobble splicing isoforms. (C) FOXM1 (Forkhead box M1) exists serial acceptor sites (AAGCAGCAGCAG) in its intron 6-7, which generates four different transcript and protein variants. The partial exon is shown in boxes and dashed lines indicate intron. Exonic nucleotides are shown in upper case letters, intronic nucleotides in lower case letters and the red letters indicate the position of tandem acceptors; blue letters indicate the position of tandem donors. Missing residues in the transcript sequences are indicated by dashes.

Mechanism of Wobble Splicing

Previous studies have shown that the high fidelity of splice site recognition involves specific networks of RNA-protein, protein-protein, and RNA-RNA interaction [4, 28, 29]. However, the details of the mechanisms for wobble splicing at tandem motifs are currently unclear. Although tandem motif is common in human genes, only a small fraction can generate wobble splicing (GTNGT: 2%; NAGNAG: 16%) [13, 15-17]. In general, we agree that different binding affinity of U1 snRNA to tandem donor sites decides 5'- wobble splicing [15, 30]. According to this theory, 5'-wobble splicing happens when one donor is sufficiently good to compete with the other donor for U1 binding. The usage of acceptor site requires more complicated than donor, since the splicing factors flexibly interact with cis-elements, such as branch point sequences, polypyrimidine tract and splice site AG. According to the linear scanning mechanism model [31-34], the spliceosome recognizes the branch point and scans downstream for the first AG. However, wobble splicing does not support this scanning model. The tandem splicing acceptor may cause trinucleotide InDel in transcripts, a phenomenon that is generated by splice-site random selection [15, 18, 30]. Using bioinformatic analysis, we observe that nucleotide preceding the AG dinucleotide may influence 3' splice site utilization, and the intronic sequence plays an important role in 3' splice site selection of the NAGNAG wobble splicing. Mutations of the region between the branch site and the NAGNAG 3' splice site indeed affected the ratio of the distal / proximal AG selection. Overall we found that regulation of the 3'-NAGNAG-based wobble splicing base on the following parameters: (i) tandem splice site (NAGNAG); (ii) nucleotide proceeding AG of the 3' tandem splice site; (iii) sequence between the BPS and the NAGNAG (including BPS and PPT)[18].

The splicing factor Slu7 have been shown to affect 3' AG selection during step II of splicing process in vitro and has been suggested to affect alternative splicing in vivo [4, 34, 35]. The correct AG is suppressed and incorrect AGs are selected as 3' splice site in the absence of hSlu7 [35]. Both Slu7 and hSlu7 act in selection of the distal 3' AG when the distance between the branch point and 3' splice site is greater than 7 or 23 nucleotides, respectively [35-37]. When the cells exposed under stress condition such as ultraviolet-C light (UV-C) and heat, function of hSlu7 was abolished by altering its cellular distribution [38]. However, we found that hSlu7 is dispensable for 3' NAGNAG-based wobble splicing. The wobble splicing patterns were not significantly changed by hSlu7 knockdown (unpublished observation). Previous studies showed that aberrant AG selection occurs in *ΔhSlu7* extracts whether the duplicate AG is located upstream or downstream of the normal AG [35]. Activation of incorrect AGs only occurs when such AGs are located within a maximal distance of ~30 nt from the branch site and a minimal distance of ~6 nt from the normal AG [35]. The fact that hSlu7 failed to alter NAGNAG-based wobble splicing may be due to the close distance (three nucleotides) between proximal and distal AG sites. Based on our observation, we believe that most of this wobble-splicing is likely due to steric hindrance from a factor bound at the surrounding tandem motif sequence. Zavolan *et al.* [9] also suggest that the wobble splicing is the result of stochastic binding of the spliceosome at the neighboring splice sites. Based on this hypothesis, Chern *et al.* recently developed a simple physical model which could predict whether splicing occurs only at one site or alternative two sites at tandem splice site [39].

Wobble Splicing Indiscriminately Selects at Tandem Motif

Alternative splicing is mediated by different nuclear splicing factors concentrations varying among tissues and cell types or by developmental stage. These can be altered by different cellular circumstances such as physiological stimuli and environmental effects. Because of the short distance between two splice sites of wobble splicing is largely excluded the possibility that splicing elements are placed between them. Therefore, splicing factors is difficultly to distinguish between proximal and distance splice sites. In 5'-wobble splicing, we do not expect much variation in splicing between tissues since U1 snRNA plays a critical role and ubiquitously expresses in high amounts [15, 30]. Interesting, the most NAGNAG-based wobble splicing is reported that the relative ratio of each wobble-splicing isoform tends to be constant among various tissues. In the study by Hiller *et al.* [13], the NAGNAG acceptors of *ITGAM*, *SMARCA4*, and *BTNL2* were reported to be tissue specific. Moreover, Tadokoro *et al.* [16] reported that the expression ratio of protein isoforms with or without a single amino acid sometimes varied extensively among tissues. Therefore, we agree with the above finding that there are many unknown factors that may mediate the ratio of 3' wobble splicing isoform-such as cell types, cell cycle, and cell-cell contacts. For example, the wobble splicing ratios of five immunologic genes, CD3ξ, CD79B, PLCγ1, CD19 and CD32B, is regulated during T and B cell activation [20]. Although most of wobble splicing is almost constant among tissues, it may generate different functional proteins which are required ubiquitously in the cells. In the case of ING4, expression of the four wobble splicing isoforms did not vary significantly in any of the cell lines examined [17, 40, 41]. Interestingly, such two wobble splicing events influences its subcellular localization, thereby modulating its ability to promote apoptosis [19].

Wobble Splicing Occurred Frequently at short-distance Tandem Splice Sites

GTNGT and NAGNAG-based wobble splicing is widespread existence in the human genes. Base on this notion, we find that the most alternative splice site usage occurs within 6nt of the dominant splice site by searching ASD database (Figure 2 and Figure 5A and B). Surprising, a high frequency of 5'-alternative splicing occur at 4nt upstream or downstream from the dominant splice site (273 cases), and 3'-alternative splicing occur at 3 nucleotides (635cases) (Figure 2 and Figure 5C). Although, a high frequency of 5' wobble splicing is located 4nt from dominant donor site, these frame-shifting transcripts would be disrupted by nonsense-mediated mRNA decay (NMD).

RNA surveillance, also known as nonsense-mediated mRNA decay (NMD), is an mRNA quality-control mechanism that degrades abnormal mRNAs such as mis-spliced mRNA transcripts[42]. By recognizing mRNAs containing premature termination codon, NMD eliminates the production of the encoded truncated protein coded by mis-spliced transcripts, which could function to the detriment of cells [43]. While most of the wrongly spliced mRNA transcripts were degraded by the NMD, the wobble spliced at $AG(N)_nAG$ or $GT(N)_nGT$ (n=1 or 4)might escape from the NMD surveillance due to the in-frame InDel (InDel of one or two amino acids) event occurred in these mRNA without generating new and premature stop codon (Figure 5). Frame-shifting tandem splice site (n=0, 2 or 3) has severe consequences for protein function because of creating different C terminus protein or mRNA degradation by NMD process.

Concluding Remarks

Recently, we have described a new concept that the tandem splice site at exon-intron boundaries may cause splicing-junction wobbling [13, 15, 16, 21, 22]. In contrast to alternative splicing in general, which may substantially affect protein structure, wobble splicing provides a mechanism to create subtle changes. These changes may subtly alter protein structure and provide a mechanism for increasing protein diversity during genome evolution. However, we also agree that particular wobble splicing cases may be regulated in certain condition, and created different function protein by changing local hydrophobicity or charge, localization of protein, or changing recognition sequences for post-translational modifications.

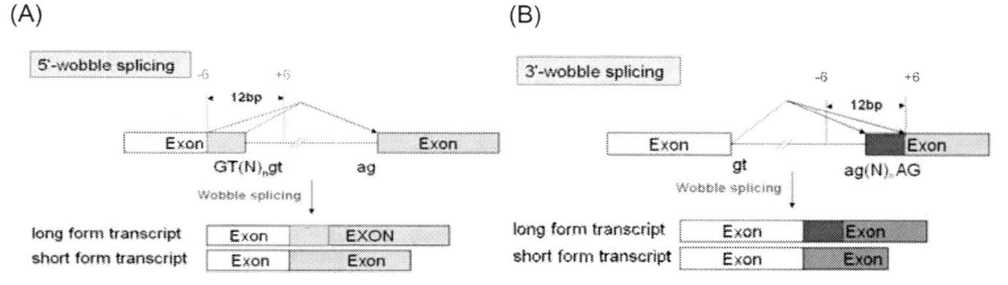

(C)

AG-toAG distance	tandem sites	number of 5' -WS	number of 3' -WS	probability effect on protein
N=0	agag or gtgt	39	52	frame shift; NMD or created new protein
N=1	agnag or gtngt	46	635	1AA InDel; increase protein diversity
N=2	agnnag or gtnngt	273	261	frame shift; NMD or created new protein
N=3	agnnnag or gtnnngt	73	168	frame shift ; NMD or created new protein
N=4	agnnnnag or gtnnnngt	76	85	2AA InDel; increase protein diversity

Figure 5. Effects of wobble splicing at short-distance tandem splicing sites on protein.
(A) and (B) 5' and 3' wobble splicing event frequently generate subtle variant transcripts by random splicing at short-tandem splice sites. (C) Depending on the distance between two splice site (GT or AG) can be distinguish into n=1,4 that preserve reading frame transcript and n=0, 2, 3, 5 that generate frame-shifting transcripts. Most frame-preserving transcripts (only subtle change in protein) contribute protein diversity. Frame-shifting transcripts are destroyed by NMD or create variant protein with a different C terminus.

There is increasing evidence that both exonic and intronic SNPs affect pre-mRNA splicing, which could alter gene expression patterns and expand protein diversity [44, 45]. Therefore, when mutations or SNPs occur around the GTNGT or NAGNAG tandem splice site may affect wobble splicing. For example, an SNP in the ABCR gene (2588G→C) is frequently found in patients with Stargardt disease 1 (STGD1), resulting in an active TAGC^{2588}AG motif, which generates two wobble splicing isoforms [46]. By searching an SNP database, we found 28.8% (127 of 441) of the EST-confirmed wobble splicing genes containing SNPs near the NAGNAG motif (~60 bp region) [18]. Experiments are still required to confirm that these SNP sites are involved in wobble splicing regulation. Since the region between the BPS and PPT is very important for wobble splicing, SNPs in this region may cause an imbalance between the wobble mRNA isoforms. Finally, we suggest that creation/deletion of a new splice site nearby dominant splice site (<6bp) through SNP or mutation might be candidates for complex human diseases.

References

[1] G. Kol, G. Lev-Maor, G. Ast, Human-mouse comparative analysis reveals that branch-site plasticity contributes to splicing regulation, *Hum Mol Genet* 14 (2005) 1559-1568.

[2] C.F. Kennedy, S.M. Berget, Pyrimidine tracts between the 5' splice site and branch point facilitate splicing and recognition of a small Drosophila intron, *Mol Cell Biol* 17 (1997) 2774-2780.

[3] K.L. Fox-Walsh, Y. Dou, B.J. Lam, S.P. Hung, P.F. Baldi, K.J. Hertel, The architecture of pre-mRNAs affects mechanisms of splice-site pairing, *Proc Natl Acad Sci U S A* 102 (2005) 16176-16181.

[4] R. Reed, Mechanisms of fidelity in pre-mRNA splicing, *Curr Opin Cell Biol* 12 (2000) 340-345.

[5] D.C. Rio, Splicing of pre-mRNA: mechanism, regulation and role in development, *Curr Opin Genet Dev* 3 (1993) 574-584.

[6] D.L. Black, Mechanisms of alternative pre-messenger RNA splicing, *Annu Rev Biochem* 72 (2003) 291-336.

[7] S. Stamm, S. Ben-Ari, I. Rafalska, Y. Tang, Z. Zhang, D. Toiber, T.A. Thanaraj, H. Soreq, Function of alternative splicing, *Gene* 344 (2005) 1-20.

[8] C.W. Sugnet, W.J. Kent, M. Ares, Jr., D. Haussler, Transcriptome and genome conservation of alternative splicing events in humans and mice, *Pacific Symposium on Biocomputing* (2004) 66-77.

[9] M. Zavolan, S. Kondo, C. Schonbach, J. Adachi, D.A. Hume, Y. Hayashizaki, T. Gaasterland, Impact of alternative initiation, splicing, and termination on the diversity of the mRNA transcripts encoded by the mouse transcriptome, *Genome Res* 13 (2003) 1290-1300.

[10] J. Stetefeld, M.A. Ruegg, Structural and functional diversity generated by alternative mRNA splicing, *Trends Biochem Sci* (2005).

[11] F. Wen, F. Li, H. Xia, X. Lu, X. Zhang, Y. Li, The impact of very short alternative splicing on protein structures and functions in the human genome, *Trends Genet* 20 (2004) 232-236.

[12] M. Akerman, Y. Mandel-Gutfreund, Alternative splicing regulation at tandem 3' splice sites, *Nucleic Acids Res* 34 (2006) 23-31.

[13] M. Hiller, K. Huse, K. Szafranski, N. Jahn, J. Hampe, S. Schreiber, R. Backofen, M. Platzer, Widespread occurrence of alternative splicing at NAGNAG acceptors contributes to proteome plasticity, *Nat Genet* 36 (2004) 1255-1257.

[14] M. Hiller, K. Huse, K. Szafranski, N. Jahn, J. Hampe, S. Schreiber, R. Backofen, M. Platzer, Single-Nucleotide Polymorphisms in NAGNAG Acceptors Are Highly

Predictive for Variations of Alternative Splicing, *Am J Hum Genet* 78 (2006) 291-302.

[15] M. Hiller, K. Huse, K. Szafranskzi, P. Rosenstiel, S. Schreiber, R. Backofen, M. Platzer, Phylogenetically widespread alternative splicing at unusual GYNGYN donors, *Genome Biol* 7 (2006) R65.

[16] K. Tadokoro, M. Yamazaki-Inoue, M. Tachibana, M. Fujishiro, K. Nagao, M. Toyoda, M. Ozaki, M. Ono, N. Miki, T. Miyashita, M. Yamada, Frequent occurrence of protein isoforms with or without a single amino acid residue by subtle alternative splicing: the case of Gln in DRPLA affects subcellular localization of the products, *J Hum Genet* 50 (2005) 382-394.

[17] K.W. Tsai, W.C. Lin, Quantitative analysis of wobble splicing indicates that it is not tissue specific, *Genomics* 88 (2006) 855-864.

[18] K.W. Tsai, W.Y. Tarn, W.C. Lin, Wobble splicing reveals the role of the branch point sequence-to-NAGNAG region in 3' tandem splice site selection, *Mol Cell Biol* 27 (2007) 5835-5848.

[19] K.W. Tsai, H.C. Tseng, W.C. Lin, Two wobble-splicing events affect ING4 protein subnuclear localization and degradation, *Experimental cell research* 314 (2008) 3130-3141.

[20] T.P. Atkinson, Y. Dai, Activation-induced changes in alternate splice acceptor site usage, *Biochem Biophys Res Commun* 358 (2007) 590-595.

[21] C. Lee, Q. Wang, Bioinformatics analysis of alternative splicing, *Brief Bioinform* 6 (2005) 23-33.

[22] C.H. Lai, L.Y. Hu, W.C. Lin, Single amino-acid InDel variants generated by alternative tandem splice-donor and -acceptor selection, *Biochem Biophys Res Commun* 342 (2006) 197-205.

[23] T.A. Thanaraj, S. Stamm, F. Clark, J.J. Riethoven, V. Le Texier, J. Muilu, ASD: the Alternative Splicing Database, *Nucleic Acids Res* 32 (2004) D64-69.

[24] K.J. Vogan, D.A. Underhill, P. Gros, An alternative splicing event in the Pax-3 paired domain identifies the linker region as a key determinant of paired domain DNA-binding activity, *Mol Cell Biol* 16 (1996) 6677-6686.

[25] P.H. Kay, M.R. Ziman, Alternate Pax7 paired box transcripts which include a trinucleotide or a hexanucleotide are generated by use of alternate 3' intronic splice sites which are not utilized in the ancestral homologue, *Gene* 230 (1999) 55-60.

[26] G. Condorelli, R. Bueno, R.J. Smith, Two alternatively spliced forms of the human insulin-like growth factor I receptor have distinct biological activities and internalization kinetics, *J Biol Chem* 269 (1994) 8510-8516.

[27] C. Rivers, A. Levy, J. Hancock, S. Lightman, M. Norman, Insertion of an amino acid in the DNA-binding domain of the glucocorticoid receptor as a result of alternative splicing, *The Journal of clinical endocrinology and metabolism* 84 (1999) 4283-4286.

[28] J.P. Staley, C. Guthrie, Mechanical devices of the spliceosome: motors, clocks, springs, and things, *Cell* 92 (1998) 315-326.

[29] R. Reed, Initial splice-site recognition and pairing during pre-mRNA splicing, *Curr Opin Genet Dev* 6 (1996) 215-220.

[30] Y. Dou, K.L. Fox-Walsh, P.F. Baldi, K.J. Hertel, Genomic splice-site analysis reveals frequent alternative splicing close to the dominant splice site, *Rna* 12 (2006) 2047-2056.

[31] C.W. Smith, T.T. Chu, B. Nadal-Ginard, Scanning and competition between AGs are involved in 3' splice site selection in mammalian introns, *Mol Cell Biol* 13 (1993) 4939-4952.

[32] S. Chen, K. Anderson, M.J. Moore, Evidence for a linear search in bimolecular 3' splice site AG selection, *Proc Natl Acad Sci U S A* 97 (2000) 593-598.

[33] C.W. Smith, E.B. Porro, J.G. Patton, B. Nadal-Ginard, Scanning from an independently specified branch point defines the 3' splice site of mammalian introns, *Nature* 342 (1989) 243-247.

[34] K. Chua, R. Reed, An upstream AG determines whether a downstream AG is selected during catalytic step II of splicing, *Mol Cell Biol* 21 (2001) 1509-1514.

[35] K. Chua, R. Reed, The RNA splicing factor hSlu7 is required for correct 3' splice-site choice, *Nature* 402 (1999) 207-210.

[36] X. Zhang, B. Schwer, Functional and physical interaction between the yeast splicing factors Slu7 and Prp18, *Nucleic Acids Res* 25 (1997) 2146-2152.

[37] A. Brys, B. Schwer, Requirement for SLU7 in yeast pre-mRNA splicing is dictated by the distance between the branchpoint and the 3' splice site, *Rna* 2 (1996) 707-717.

[38] N. Shomron, M. Alberstein, M. Reznik, G. Ast, Stress alters the subcellular distribution of hSlu7 and thus modulates alternative splicing, *J Cell Sci* 118 (2005) 1151-1159.

[39] T.M. Chern, E. van Nimwegen, C. Kai, J. Kawai, P. Carninci, Y. Hayashizaki, M. Zavolan, A simple physical model predicts small exon length variations, *PLoS Genet* 2 (2006) e45.

[40] J.C. Shen, M. Unoki, D. Ythier, A. Duperray, L. Varticovski, K. Kumamoto, R. Pedeux, C.C. Harris, Inhibitor of growth 4 suppresses cell spreading and cell migration by interacting with a novel binding partner, liprin alpha1, *Cancer Res* 67 (2007) 2552-2558.

[41] M. Unoki, J.C. Shen, Z.M. Zheng, C.C. Harris, Novel Splice Variants of ING4 and Their Possible Roles in the Regulation of Cell Growth and Motility, *J Biol Chem* 281 (2006) 34677-34686.

[42] L.E. Maquat, Nonsense-mediated mRNA decay: splicing, translation and mRNP dynamics, *Nat Rev Mol Cell Biol* 5 (2004) 89-99.

[43] L.F. Lareau, R.E. Green, R.S. Bhatnagar, S.E. Brenner, The evolving roles of alternative splicing, *Curr Opin Struct Biol* 14 (2004) 273-282.

[44] L. Cartegni, S.L. Chew, A.R. Krainer, Listening to silence and understanding nonsense: exonic mutations that affect splicing, *Nat Rev Genet* 3 (2002) 285-298.

[45] N. Lopez-Bigas, B. Audit, C. Ouzounis, G. Parra, R. Guigo, Are splicing mutations the most frequent cause of hereditary disease?, *FEBS Lett* 579 (2005) 1900-1903.

[46] A. Maugeri, M.A. van Driel, D.J. van de Pol, B.J. Klevering, F.J. van Haren, N. Tijmes, A.A. Bergen, K. Rohrschneider, A. Blankenagel, A.J. Pinckers, N. Dahl, H.G. Brunner, A.F. Deutman, C.B. Hoyng, F.P. Cremers, The 2588 G-->C mutation in the ABCR gene is a mild frequent founder mutation in the Western

European population and allows the classification of ABCR mutations in patients with Stargardt disease, *Am J Hum Genet* 64 (1999) 1024-1035.

In: The Human Genome: Features, Variations...
Editor: Akio Matsumoto and Mai Nakano
ISBN: 978-1-60741-695-1

Chapter 7

Alternative Splicing Transcripts Affected by Junction Tandem Repeats in the Human Genome

Chun-Hung Lai[1] and Wen-chang Lin[1,2,*]
[1]Institute of Biomedical Sciences, Academia Sinica, Taipei, Taiwan, Republic of China;
[2]Institute of Biomedical Informatics, School of Medicine, National Yang-Ming University, Taipei, Taiwan, Republic of China

Abstract

Subtle alternative splicing at short distance is defined as the selection between two or more adjacent splicing signals during mRNA processing. The co-existence of expressed sequences with such minute distance differences is typified as Wobble Splicing, which implies the ambiguous recognition of splicing complexes in selection between the nearby splicing sites. Successive tandem splice acceptors arrangement, such as NAGNAG, is the most common wobble splicing type in eukaryotic genomes. In addition to the short distance tandem splicing sites previously reported, DNA segmental duplication or multiplication also has the potential to duplicate the splicing signals *in situ*, once it located at the boundaries between intron and exon. Herein, junctional tandem repetitive sequences in human are mapped and investigated in this study. The splicing site selection among the homologous duplicated splicing sequences was also detected and characterized. Junctions with duplicated 5' splicing signatures are more prevalent than 3' ones by summation of total numbers of splicing sites observed. The distance between the duplicated repetitive sequences seems to play a critical role in the alternative splicing site selection as observed by ESTs. This reveals the additional control mechanism of splicing acceptor site selection, which contain branch site sequence and polypyrimidine tract aside from the minimal dinucleotide -AG signal.

* *Corresponding author: Dr. Wen-chang Lin, Institute of Biomedical Sciences, Academia Sinica, Taipei 115, Taiwan. Phone: +(886)-2-2652-3967; Fax: +(886)-2-2782-7654;
e-mail: wenlin@ibms.sinica.edu.tw.

Keywords: Alternative splicing, tandem repeats

Introdcution

Alternative Splicing in Transcription

In contrast to prokaryotes, the mRNA transcripts of multi-cellular eukaryotic genes are post-processed after transcription [1, 2, 3]. Generally, introns are trimmed away, and exons are ligated together as "meaningful" message strings for the molecular translation machinery complex. Almost over 90% of genes could generate more than one form of the processed mRNAs [2], they are produced through the alternative inclusion or exclusion of particular exon(s) or the usage of different splice sites, which are defined as alternative splicing.

Alternative splicing contributes to the molecular diversity and the gene regulation of genome functions. Categories of alternative splicing classes include alternative exon choice; in some cases they are mutually exclusive; while in some cases the intron is retained instead of exon selection. In addition, alternative splicing donor or acceptor sites selection remains as an important part of alternative splicing events characterized to date.

The basic features of splicing donor and acceptor sites include the nucleotide sequence of the splice sites, the cognitive di-nucleotide GT and AG as 5' splice site (5'ss) and 3' splice site (3'ss) which are utilized for over 90% of the identified splicing junctions [4, 5]. Theoretically, the successive arrangement such as GTGT or AGAG could be the minimal recognizable alternative splicing signals, and some of these alternatively processed variants were detected in expressed forms [6]. However, due to the nucleotide motif occurrence and ORF frame constraint during genome evolution, NAGNAG and GTNNGT generate the most frequent transcripts observed than other splicing di-nucleotide signals observed in recent studies [6, 7, 8, 9].

Wobble Splicing at Tandem Splice Site Sequences

In previous studies, such subtle alternative splicing within short distance is consequently termed Wobble Splice. It is due to the fact that only few nucleotide length-difference of these splicing variants and that similar distribution ratio of these variants in most tissues examined [9, 10]. However, there are limited cases with reported tissue-specific splicing patterns [7, 8].

The precise wobble splicing regulation is largely unknown. Accordingly, due to the adjacent occurrences of splicing signal sequences at the splicing junction, RNA splicing complexes (the spliceosome) [11, 12], could not be able to distinguish them precisely [6, 9]. This will then create the subtly diverse splicing variants with only few base pair in difference. From previous studies, the wobble spliced isoform generation and distribution is the net result of complex factors influenced by surrounding adjacent nucleotide(s), polypyrimidine tract sequence and intron length [13], and other unknown *cis* or *trans* regulatory factors [14].

The origin of such tandem wobble splicing sequences is also not clear yet. It is likely that sequence duplication could be the evolution forces of creating wobble splicing junctions.

DNA sequence duplication (or multiplication) has the potential to create these identical or similar "tandem splice sites" required for wobble splicing. Tandem sequence sites at intron-exon junction could therefore serve as choices for splicing complex since they are composed of similar sequence patterns at relatively close proximity [15]. It is speculated that once the junction sequences were duplicated, the RNA splicing complex could have difficulties in distinguishing the "original" and the adjacent "similar" alternative signal(s). This might be the source of wobble splicing events.

On the other hand, repetitive element duplication could be another source of wobble splicing sites in the genome. Herein, we investigated the possibility that the biological roles of repetitive sequence elements in alternative splicing regulations. Tandem repeat sequences (TRs) were searched [16, 17] through data-mining, and the TRs located at intron-exon boundaries of human genes were further inspected.

Results

Tandem Repeats in Wobble Splicing Generation

We performed the classification of the TRs according to the genomic location by the grouping and sorting functions provided by the Tandem Repeats database. Tandem Repeats Database (TRDB) organized and characterized the human tandem repeat sequences in genome [17]. TRs within gene regions were specifically divided out from whole genome TRs (WG-TRs), and sub-divided into intronic TRs (I-TRs), exonic TRs (E-TRs) and splice junction TRs (J-TRs). Using the Tandem Repeats Finder program [16], annotations of the TRs identified and characterized were output and examined. Regions spanning the intron-exon boundaries with at least one nucleotide apart were selected and further interrogated with UCSC genome browser [18] to identify the expression sequence tags (ESTs) covering these putative wobble splicing transcripts.

Among 947,696 regions containing characterized TRs of the human genome, 328,104 (~34.62%) are located in the intra-genic region. Among these intra-genic TRs, I-TRs stand for most of them, with 318,199 (96.98%; about 33.58% of total TRs) loci. On the other hand, 8,229 intra-genic TRs (2.5%; about 0.87% of total TRs) are attributed as E-TRs. Except for the repetitive sequences situated at the defined gene regions, there are 1,676 TRs which are located at the boundaries between intron and exon (0.5%; about 0.17% of total TRs), they span over the junction either of the 5'ss or 3'ss. Hence, they are termed J-TRs and be involved in the formation of wobble splicing. They may become whole or part of the splicing signal sequences that could be recognized and processed by the splicing complex.

These repetitive sequences arrayed in tandem are classified by length with their core sequence units as microsatellites and minisatellites [19]. Typically, repeating element unit runs from mono-, di-, tri-... nucleotides are considered to be microsatellite; repeating runs of ten or more nucleotides are attributed to be minisatellites. These two categories of satellite DNA segments of E-TRs and I-TRs are quite different in their distribution pattern (Figure 1A). TRs with pattern size over 10 are relatively abundant than that of below 10 bps TRs, therefore, short TRs are rare in splicing junctions.

The distribution pattern of repeat lengths of J-TRs and E-TRs reveals the relatively abundant occurrence of the pattern length by the multiples of three (figure 1A, lower panels). This reflects that DNA nucleotide size change was also under the regulated control and evolution constraint, nucleotide insertion or deletion change by the number of multiples of three could have the lowest impact of the translated protein product [20].

In order to look into the repetition extent of the TRs, their copy number (or phase number) distribution was plotted (Fig 1B). Most of the repeats have a copy number around one to three, majority of them has two, regardless of the TRs' genomic location. This may reflect the duplication nature of DNA during replication process, for example, DNA polymerase slippage in replication [21]. DNA repeats would have accumulated changes over the evolution time, the variations and insertions or deletions of the nucleotides may degenerate over time and change the TR element features, such as copy numbers and sequences.

A

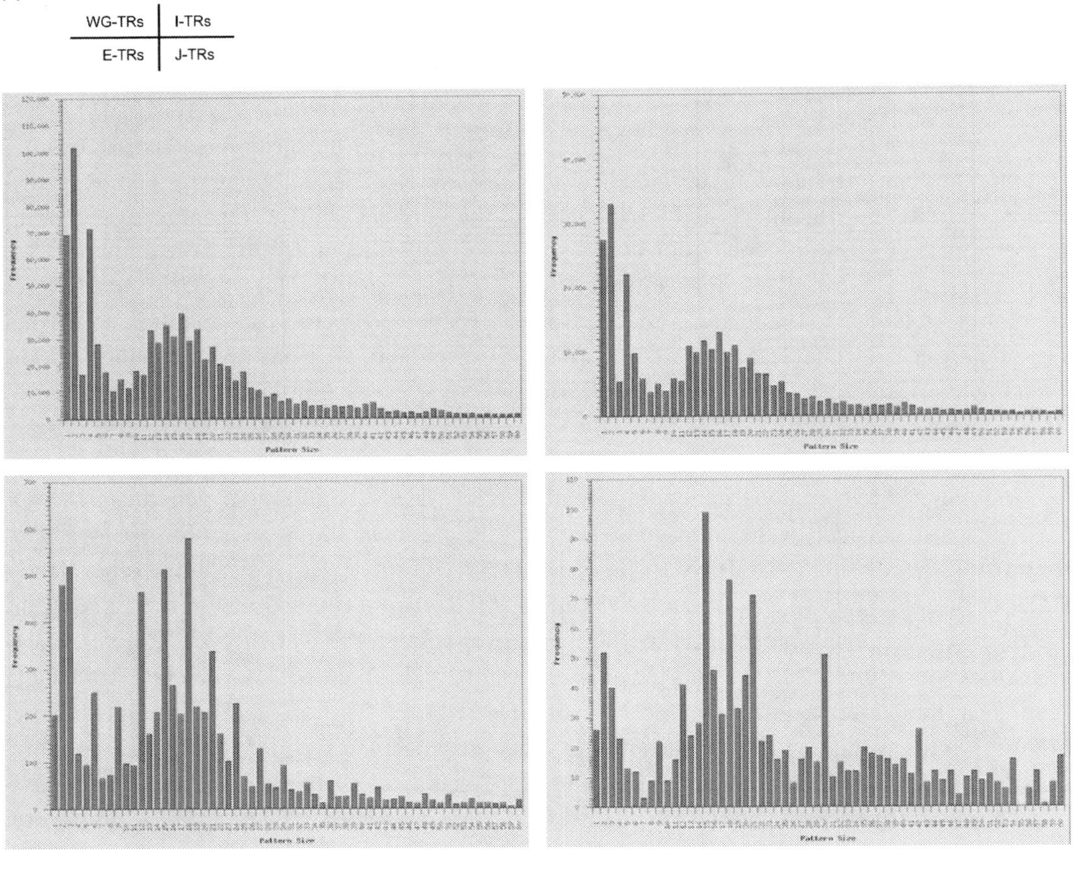

Figure 1. (Continued on next page.)

B

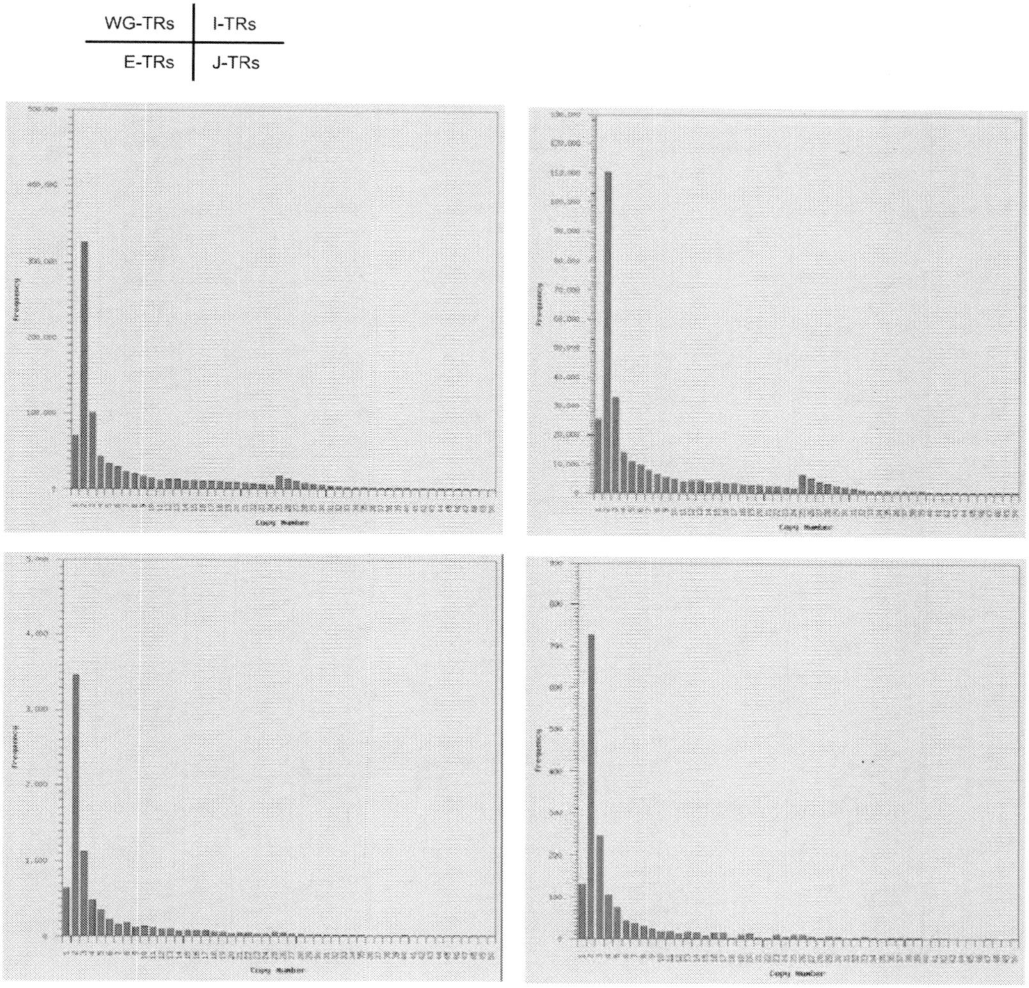

Figure 1. Pattern size and copy number distribution of human TRs. TRs within gene are sub-grouped as I-, E-, J- TRs according to their location at intron, exon, or at the junction between intron and exon. The horizontal axis represents the pattern size in (A) or the copy number in (B) of each TR defined by Tandem Repeats Finder [16]. The pattern size range is shown from mono to sixty nucleotide repeats in TR period size, and the copy number is from one to fifty in integer of period number. The vertical axis shows occurrence of the TR region. All plots are performed and output from TRDB [17].

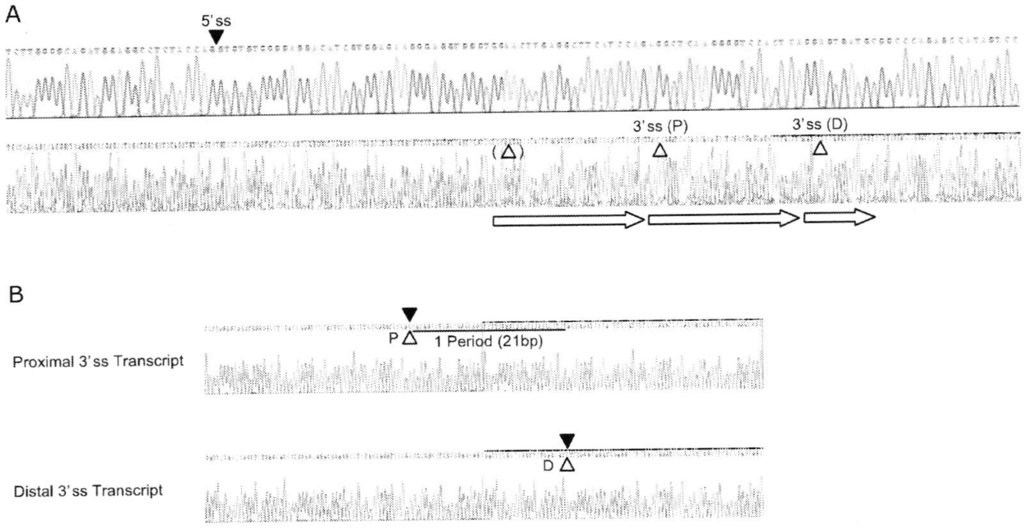

Type	Clones	ESTs	ASD
P	33 (57%)	15 (58%)	6 (55%)
D	25 (43%)	11 (42%)	5 (45%)

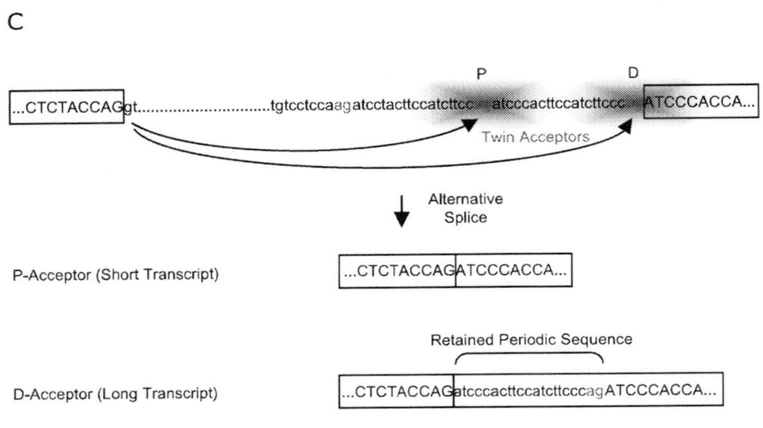

Figure 2. Alternative splicing on tandem repetitive splice acceptors of NCKIPSD. (A) Chromatograms show the genomic sequence from partial exons 3 and 4, and the intronic sequence in between. 5'ss is marked with a black triangle and 3'ss with a triangle. The 3'ss situated relatively in proximity or distant to 5'ss is denoted with "P" or "D". The triangle in parentheses shows the most proximal proposed 3'ss site in the TR tract, though no this type of transcript detected to date. Arrows illustrate the TR tract that is consisted with a 21 bp nucleotide repeating 2.4 times. (B) Sequences of the splicing variants by utilization of P or D acceptor. P and D transcripts are characterized with exactly 1 copy difference in the expressed J-TR region. The percentage of these isoforms is obtained either by cDNA-clone sequencing, and ESTs counting from UCSC genome browser and the alternative splice database (ASD). The genomic and the total RNA for reverse transcription were prepared from KatoIII cell line. (C) Schematic illustration of Wobble Splicing acts on J-TRs. Recognizable splice acceptors, denoted with red "ag", are arrayed in tandem where alternative processing of the mRNA could occur. The denser shadow area around the P site shows its stronger splice site choice for splicing than D site for this example.

Table 1. J-TRs located at the splice donor and acceptor sites with aligned ESTs isoforms in the human genome

Repeat ID[a]	Pattern Size[b]	Copy Number	Gene Name	ESTs Isoforms [c]	Syncopated Pattern of Splice Sites [d]
5'ss					
117152666	37	4	DECR2	32:2	CTCGTGTGCTCTATGAGAAGTTCTTCCGGgt gggtgcctcgtgcgctctgtgagaagttcttccg g][gt gggtgcctcgtgcgctctgtgagaagttcttccgg//gtgggtgccttgtgcgctctgtgagaagttcttccgg// gtgggtg
117803301	49(48)	2.3	SAMD11	5:8	CCTCCCCAGCCACG][GT GAGGACCCACCCTGGCATGATCCCCCTCATCACCT CCCCAGCCACGgt gaggacccaccctggcatgatctcccctcatcacctccccagccac
117675703	10(9,10,10)	4	MGC11257	2:3:5:34	AG][GT GGGCGAC][GT GGGCGCAG][GT GGGCGCAGgt aggcgc
117153382	16	1.9	UNKL	1:12	AG][GT GCGCCCCCCGCCAGgt gcgtcccccgc
117083793	37	2.2	SCARF1	11:1	GTGTGGCCAGCAGgt gggtctcgtgggagtctgatgcgtgtggccagcag][gt gggtctcgtgggagtc tgatgcgggtggc
117413204	57(40,16)	3.1	CLN8	67:3:3[e]	TGAGTGGCAGCCCAGgt gagcgctcagggagcccgggtgagggccgggaacccag][gt gagggg cagcccag][gt gagcgcgcggggtgcccgggttaggggccgggaacccaggtgaggggcagcccag//gtg agcgcgcgtggagcccaggcgaggggggctgggaagcccgggtgagggg
117676656	14	3.5	CHST12	38:11	GGGCGCGAGgt gaggggcgcgag][gt gagggtcgcgag//gtgaggggcgcg
117620483	29	2	SH3BP2	46:1	TCCGAGGAGGAGCGCAAGgt gactgggggtccgaggacgagtgcaag][gt gactgggggt
117244573	70	2.1	ITFG2	19:4	CCACAGCCTGCTGCAGGAGCTGGGCGTGGgt gagtcccagaaaagccagtggcccgagt gcccaggagacccacagcatgctgcaggagctgggcgtgg][gt gagtcccagaaaagccagtggcccgggt gcccaggagacccacagc
117084850	35	2.1	ATP2A3	14:1	TGCAGTACGAGgt gggtgcaggagccgattctccctgcagtacgag][gt gggtgcaggagccaagtctc cctgcag
117245973	45	3	PEX5[f]	114:6	GCCTCTGAGGCAgt gagtgttcttgaggtggaaagcccaggtgcagcctctgaggca][gt gagtgttctt gaggtggaaagcccaggtgcagcctctgaggca//gtgagtgttcttgaggtggaaagcccaggtg
117033933	69(138)	7.6	CD209	1:2	GATCTACCAGAACCTGACCCAGCTTAAAGCTGCAGTGG//GTGAGCTCTCAGA GAAATCCAAGCTGCAGGAGATCTACCAGGAGCTGACCCAGCTGAAGGCTGCA GTGG//GTGAGCTTCCAGAGAAATCTAAGCTGCAGGAGATCTACCAGGAGCTG ACCCGGCTGAAGGCTGCAGTGG//GTGAGCTTCCAGAGAAATCTAAGCTGCAG GAGATCTACCAGGAGCTGACCTGGCTGAAGGCTGCAGTGG//GTGAGCTTCC AGAGAAATCTAAGATGCAGGAGATCTACCAGGAGCTGACTCGGCTGAAGGCT GCAGTGG//GTGAGCTTCCAGAGAAATCTAAGCAGCAGGAGATCTACCAGGAG CTGACCCGGCTGAAGGCTGCAGTGG][GT GAGCTTCCAGAGAAATCTAAGCA GCAGGAGATCTACCAGGAGCTGACCCGGCTGAAGGCTGCAGTGG//GTGAGC TTCCAGAGAAATCTAAGCAGCAGGAGATCTACCAGGAGCTGACCCAGCTGAA GGCTGCAGTGGgt gag
117807075	21	2	PER3	4:2	CGAATGGTGGTGgt gagtcagcgaatggtggtg][gt gagtcagc
116961795	18	2.2	SEC23B	142:104	CGGCTCTAGCTAGgt gagcggctccggccag][gt gagcgg
117738062	39	2.1	ITSN2	3:3	GAAAGCTGAGGAGAAACAAC][GT AAGGATAAGGATACTTTGAAAGCTGAGGA GAAAAAACgt aaggataaggatactttga
117815344	12	27	IFI6	126:23:3	GCAGgt gagaatgcgg][gt aaggatgcag][gt aaggggacag//gtaagggtgcag//gtaaggatgcaa/ /gtaagggtgcag//gtaaggatgcgg//gtaaggatgcag//gtaaagatgcgg//gtaaggatgcag//gtaagg atgcgg//gtaaggatgcag//gtaagggtgcag//gtaaggatgcgg//gtaaggatgcag//gtaaggatgcgg/ /gtaaggatgcag//gtaaggatgcag//gtaagggtgcgg//gtaaggatgcgg//gtaaggatgcag//gtaagg atgtgg//gtaaggatgcag//gtaagggtgcag//gtaagggaacag//ataagggtgcag//gtaagg
117117796	13	1.9	MTMR15	13:6	CCCAGgt gcgcgtcccag][gt gcgcg
117468072	22	2	NRM	83:9	GCCCTGGCCTTGCAGgt atgaggccctggccttgcag][gt atga
117690001	18	1.9	IGFBP3	217:4	GCCGCCAGCTCCAGgt gagccgcccgcgccag][gt g
116956204	60	6.9	PCNT	2:1	CGGCGGCAAGgt gtggggagggggaaggcgcgaggtcccccgggagaggctggacacgcggca gcaag][gt gtggggagcggggaaggcacgaggcccacccgggagaggctggacacgcggcagcaag//gt gtggggagcggggaaggcacgaggcccacccgggagaggctggacacgcggcagcaag//gtgtggggag cggggaaggcacgaggcccacccgggagaggctggacacgcagcagcaag//gtgtggggagggggaagg cgtgaggccccccctgggagaggctggacacgcggcagcaag//gtgtggggaacggggaaggcacgaggcc cacccgggagaggctggacacgcggcagcaag//gtgtggggagggggaaggcgtgaggccccccctggga gaggctgg
116956279	16	2.3	S100B	1:177	GAGACCAGGAAGGG][GT GAGACAAGGAAGAGgt gaga
117576642	38(40)	9.8	HYAL1	1:2	ccccgtctgggaa//gtgaggagcgtctctgcctggccgcccatcgtctgggat//gtgaggagcccctctgcctgg ctgcccagtctggaaa//gtgaggagcgtctctgcccagccgccatcccatctaggaa//gtgagcagcgcctcttc ccggccgccatcccatctaggaa//GTGCGGAGCCTCTCTGCCCGGCCACCCATCGTCT GAGAT//GTGGGGAGCGCCTCTGCCCTGTCGCCCCGTCCGGGAT//GTGAGGA GCGTCTCTCCCTGGCCGCCCCGTCTGAGAA][GT GAGGAGCCCCTCCGCCCA GCAGCCGCCCTGTCTGAGAAgt gaggagcccctccgcccggcagccaccctgtctgggaa//gtg aggagggtctccgccc
117304857	16	3.6	LPXN	132:1	TGAGATGCAGGCCAAGgt gagtgcaggccaag][gt gagtgcaggccaag//gtgagtgcag

Table 1. (Continued)

Repeat ID[a]	Pattern Size[b]	Copy Number	Gene Name	ESTs Isoforms [c]	Syncopated Pattern of Splice Sites [d]
117748341	15	2	FANCL	41:7	CCTCAGgt aaattctcctcag][gt aaattct
117076088	30(3,27)	2	SERPINB13	17:1:10[g]	AAGAATAAAGGCTGAAGAAAAAGAG][GT Ggt aagaataaaggctgaaggaaaagag][gt gg
116977331	18	2.6	CHRNA4	2:1	GTCCTCTACAACAAgt gagtcctctacaacaa][gt gagtcctctaca
117308341	37	4.6	TCIRG1	22:1	TGCAGGAGgt gggtgcccccggccttccggaggcgggtgtaggag][gt gggtgcccccggcctcccggaggtgggtgcaggag//gtgggtgccctggcctggcctccaggaggtgggtgcaggag//gtgggtgccctggcctggcctccaggaggtgggtgcaggag//gtgggtgcccccggcc
117309920	23	2.8	DHCR7	8:245:1	AAGGTGGGCGCAG][GT GAGGGGCCGAGGTGTGCGCAGgt gagtggccaaggtgtgcgcag][gt gag
117701684	18	1.9	SRCRB4D	5:2	AGgt gaggctgacagcgaag][gt gaggctgacagcg
117113183	40	1.9	RAC3	19:1	CTTCCCCGGAGAGTACATCCCCACCGTgt gagtgtgggggcttcccgggagagcacaggccctccgt][gt gagtgtgg
117113451	21(42)	11	FLJ22222	41:1	GCGCACGTGGACAAGgt gagcgcgcccgtgggcagg//gtgagcgtgcccgtggacaag][gt gagcgtgcccgtgggcacg//gtgagtgcgcacgtgggcagg//gtgagcgcgctcgtggacaag//gtgagcgtgcccgtgggcagg//gtgagtgcgtacgtgggcagg//gtgagtacgcacgtgggcagg//gtgagcgcgcatgtggacagg//gtgagcgcgcatgtggtcaag//gtgag
117437468	18	2	MRPS28	18:39	CTACAGAAG][GT GGAGCCCCTACAGAAGgt ggagccc
117396808	38	2	HSD17B3	12:2	GTCATCATCCAGgt gaggtgggcagctgtggagtccttgtcatcatccag][gt gaggtgggcggctgtgagtccttg
117708637	28(27)	2.3	ZNF498	4:21	AGCGG][GT GAGAGGCCCTTCAGGGCCGCGGCGGgt gagagcctcttcagggccgcagcgg//gtg
117362720	36	3.1	AS3MT	2:18	CCTGAGTCGCAGGCCGAGGAGACA//GTGAGTGCGCGCCCTGAGTCGCAGGCCGAGGAGACA][GT GAGTGCGCGCCCTGAGTCGCAGGCCGAGGAGACAgt gagtgcgcgccctg
117277879	36	9.9	DDX54	87:1	Ggt gagtgatcgagtgagccgcccctccatgctcggg][gt gagtgaccgagtgagctgctcctccgtgcctggg//gtgagtgatcgagtgagccgcccctccctccatgcccggg//gtgagtgagtgagtgagccgcccctccatgcccggg//gtgagtgatcgagtgagctgcccctccaggcccagg//gtgagtgagagagtgagctgcccctccatgcctggg//gtgagtgagcgagtgagcctcccctacctccattccctgaa//gtgagtgagcgagtgagccgcccctccattccctggg//gtgagtgagcgagtgagccacccctccaggcccagg//gtgagtgatcgagtgagctgccctccatgcc
117768261	14	1.9	IWS1	45:1	GAACAGgt atgagggaacag][gt atgag
117373144	34	2	MTG1	97:1	CACCTCAGGAAAGgt actggcgcgtgcggctgatcacctcaggaaag][gt actggcgcgtgtggctgat
117497609	24	3.1	TNFAIP3	26:15:21	CCCGGAGAGgt aaccgccgcgcctcccggagag][gt aaccgccgcgcctcccggagag][gt aaccgccgcgcctccc
117560675	15	2.4	FGFR4	22:4	GGAGGAGCCAG][GT GAGGAGGAGCCAGgt gagcagga
117785869	15	2.3	MYO1B	1:12	GCGCCCGGCGAG][GT AGCGCCAGGCGAGgt agcgc
3'ss					
117803857	42	2	MIB2	22:27	ccctctgcccacag GTCCCGAGCAGCCCGGCCCACCATGGACCCCTCTGCCCACAG][GTCCCGAGCAGCCCCGCCCAACATGGACCC
117374524	17	2	RCL1	2:105	ttttcttcacag][attttttttcttcac agATTTT
117245645	60(109)	2.7	SCNN1A	55:2	tccactccggggctcatgaag GGGAACAAGCTGGAGGAGCAGGACCCTAGACCTCTGCAGCCCATACCAGGTCTCATGGAG//GGGAACAAGCTGGAGGAGCAGGACTCTAGCCCTCCACAGTCCACTCCAG][GGCTCATGAAG//GGGAACAAGCGTGAGGAGCAGG
116929343	19	2.1	HTF9C	1:36	cacttcccttgtgacag][gccacttcccttgtgacag GCC
116950275	54(51)	2	SON	26:28	attgtatgtttttcttgaatag GTATTGAGAAATGGAGCCCTTACCAGCCCAATTGTATGTTTTTCTTGAATAG][GTATTGATAAATGGAAGCGCTTACCAGCCCA
116950282	119(116)	2	CRYZL1	32:3	atttctaatttcattccttctcag ACCTCAGTTGGATGAACCCATTCCACTGTATGAGGCAAAAGTTTCCATGGAAGCTGTTCAGAAAAATCAAGGAAGAAAAAAGCAAGTTGTTCAATTTTAATTTTCTTCTTTCTCAG][ACCTCAGTCGGATGAACATATTCCAGTATTTGAAGCCAGAATTTTCTTTGGAAATTGTTGAGAAAAACCAAGGAAGATAAAACAAGTTG
117575864	21	2.4	NCKIPSD	15:11	ag//atcctacttccatcttccag][atcccacttccatcttccc agATCCCAC
117437660	36(72)	3.8	ZBTB10	1:1	ag GTACTTCACATGATTTCAAGTATGGTTTGATGCCTG//GTCCTTCAAATGATTTCAAGTATGGATTGATACCAG][GTACTTCAAATGATTTCAAGTATGGATTGATACCAG//GTGCTTCAAATGATTTCAAGTATGGATT

Repeat ID[a]	Pattern Size[b]	Copy Number	Gene Name	ESTs Isoforms [c]	Syncopated Pattern of Splice Sites [d]
117185722	53	3	ACSF3	1:16	taaccagagcccccttcctcag//aggacaccgtggtgtttaaggacggccagttaaccagagccccttttcctcag][gggacaccgtggtgtttaaggatggccagttaaccagagccccttttcctcagGGGACACCGTGGTGTTTAAGGATGGCCAGT

[a] Each TR characterized by Tandem Repeats Finder is numbered, the data can be accessed from TRDB.
[b] Pattern size shows the length of the repeat unit, due to the possible insertion or deletion of nucleotide(s) in each period, the length difference of the transcript isoforms is not necessarily the pattern size. The number in parentheses indicates the observed period length differences.
[c] The order shown is in accordance with the postulated utilization of the splice site that generates the detected variants from 5' to 3'. For instance, four isoforms are produced by the tandem splicing array of MGC11257, the number of matching ESTs is 2:3:5 which use the intra-exonic splice sites in succession from 5' to 3', while 34 ESTs are assigned to the 3'-most splicing and uses the reference splice site. Another example is, for NCKIPSD, 15 ESTs best matched to the second postulated 3'ss and 11 to the reference site by transcription direction from 5' to 3'.
[d] Exon sequence is typed in upper-case, and intron is in lower-case. Sequence underlined in HYAL1 gene indicates the outside transcript region where its TR extends. The periods are syncopated based on the splicing di-nucleotides, "gt" for 5'ss, and "ag" for 3'ss. Period copies are divided by double slash except for the reference junction, once the solid transcript matches to any of the candidate splicing site, the di-nucleotide was bold-typed and the double slash was changed to a pair of brackets.
[e] A second splice donor exists and interrupts the 57-nucleotide splicing pattern.
[f] A deletion-insertion-polymorphism (DIP) record in the TR region, rs:11269713, with a 45 bp indel.
[g] The tri-nucleotide wobble splice acts on GTGgta interrupts the 30-nucleotide splicing pattern.

Alternative Wobble Splice Site Selection in J-TRs

To further investigate the gene expression region where J-TRs resides, ESTs records were examined for all J-TRs covering regions using UCSC genome browser. In total, 41 5'ss and 9 3'ss J-TRs were investigated with alternatively spliced variants (Table 1). Most of these J-TRs are consisted of around two copies of the core sequence units; TR in IFI6 has the maximal repeating number of 27 by a 12 bps pattern. The expression ratio between the transcripts shows that the tandem splice site proximal to the exon of J-TR would be a stronger choice for splicing. Especially in the 41 5'ss TRs, 28 (~68%) have short ESTs transcripts over long ones, this is consistent with the splice-site dominance observation conducted by Yimeng et al [20].

ESTs isoforms aligned to a single J-TR tract are mostly differed in one copy length of that repeating unit, where the conventional splicing di-nucleotide could be found and would be served for the sites for splicing. More short distance wobble splicing variants were also detected by ESTs variants of CLN8 and SERPINB13 genes, TR at SCNN1A gene also could generate a transcript isoform which is spliced at a shifted distance (Table 1).

In order to verify the alternative splicing of J-TRs, the genomic and the expressed sequences of the TR region of the gene were cloned and sequence examined. As illustrated by J-TR junction sequence confirmation of NCKIPSD gene (Figure 2), two major forms of transcripts differed with 21 nucleotides are detected. The length difference is exactly the copy length of the duplicated (or multiplicated) region. Because the similarity between the tandem-placed splicing signals would create the ambiguity for the splicing factors recognition.

In summary, tandem repetitive sequences are distributed widely and abundantly in the human genome, however, most of them are not expressed. We characterize that a fraction of the TRs may act as tandem splicing sequence tracts for either of the splicing donor or acceptor sites. The isoforms of the expressed J-TRs are thus characterized as the novel processed products of wobble splicing.

References

[1] Modrek, B. & Lee, C. (2002). A genomic view of alternative splicing. *Nat Genet., 30*, 13-19.

[2] Pan, Q., Shai, O., Lee, L .J., Frey, B. J. & Blencowe, B. J. (2008). Deep surveying of alternative splicing complexity in the human transcriptome by high-throughput sequencing. *Nat Genet., 40*, 1413-1415.

[3] Barbazuk, W. B., Fu, Y. & McGinnis, K. M. (2008). Genome-wide analyses of alternative splicing in plants: opportunities and challenges. *Genome Res., 18*, 1381-1392.

[4] Sheth, N., Roca, X., Hastings, M. L., Roeder, T., Krainer, A. R. & Sachidanandam, R. (2006). Comprehensive splice-site analysis using comparative genomics. *Nucleic Acids Res., 34*, 3955-3967.

[5] Burset, M., Seledtsov, I. A. & Solovyev, V. V. (2000). Analysis of canonical and non-canonical splice sites in mammalian genomes. *Nucleic Acids Res., 28*, 4364-4375.

[6] Chern, T. M., van Nimwegen, E., Kai, C., Kawai, J., Carninci, P., Hayashizaki, Y. & Zavolan, M. (2006). A simple physical model predicts small exon length variations. *PLoS Genet., 2*, e45.

[7] Hiller, M., Huse, K., Szafranski, K., Jahn, N., Hampe, J., Schreiber, S., Backofen, R. & Platzer, M. (2004). Widespread occurrence of alternative splicing at NAGNAG acceptors contributes to proteome plasticity. *Nat Genet., 36*, 1255-1257.

[8] Tadokoro, K., Yamazaki-Inoue, M., Tachibana, M., Fujishiro, M., Nagao, K., Toyoda, M., Ozaki, M., Ono, M., Miki, N., Miyashita, T. & Yamada, M. (2005). Frequent occurrence of protein isoforms with or without a single amino acid residue by subtle alternative splicing: the case of Gln in DRPLA affects subcellular localization of the products. *J Hum Genet., 50*, 382-394.

[9] Lai, C. H., Hu, L. Y., & Lin, W. C. (2006). Single amino-acid InDel variants generated by alternative tandem splice-donor and -acceptor selection. *Biochem Biophys Res Commun. 342*, 197-205.

[10] Tsai, K. W. & Lin, W. C. (2006). Quantitative analysis of wobble splicing indicates that it is not tissue specific. *Genomics., 88*, 855-864.

[11] Valadkhan, S. (2007). The spliceosome: a ribozyme at heart? *Biol Chem., 388*, 693-697.

[12] Schellenberg, M. J., Ritchie, D. B. & MacMillan, A. M. (2008). Pre-mRNA splicing: a complex picture in higher definition. *Trends Biochem Sci., 33*, 243-246.

[13] Tsai, K. W., Tarn, W. Y. & Lin, W. C. (2007). Wobble splicing reveals the role of the branch point sequence-to-NAGNAG region in 3' tandem splice site selection. *Mol Cell Biol., 27*, 5835-5848.

[14] Wang, Z. & Burge, C. B. (2008). Splicing regulation: from a parts list of regulatory elements to an integrated splicing code. *RNA, 14*, 802-813.

[15] Turri, M. G., Cuin, K. A. & Porter, A. C. (1995). Characterisation of a novel minisatellite that provides multiple splice donor sites in an interferon-induced transcript. *Nucleic Acids Res., 23*, 1854-1861.

[16] Benson, G. (1999). Tandem repeats finder: a program to analyzed DNA sequences. *Nucleic Acids Res., 27*, 573–580.

[17] Gelfand, Y., Rodriguez, A. & Benson, G. (2007). TRDB--the Tandem Repeats Database. *Nucleic Acids Res., 35*, (Database issue) D80-D87.

[18] Kent, W. J., Sugnet, C. W., Furey, T. S., Roskin, K. M., Pringle, T. H., Zahler, A. M. & Haussler, D. (2002). The human genome browser at UCSC. *Genome Res., 12*, 996-1006.

[19] Debrauwere, H., Gendrel, C. G., Lechat, S. & Dutreix, M. (1997). Differences and similarities between various tandem repeat sequences: minisatellites and microsatellites. *Biochimie., 79*, 577-586.

[20] Dou, Y., Fox-Walsh, K. L., Baldi, P. F. & Hertel, K. J. (2006). Genomic splice-site analysis reveals frequent alternative splicing close to the dominant splice site. *RNA. 12*, 2047-2056.

[21] Levinson, G. & Gutman G. A. (1987). Slipped-strand mispairing: a major mechanism for DNA sequence evolution. *Mol Biol Evol., 4*, 203-221.

In: The Human Genome: Features, Variations...
Editor: Akio Matsumoto and Mai Nakano
ISBN: 978-1-60741-695-1

Chapter 8

Genetic Susceptibility to Complex Traits: Moving Towards Informed Analysis of Whole-Genome Screens

Michael R. Green[1,2], Emily Camilleri[1], Maher K. Gandhi[2,3] and Lyn R. Griffiths[1,2*]

[1]Genomics Research Centre, Griffith Institute of Health & Medical Research, Griffith University, Southport, Queensland, Australia;

[2]Griffith Medical Research College, A joint initiative of Griffith University and the Queensland Institute of Medical Research, Herston, Queensland, Australia;

[3]Clinical Immunohaematology Laboratory, Queensland Institute of Medical Research, Herston, Queensland, Australia

Abstract

Susceptibility to complex traits, by definition, involves aetiological polymorphisms at multiple genetic loci combined with variable contributions by environmental factors. However, the approaches taken to identifying genetic loci implicated in susceptibility to complex traits frequently overlooks the compounding contribution of multiple loci in favour of highlighting a single gene solely responsible for predisposition. It is only in a small minority of cases that this has resulted in clear disease heritability associated with polymorphisms in a single gene. More often, this approach has led to an accumulation of single-gene associations with minor contributions to disease susceptibility. As the genomic era advances and genome-wide screens become higher in resolution and throughput, the need for simultaneous consideration of multiple loci is becoming more important. With special reference to non-Hodgkin's lymphoma (NHL), this chapter will overview the current progress made in elucidating genetic polymorphisms associated with disease susceptibility. We also present novel data from a high-resolution single nucleotide polymorphism (SNP) microarray screen for susceptibility loci that are

* Corresponding author: l.griffiths@griffith.edu.au

involved in NHL. Using an 'informed approach', the findings are highlighted within the context of cellular pathways, and provide insight and new ideas for methods of analysis for genome-wide screens for susceptibility.

Introduction

Disease susceptibility is conferred by both environment and genetic risk factors. In many cases, the genetic contribution to disease susceptibility is minor in comparison to the environmental risk factors. Furthermore, the genetic contribution to disease generally consists of a low number of alleles that confer high-risk in addition to a high number of alleles that confer low-risk (Figure 1). The aim of many genetic susceptibility studies is to elucidate high-risk alleles that show an outstanding association with a disease phenotype. However, within complex diseases, focussing on determining high-risk alleles and thereby overlooking the alleles with minor contributions to disease risk may overlook the majority of the genetic components of the disease. With the advent of high-resolution single nucleotide polymorphism (SNP) microarrays, that are capable of simultaneously genotyping hundreds of thousands of polymorphic markers, the ability to scan the genome for minor but compounding contributors to disease susceptibility has become achievable. Here, we will provide an overview of the evolution of approaches to determining genetic susceptibility, their application and successes in non-Hodgkin's Lymphoma (NHL), and our impressions of the future of genome-wide susceptibility scans.

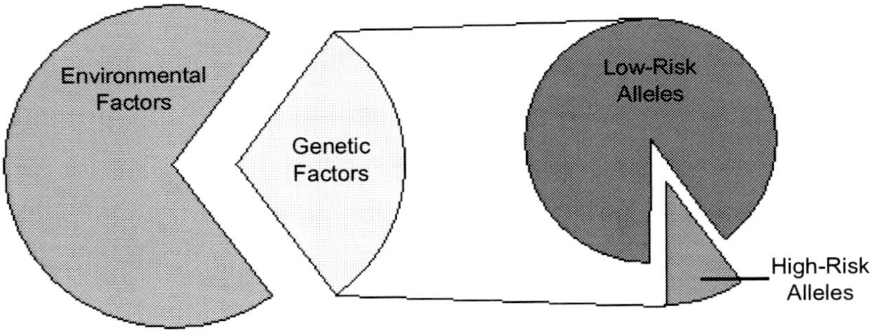

Figure 1. Illustration of the genetic component of disease, showing that the genetic component may be a small part of the overall aetiology, and high-risk alleles are a small part of the genetic basis for susceptibility.

Evolution of Genome-Wide Approaches for Determining Susceptibility

At a time in which knowledge of polymorphisms within the human genome is evolving at an ever increasing rate, genetic studies for determining susceptibility to disease have also been forced to adapt. Family-based linkage studies have been successful in determining genetic loci associated with diseases such as Huntington's, Alzheimer's and familial breast cancer. These studies are based upon the co-inheritance of genetic markers (polymorphisms) and phenotypes through families over several generations. However, this approach has not been as successful in determining susceptibility to non-mendelian complex traits due to the inability to detect inheritance patterns within family pedigrees. This has led to the growing popularity of association studies, wherein the contribution of genetic loci to disease susceptibility is determined by comparing allele frequencies of genetic polymorphisms between affected cases and unaffected controls, resulting in higher power and genetic resolution than that obtained from linkage studies [Risch and Merikangas, 1996]. Controls for these studies can either be family-based, or be made up of a cohort of individuals derived from a population. Family-based studies use unaffected relatives (for example siblings or cousins) as controls, or can alternatively employ pseudosiblings formed from the set of genotypes that are not passed down from the parents to an affected offspring [Thomson, 1995]. These studies have the advantage of overcoming population stratification; a phenomenon where ethnic differences in both allele frequency and disease risk can cause an accumulation of type I errors (i.e. false associations). However, the difficulty in recruiting cases and their relatives is a major limiting factor for these studies.

Association studies employing cohorts of a population as case and control subjects are now the most popular format for genome-wide investigations. These studies come in two forms: (i) hypothesis-based investigations, or (ii) hypothesis-generating investigations. The hypothesis-based investigations are directed towards identifying associations between a disease and polymorphisms within candidate genes for which there is usually functional evidence for a role in the disease in question. This approach has been utilised to determine many genetic risk factors for NHL (see below), but is limited towards studies of genes implicated in well-defined etiological mechanisms of the disease. In order to overcome this limitation, and allow investigation of diseases with varied or poorly-defined aetiologies, hypothesis-generating approaches can be employed. These studies use genome-wide screens in order to find differences between case and control populations, and then generate hypotheses based upon the findings. A platform allowing high-throughput whole-genome screens for these hypothesis-generating studies is SNP microarrays.

SNPs are both abundant and spread evenly across the genome, making them excellent markers for high-resolution scans for genetic susceptibility loci. Oligonucleotide microarray-based interrogation of these markers allows high-throughput analysis with ever increasing coverage and decreasing cost. New generation arrays allow for much higher resolution than that attainable with other polymorphic markers such as microsattelites, and therefore allow the elucidation of more narrowly defined loci associated with disease susceptibility. The interrogation of large numbers of samples, which are needed in order to provide significant power to associations, can also be further facilitated through the use of DNA pooling. This method pools genomic DNA from cases and controls and extrapolates allele frequencies for each SNP using the relative probe fluorescence intensities from the SNP microarrays. By performing multiple replicates of the prepared pools, inferred allele frequencies can be remarkably accurate [Bansal et al., 2002; Docherty et al., 2007], and allow a first-pass screen of the genome to be performed at low cost. Once allele frequencies have been derived for SNPs, differences in allele representation between case and control cohorts can be used to provide associations between genotypic variants and disease susceptibility. These polymorphisms may be either directly associated with disease susceptibility due to a functional role in the expression or coding of a gene, or indirectly associated with susceptibility due to linkage disequilibrium with a directly associated allele. Alleles conferring a large relative risk can lead to large differences in allele frequency between case and control cohorts, whereas alleles conferring a small relative risk demonstrate more modest differences between cohorts [Carlson et al., 2004]. Unfortunately, it is the objective of most genome-wide association studies to direct their attention towards alleles that are most differentially represented between case and control cohorts. This effectively overlooks the contribution of low-risk alleles that make up the majority of the genetic component of disease susceptibility.

Genetic Susceptibility to Non-Hodgkin's Lymphoma

A genetic basis for susceptibility to NHL has become evident due to the strong aggregation of lymphomas in otherwise healthy families [Wang et al., 2007]. The most pronounced risk factor for developing NHL is strong immunodeficiency due to disorders such as Ataxia Talangiectasia (AT), Bloom's syndrome or Nijmegen breakage syndrome (NBS). In these diseases, susceptibility to NHL is conferred by defects within pathways that respond to DNA damage [Tran et al., 2008]. This observation has lead to an accumulation of hypothesis-based studies that have found numerous associations between NHL and polymorphisms in the DNA repair system. Furthermore, susceptibility has also been linked to loci associated with folate and one-carbon metabolism, as well as immune-responses.

Variants within DNA Repair Genes

Genes involved in all facets of DNA-repair have had variable associations with disease susceptibility. This includes genes involved in lymphocyte developmental processes such as the recombination of immunoglobulin and T-cell receptor variable, diversity and joining regions, and the associated non-homologous end-joining repair pathway (RAG1 [Hill et al., 2006], LIG4 [Hill et al., 2006] and H2AFX [Novik et al., 2007]). However, genes involved in other mechanisms of DNA repair that are not necessary for lymphocyte development have also been associated with susceptibility, such as homologous recombination (BRCA1 [Shen et al., 2006], BRCA2 [Hill et al., 2006], WRN [Hill et al., 2006; Shen et al., 2006] and XRCC3 [Ekstrom-Smedby et al., 2006a; Shen et al., 2006]), nucleotide excision repair (ERCC2 [Ekstrom-Smedby et al., 2006a], ERCC5 [Shen et al., 2006]), base excision repair (XRCC1 [Ekstrom-Smedby et al., 2006a; Shen et al., 2007]), direct repair (MGMT [Hill et al., 2006; Shen et al., 2007]), and mismatch repair (MSH2 [Paz-y-mino et al., 2003; Hishida et al., 2003a]).

Variants within Folate Related Genes

The folate and one-carbon metabolism pathway is also linked to DNA-repair through the role of 5,10-methylenetetrahydrofolate reductase (MTHFR) in catalyzing the dissociation of thymine-thymine dimers. However, more important for lymphoma risk is the role if this pathway in regulating DNA methylation by affecting the amount of available methionine, and in controlling the balance of deoxynucleotide synthesis, particularly uracil which can lead DSB formation when misincorporated into DNA [Skibola et al., 2004]. Several genes within this pathway have been associated with lymphoma risk, including MTHFR [Matsuo et al., 2001; Matsuo et al., 2004; Niclot et al., 2006; Rudd et al., 2004], methionine synthase [Matsuo et al., 2001; Lincz et al., 2003; Niclot et al., 2006], serine hydroxymethyltransferase [Hishida et al., 2003b], thymidylate synthase [Hishida et al., 2003b; Skibola et al., 2004; Niclot et al., 2006], and folylpolyglutamate synthase [Lim et al., 2007]. However, these associations are heavily confounded by variation in the dietary intake of folate and methionine and multiple co-factors for their metabolism [Lim et al., 2007].

Immune Related Gene Variants

Although the association between immune-gene polymorphisms and lymphoma in healthy individuals is not as pronounced as the association with immunodeficiency, there have been several findings linking these polymorphisms with both lymphoma susceptibility and disease outcome. For example, in Hodgkin's lymphoma there are strong links between polymorphisms with the human leukocytes antigen (HLA) class I and class II regions due to their role in mediating responses to the causative pathogen, Epstein-Barr virus (EBV). Although EBV does not play as prominent of a role in NHL aetiology, these same polymorphisms have also been associated with NHL susceptibility [Houlston et al., 2003; Machulla et al., 2001]. A polymorphism within the TNFα gene, localized to the HLA class III gene cluster, has also been associated with lymphoma risk [Rothman et al., 2005], as have the genes for interleukin-10 (IL-10) [Guzowski et al., 2005; Rothman et al., 2005] and the alpha sub-unit of the IL-10 receptor (IL-10RA) [Nieters et al., 2006] which reduce TNFα expression. Interestingly, a recent study by Wang *et al.* (2007b) has provided evidence that lymphoma predisposition associated with autoimmune disease is restricted to those individuals carrying a variant allele of either of the implicated TNFα or IL-10 polymorphisms. Furthermore, TNFα and IL-10 gene polymorphisms have also been linked with patient outcome, demonstrating that their role is not restricted to disease genesis [Warzocha et al., 1998; Lech-Maranda et al., 2004].

Inconsistencies in Previous Associations

Despite the accumulation of positive associations with NHL, many of these investigations were hypothesis-based and their replication is disappointingly low (Table 1). This is contributed by factors associated with study design, including low statistical power, multiple hypothesis testing and population sub-structure; but it is also a factor of gene-environment and gene-gene interactions. For example, as well as the gene-environment interactions previously highlighted between HLA polymorphisms and EBV infection, the association between XRCC3 polymorphisms and the risk of developing Follicular Lymphoma are exacerbated in individuals that smoke [Ekstrom-Smedby et al., 2006]. Gene-gene interactions have also been shown in NHL between the DNA repair genes RAG1 and LIG4 [Hill et al., 2006], as well as the TYMS, MTR and MTHFR genes that function in folate metabolism; associations with two polymorphisms within the TYMS only gained significance when analysed in tandem with polymorphisms in the MTR and MTHFR genes [Niclot et al., 2006].

Informed Analysis of Genome-Wide SNP Microarray Data

The large number of polymorphisms associated with a moderately increased risk of developing NHL, as well as the large amount of inconsistency among independent studies, should demonstrate that complex disorders such as NHL are the result of susceptibility alleles within multiple genes that each have a minor contribution to disease onset. That is, these alleles may be neither necessary nor sufficient to individually cause disease, but may act in concert with other alleles to cause disease when a threshold of susceptibility is reached [Johnson and Todd, 2000]. Therefore, modern genetic mapping of complex diseases should allow for simultaneous consideration of multiple unlinked loci [Cox et al., 1999].

The consideration of multiple loci in microarray data has been addressed by the development of Gene Set Enrichment Analysis (GSEA) platforms which allow the interpretation of aberrations within the context of cell signalling pathways or gene ontology [Subramanian et al., 2005]. However, this approach thus far has been restricted to gene expression microarray analysis. Therefore, the wealth of information provided by genome-wide SNP microarray studies has been only relatively crudely interpreted in terms of deriving genetic aetiology. To address this issue and look at a better way of defining genetic contributors to disease, here we describe a simple method for GSEA of SNP microarray data. In order to illustrate the potential of this approach to analyse whole-genome screens for genetic susceptibility, we have used an example data-set from Affymetrix 250K Sty SNP microarrays performed on pooled genomic DNA of NHL cases (n=100) and controls (n=100). These data will allow us to provide an example of how whole-genome association study results can be interpreted within the context of cellular pathways in order to elucidate genes with a compounding contribution to disease susceptibility.

Methodology for Informed Analysis of Genome-Wide Susceptibility Screens

This method employs the calculatation relative allele signal (RAS) values from probe intensity data of cases and controls, followed by the application of thresholds to find informative SNPs, and the use of a web-based GSEA platform to derive meaningful cellular-pathway based associations. Probe intensity data are first cleaned by removing SNPs with failed calls in any of the first 3 quartets in order to eliminate outliers. The RAS values and delta-RAS values (ΔRAS), representing the difference in allele frequency between the case and control pools, are then calculated for every SNP. ΔRAS values fit a normal distribution (Figure 2), allowing simple thresholds to be applied to the tails in order to define informative SNPs. The selection of informative SNPs is therefore based upon the relative distribution of their ΔRAS values compared to other markers, rather than statistical stringencies. This allows the selection of low-risk as well as high-risk alleles in order to provide a platform for comprehensive analysis of compounding contributors to disease susceptibility.

Annotation data accompanying the Affymetrix SNP arrays are then used in order to determine genes associated with informative SNPs. It is this level of the analysis at which a threshold is applied in order to control for type I errors. The use of Affymetrix SNP microarrays in this method allows for the presence of multiple polymorphisms associated with most genes. Utilising only those genes with multiple low-stringency associations allows the elimination of numerous false-positive associations. Genes of interest (GOI) are therefore defined as those with multiple (≥3) informative SNPs showing association with the trait. GOIs elucidated in our data included genes with well-defined roles in NHL pathogenesis, such as BCL2 [Hanada et al., 1993; Kramer et al., 1998], FOXP1 [Barrans et al., 2005; Brown et al., 2005; Brown et al., 2007] and SWAP70 [Shipp et al., 2002]. There were also a host of cell-cycle regulatory genes (eg. CCND1, CCND2, CDK6 and CDK8), genes involved in cytokine signalling (eg. IL6, IL12B, CXCR4, and IL10RA), DNA repair genes (eg. MGMT, ERCC4, MSH2, RAG2 and XRCC4), growth factors and their receptors (eg. FGF1, FGFR2, VEGFC, and PDGFC), and well as genes involved in signal transduction from growth factor and interleukin receptors (eg. GRB2, FYN and JAK2).

Table 1. Examples of contradictory findings regarding genetic susceptibility in NHL.

Gene	Positive Associations	Negative Associations
RAG1	Hill et al., 2006	Shen et al., 2007 Scott et al., 2007
BRCA2	Hill et al., 2006	Scott et al., 2007
XRCC1	Hill et al., 2006 Shen et al., 2006	Ekstrom-Smedby et al., 2006
MSH2	Paz-y-mino et al., 2003	Hishida et al., 2003a

Table 1. (Continued)

Gene	Positive Associations	Negative Associations
TYMS	Skibola et al., 2004	Lim et al., 2007
	Niclot et al., 2006	
SHMT1	Hishida et al., 2003b	Niclot et al., 2006
MTHFR	Skibola et al., 2004	Lim et al., 2007
	Matsuo et al., 2001	Gemmati et al., 2004
	Niclot et al., 2006	Lincz et al., 2003

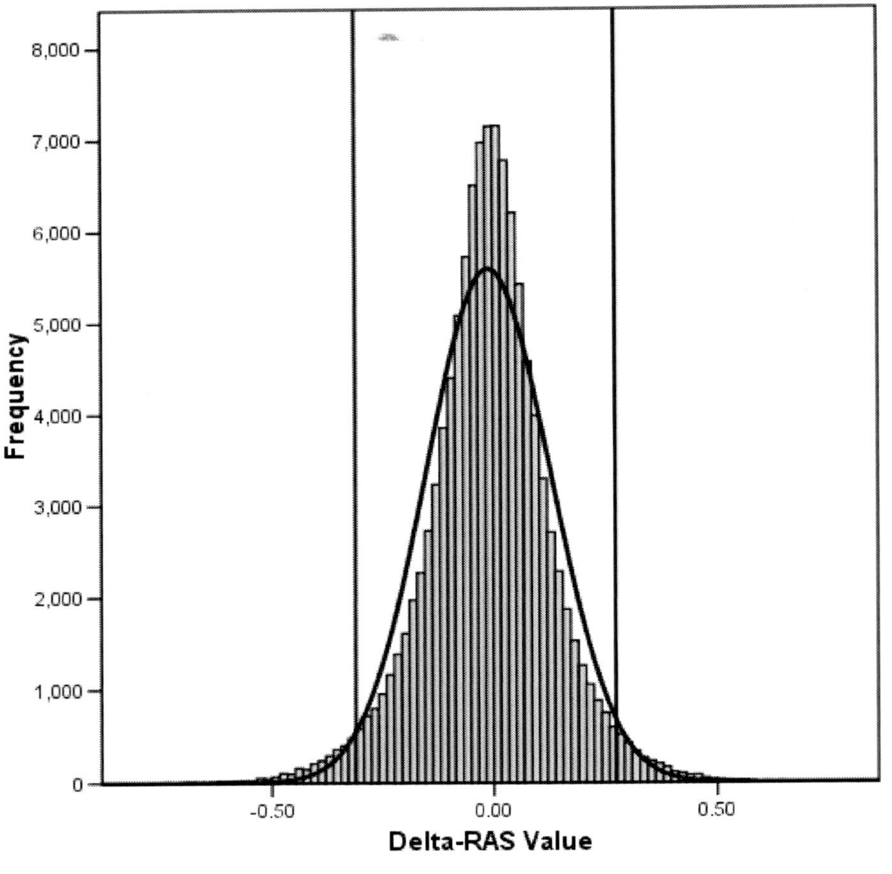

Figure 2. Delta values generated from the differences in case and control allele frequencies showing a normal distribution to which simple thresholds can be applied.

In order to make sense of these long lists of GOIs, they can be organised by cell signalling pathways using a freely available web-based GSEA platforms such as WebGestalt [Zhang et al., 2005]. These utilities group candidate genes based upon pre-defined associations with KEGG pathways, BioCarta pathways or gene ontology. GOI resident in the same cellular pathways may have a compounding contribution to disease susceptibility and therefore should be considered in unison in association studies. However, there may also be interaction between pathways due to pathway crosstalk or pathways containing common constituents. These pathway associations can therefore be represented graphically using nodes consisting of cellular pathways, and straight edges linking nodes that are mutually dependent due to crosstalk or common constituents. Pathways that are central to disease susceptibility also fall centrally within graphical representations of the pathway data. For example, using this method, we associated several cell signalling pathways with susceptibility to NHL, each containing between 2 and 27 mutually inclusive constituents. Several of these pathways have important roles in the cellular physiology of B-cells, the cell-of-origin for the subtypes of NHL investigated here.

Results of Pilot Study into NHL Susceptibility

Visualisation of GSEA results by linking pathways by common components elucidated three central pathways involved in cytokine signalling and the transduction of signals through the MAPK and JAK-STAT pathways. The transduction of signals from cytokine receptors is critical for the normal development of lymphocytes. But these normal cellular processes have also been described to have a pathogenic role in NHL. The expression of several cytokines has been linked with NHL pathogenesis [Fayad et al., 2001; Hulkkonen et al., 2000; Jones et al., 2002], including IL6 which was highlighted as a candidate in this investigation and contains polymorphisms within its 5' flanking region that alter its expression [Hulkkonen et al., 2000]. Signals from cytokine receptor are transduced through the MAPK and JAK-STAT pathways resulting in the activation of transcription factors (Figure 3). This resultant gene transcription can result in the initiation of processes such as proliferation or inhibition of processes such as apoptosis [Imada and Leonard, 2000; Treisman, 1996].

The interaction between 50 GOIs within the three pathways presents a situation where aberrations at different levels of the signalling cascade can have a compounding effect on NHL susceptibility. Therefore, the combinations of low-risk polymorphisms that effect the function or expression of cytokines, cytokine receptors, scaffolding proteins, signal transduction molecules, transcription factors and inducers or inhibitors thereof, could result in an accumulated high-risk of NHL. The statistical analysis of these multi-locus models is beyond the scope of this chapter. However, it should be noted that the formation of interaction networks within and between the central pathways resembles that of Bayesian networks [Rodin and Boerwinkle, 2005]. Using Bayesian model averaging, it is therefore possible to select a model or set of models that explain the complex relationship between genetic risk factors [Fridley, 2008]. These analyses can be performed using the R library RMA (http://cran.r-project.org/src/contrib/ descriptions/BMA.html) and are reviewed in Hoeting et al. (1999).

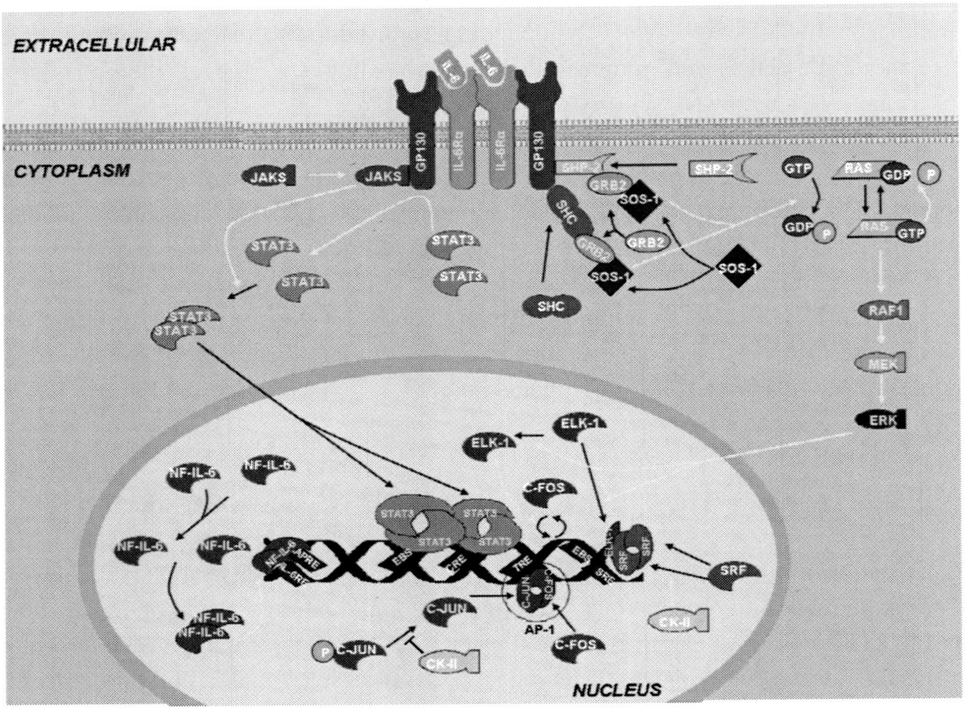

Figure 3. An example of signal transduction from a cytokine receptor, in this case the IL6 receptor, through the JAK-STAT (left) and MAPK (right) signalling pathways.

Conclusion

Studies to derive genetic features associated with susceptibility to NHL have generated a plethora of associations accompanied by numerous inconsistencies. These studies have employed strict statistical stringencies to their investigations in the hope of deriving a single polymorphism or small number of polymorphisms associated with a high risk of NHL. However, the cryptic inheritance patterns of this disease highlight the complexity of its genetic basis, lending proof to the hypothesis that susceptibility is conferred by a large number of interacting and compounding low-risk alleles. In order to address this and provide a means of elucidating interacting low-risk alleles, we have described a method of informed gene pathway analysis for genome-wide susceptibility screens. Using this approach within a pilot study we have implicated 50 genes overlapping three cell signalling pathways that have a plausible role in NHL disease etiology. This provides support for the use of our method of informed analysis for the elucidation of compounding contributors to susceptibility of complex traits such as NHL.

References

Bansal, A., van den Boom, D., Kammerer, S., Honisch, C., Adam, G., Cantor, C., Kleyn, P. & Braun, A. (2002), Association testing by DNA pooling: An effective initial screen. *Proc. Natl. Acad. Sci.*, *99*, 16871-16874

Barrans, S., Fenton, J., Banham, A., Owen, R. & Jack, A. (2004). Strong expression of FOXP1 identifies a distinct subset of diffuse large B-cell lymphoma (DLBCL) patients with poor outcome. *Blood*, *104*, 2933-2935

Brown, P., Ashe, S., Leich, E., Burek, C., Barrans, S., Fenton, J., et al. (2008), Potentially oncogenic B-cell activation induced smaller isoforms of FOXP1 are highly expressed in the activated B-cell-like subtype of DLBCL. *Blood*, *111*, 2816-2824

Brown, P., Marafioti, T., Kusec, R. & Banham, A. (2005). The FOXP1 transcription factor is expressed in the majority of follicular lymphomas but is rarely expressed in classical and lymphocyte predominant Hodgkin's lymphoma. *J. Mol. Histol.*, *36*, 249-256

Carlson, C., Eberle, M., Kruglyak, L. & Nickerson, D. (2004). Mapping complex disease loci in whole-genome association studies. *Nature*, *429*, 446-452

Cox, N., Frigge, M., Nicolae, D., Concannon, P., Hanis, C., Bell, G. & Kong, A. (1999). Loci on chromosome 2 (NIDDM1) and 15 interact to increase susceptibility to diabetes in Mexican Americans. *Nat. Genet.*, *21*, 213-215

Diepstra, A., Niens, M., Vellenga, E., van Imhoff, G., Nolte, I., Schaapveld, M. et al. (2005). Association with HLA class I in Epstein-Barr-virus-positive and with HLA class III in Epstein-Barr-virus-negative Hodgkin's lymphoma. *Lancet*, *365*, 2216-2224

Docherty, S., Butcher, L., Schalkwyk, L. & Plomin, R. (2007). Applicability of DNA pools on 500 K SNP microarrays for cost-effective initial screens in genomewide association studies. *BMC Genomics*, *8*, 214

Ekstrom-Smedby, K., Hjalgrim, H., Askling, J., Chang, E., Gregersen, H., Porwit-MacDonald, A. et al. (2006b). Autoimmune and chronic inflammatory disorders and risk of non-Hodgkin's lymphoma by subtype. *J. Natl. Cancer I.*, *98*, 51-60

Ekstrom-Smedby, K., Lindgren, C., Hjalgrim, H., Humphreys, K., Schollkopf, C., Chang, E. et al. (2006). Variation in DNA repair genes ERCC2, XRCC1, and XRCC3 and risk of Follicular Lymphoma. *Cancer Epidemiol. Biomarkers Prev.*, *15*, 258-265

Ellis, N. and German, J. (1996). Molecular genetics of Bloom's syndrome. *Hum. Mol. Genet.*, *5*, 1457-1463

Fayad, L., Keating, M., Reuben, J., O'Brien, S., Lee, B., Lerner, S. *et al.* (2001), Interleukin-6 and interleukin-10 levels in chronic lymphocytic leukaemia: correlation with phenotypic characteristics and outcome. *Blood*, *97*, 256-263

Gemmati, D., Ongaro, A., Scapoli, G., Della Porta, M., Tognazzo, S. & Serino M. (2004). Common gene polymorphisms in the metabolic folate and methylation pathway and risk of acute lymphoblastic leukaemia and non-Hodgkin's lymphoma. *Cancer Epidem. Biomar.*, *13*, 787-794

Guzowski, D., Chandrasekaran, A., Gawel, C., Palma, J., Koenig, J., Wang, X. et al. (2005), Analysis of single nucleotide polymorphisms in the promoter region of IL-10 by denaturing high-performance liquid chromatography. *J. Biomol. Tech.*, *16*, 154-166

Hanada, M., Delia, D., Aiello, A., Stadtmauer, E. & Reed, J. (1993). Bcl-2 gene hypomethylation and high-level expression in B-cell chronic lymphocytic leukemia. *Blood*, *82*, 1820-1828

Hill, D., Wang, S., Cerhan, J., Davis, S., Cozen, W., Severson, R. et al. (2006), Risk of non-Hodgkin's lymphoma (NHL) in relation to germline variation in DNA repair and related genes. *Blood*, *108*, 3161-3167

Hishida, A., Matsuo, K., Hamajima, N., Ito, H., Ogura, M., Kagami, Y. et al. (2003a). Polymorphism in the hMSH2 gene (gIVS 12-6T→C) and risk of non-Hodgkin's lymphoma in a Japanese population. *Cancer Genet. Cytogen.*, *147*, 71-74

Hishida, A., Matsuo, K., Hamajima, N., Ito, H., Ogura, M., Kagami, Y. et al. (2003b). Association between polymorphisms in the thymidilate synthase and serine hyodroxymethyltransferase genes and susceptibility to malignant lymphoma. *Haematologica*, *88* 156-166

Hoeting, J., Madigan, D., Raftery, A. & Volinsky, C. (1999). Bayesian model averaging; a tutorial (with discussion). *Stat. Sc.*, *14*, 382-401

Houlston, R., Sellick, G., Yuille, M., Matutes, E. & Catovsky, D. (2003). Causation of chronic lymphocytic leukemia – insights from familial disease. *Leuk. Res.*, *27*, 871-876

Hulkkonen, J., Vilpo, J., Vilpo, L., Koski, T. & Hurme M. (2000). Interleukin-1 β, interleukin-1 receptor agonist and interleukin-6 plasma levels and cytokine gene polymorphisms in chronic lymphocytic leukaemia: correlation with prognostic parameters. *Haematologica*, *85*, 600-606

Imada, K. & Leonard, W. (2000). The Jak-STAT pathway. *Mol. Immunol.*, *37*, 1-11

Johnson, G. & Todd, J. (2000). Strategies in complex disease mapping. *Curr. Opin, Genet. Dev.*, *10*, 330-334

Jones, E., Pringle, J., Angel, C. & Rees, R. (2002). Th1/Th2 cytokine expression and its relationship with tumor growth in B cell non-hodgkin's lymphoma (NHL). *Leukemia Lymphoma*, *43*, 1313-1321

Kramer, M., Hermans, J., Wijburg, E., Philippo, K., Geelen, E., van Krieken, J. *et al.* (1998). Clinical relevance of BCL2, BCL6 and MYC rearrangements in diffuse large B-cell lymphoma. *Blood*, *92* 3152-3162

Lech-Maranda, E., Baseggio, L., Bienvenu, J., Charlot, C., Berger, F., Rigal, D. et al. (2004). Interleukin-10 gene promoter polymorphisms influence the clinical outcome of diffuse large B-cell lymphoma. *Blood*, *103*, 3529-3534

Lim, U., Wang, S., Hartge, P., Cozen, W., Keleman, L., Chanock, S. et al. (2007). Gene-nutrient interactions among determinants of folate and one-carbon metabolism on the risk of non-Hodgkin's lymphoma: NCI-SEER case-control study. *Blood*, *109*, 3050-3059

Lincz, L., Scorgie, F., Kerridge, I., Potts, R., Spencer, A. & Enno, A. (2003) Methionine synthase genetic polymorphism MS A2756G alters susceptibility to follicular but not diffuse large B-cell non-Hodgkin's lymphoma or multiple myeloma. *Brit. J. Haematol.*, *120*, 1051-1054

Machulla, H., Muller, L., Schaaf, A., Kujat, G., Schonermarck, U. & Langner, J. (2001). Association of chronic lymphocytic leukemia with specific alleles of the HLA-DR4:DR53:DQ8 haplotype in German patients. *Int. J. Cancer*, *92*, 203-207

Matsuo, K., Hamajima, N., Suzuki, R., Ogura, M., Kagami, Y., Taji, H. et al. (2004). Methylenetetrahydrofolate reductase gene (MTHFR) polymorphisms and reduced risk of malignant lymphoma. *Am. J. Hematol.*, *77*, 351-357

Matsuo, K., Suzuki, R., Hamajima, N., Ogura, M., Kagami, Y., Taji, H. et al. (2001). Association between polymorphisms of folate- and methionine-metabolising enzymes and susceptibility to malignant lymphoma. *Blood*, *97*, 3205-3209

Niclot, S., Pruvot, Q., Besson, C., Savoy, D., Macintyre, E., Salles, G. et al. (2006) Implication of the folate-methionine metabolism pathways in susceptibility to follicular lymphoma. *Blood*, *108*, 278-285

Niens, M., Jarrett, R., Hepkema, B., Nolte, I., Diepstra, A., Platteel, M. et al. (2007). HLA-A*02 is associated with reduced risk and HLA-A*01 with an increased risk of developing EBV+ Hodgkin's lymphoma. *Blood*, *110*, 3310-3315

Nieters, A., Beckmann, L., Deeg, E. & Becker, N. (2006). Gene polymorphisms in toll-like receptors, interleukin-10, and interleukin-10 receptor alpha and lymphoma risk. *Genes Immun.*, *7*, 615-624

Novik, K., Spinelli, J., MacArthur, A., Shumansky, K., Sipahimalani, P., Leach, S., Lai, A., Connors, J., Gascoyne, R., Gallagher, R. & Brooks-Wilson, A. (2007). Genetic variation in H2AFX contributes to risk of non-Hodgkin's lymphoma. *Cancer Epidemiol. Biomarks Prev.*, *16* 1098-1105

Paz-y-mino, C., Fiallo, B., Morrillo, S., Acosta, A., Gimenez, P., Ocampo, L. & Leone, P. (2003). Analysis of the polymorphism [gIVS12-6T>C] in the hMSH2 gene in lymphoma and leukaemia. *Leukemia Lymphoma*, *44*, 505-508

Rothman, N., Skibola, C., Wang, S., Morgan, G., Lan, Q., Smith, M. et al. (2005). Genetic variation in TNF and IL-10 and risk of non-Hodgkin's lymphoma: a report from the interlymph consortium. *Lancet Oncol.*, *7*, 27-38

Sancar, A., Lindsey-Boltz, L., Unsal-Kacmaz, K., Linn, S. (2004). Molecular mechanisms of mammalian DNA repair and the DNA damage checkpoints. *Annu. Rev. Biochem.*, *73*, 39-85

Scott, K., Adamson, P., Barrans, S., Worrillow, L., Willett, E. & Allan, J. (2007). RAG1 and BRCA2 polymorphisms in non-Hodgkin's lymphoma. *Blood*, *109*, 5522-5523

Shen, M., Zheng, T., Lan, Q., Zhang, Y., Zahm, S., Wang, S. et al. (2006). Polymorphisms in DNA repair genes and risk of non-Hodgkin's lymphoma among women in Connecticut. *Hum. Genet.*, *119*, 659-668

Shipp, M., Ross, K., Tamayo, P., Weng, A., Kutok, J., Aguiar, R., Gaasenbeek, M., Angelo, M., Reich, M., Punkus, G., Ray, T., Koval, M., Last, K., Norton, A., Lister, A., Mesirov, J., Neuberg, D., Lander, E., Aster, J. & Golub, T. (2002). Diffuse large B-cell lymphoma outcome prediction by gene expression profiling and supervised machine learning. *Nature Med.*, *8*, 68-74

Skibola, C., Forrest, M., Coppede, F., Agana, L., Hubbard, H., Smith, M. et al. (2004). Polymorphisms and haplotypes in folate metabolising genes and risk of non-Hodgkin's lymphoma. *Blood*, *104*, 2155-2162

Subramanian, A., Tamayo, P., Mootha, V., Mukherjee, S., Ebert, B., Gillette, M. *et al.* (2005). Gene set enrichment analysis: a knowledge-based approach for interpreting genome-wide expression profiles. *Proc. Natl. Acad. Sci.*, *102*, 15545-15550

Tran, H., Nourse, N., Hall, S., Green, M., Griffiths, L. & Gandhi, M. (2008). Immunodeficiency-associated lymphomas. *Blood Rev.*, *22*, 261-268

Treisman, R. (1996). Regulation of transcription by MAP kinase cascades. *Curr. Opin. Cell Biol.*, *8*, 205-215

Varon, R., Vissinga, C., Platzer, M., Cerosaletti, K., Chrzanowska, K., Saar, K. et al. (1998). Nibrin, a novel double-strand break repair protein, is mutated in Nijmegen breakage syndrome. *Cell*, 93: 467-476

Wang, S., Cozen, W., Cerhan, J., Colt, J., Morton, L., Engels, E. et al. (2007b). `Immune mechanisms in non-Hodgkin's lymphoma: joint effects of the TNF G308A and IL10 T3575A polymorphisms with non-Hodgkin's lymphoma risk factors. *Cancer Res.*, *67*, 5042-5053

Wang, S., Slager, S., Brennan, P., Holly, E., De Sanjose, S., Bernstein, L. et al. (2007). Family history of hematopoietic malignancies and risk of non-Hodgkin lymphoma (NHL): a pooled analysis of 10 211 cases and 11 905 controls from the International Lymphoma Epidemiology Consortium (InterLymph). *Blood*, *109*, 3479-3488

Warzocha, K., Ribeiro, P., Bienvenu, J., Roy, P., Charlot, C., Rigal, D. et al. (1998). Polymorphisms within the tumor necrosis factor locus influence non-hodgkin's lymphoma outcome. *Blood*, *10*, 3574-3581

Zhang, B., Kirov, S. & Snoddy, J. (2005). WebGestalt: an integrated system for exploring gene sets in various biological contexts. *Nucleic Acids Res.*, *33*, W741-W74

In: The Human Genome: Features, Variations... ISBN: 978-1-60741-695-1
Editor: Akio Matsumoto and Mai Nakano

Chapter 9

The Personal Genome: Science and Beyond

Kung-Hao Liang and Hua-Mei Chang
Vita Genomics Inc.
7F, No.6, Sec.1, Jungshing Rd., Wugu Shiang, Taipei County, 248 Taiwan

Abstract

The twenty-first century is the era of the personal genome, enabled by a trilogy of scientific achievements: the Human Genome Project, the international HapMap project, and large-scale genome-wide association studies. Based on these achievements, service providers emerge to offer ordinary people an unprecedented opportunity to view their genetic heritage. These services usually include personal ancestry analysis and lifetime risk estimations for various common complex diseases. The era of the personal genome will have a great impact on many aspects of life. It promises a better health-care system featuring preventive and personalized medicine. In this chapter, we introduce multiple aspects of the personal genome, including the science, technology, applications and concerns.

Introduction

We have witnessed in the past decade an escalation in the scientific advancement of the human genome, the entire inherited blueprint of a human body passed down from generation to generation. The human genome not only promised a better health-care system featuring preventive and personalized medicine, but also made broad and profound impacts on basic human biology, personal identity, ancestry, forensic applications, global business and economy, as well as our own perception of humanity. Many impacts were previously unforeseeable.

Science and Technology

A trilogy of milestones was achieved by publicly and privately funded projects pertaining to the human genome. The human genome is made of deoxyribonucleic acid (DNA) molecules in a double-strand helix structure. DNA is a long sequence of nucleotides with four basic elements represented by the four alphabet letters A, T, G and C. The entire human genome comprises approximately 3 billion nucleotide bases. The celebrated Human Genome Project (HGP), the first of the trilogy, is run by an alliance of international scientists. Its goal is to produce sequences of alphabet codes representing the nucleotide bases in the genome. This project started in 1990, but it did not reach full speed until the late 1990s when the privately funded genomics corporation, Celera, joined the competition to achieve the same goal [Waterson, 2002]. The reference sequences were jointly achieved by the two groups together in 2003. With the reference sequences available, every gene can be accurately marked on a position in the genome. In addition, previously unknown genes may also be predicted based on the reference sequences.

Most parts of the human genome are identical for every individual; the differences are called variants. The single nucleotide polymorphism (SNP) is a very important class of variants. The HGP analyzed only a handful of persons, aiming to provide a common set of reference sequences representing *Homo sapiens*. The subsequent major task after the HGP has been to reveal the personal DNA variations and their corresponding frequencies in a population. Of course, this involves the analysis of more people. The International HapMap project, the second of the trilogy, was thus initiated in 2002. Phases I and II of this project together achieved a genome-wide scan of a monumental 3.1 million SNPs for 270 persons of European, African, Chinese and Japanese descent [HapMap, 2007]. The scanned data proved to be one of the most valuable resources for scientists around the globe with multiple uses, including the study of relationships between adjacent SNPs (known as linkage disequilibriums), the selection of representative SNPs (known as tag SNPs) for a genomic region, and the comparison of differences of allele frequencies in different ethnic groups.

The HapMap project is designed based on the common-disease-common-variant (CDCV) hypothesis, considering variants of higher variant frequencies (i.e., common variants) as major contributors to disease mechanisms [Rotimi 2004]. Therefore, this project focused on decoding those common variants with frequencies higher than 1%. Variant frequencies are different in different ethnic groups, thus four representative groups were selected at Phases I and II, and the number of groups expanded to 11 in Phase III.

The third of the trilogy is a series of genome-wide association studies for capturing the underlying relationship between genetic variants and disease mechanisms. Many genetics-phenotype associations have been established and validated since 2007, unraveling part of the mysteries of illness. The diseases investigated, to name a few, include coronary artery disease [Samani, 2007], type 2 diabetes [Sladek, 2007], prostate cancer [Thomas, 2008], systemic lupus erythematosus [SLEGEN, 2008] and many other common-complex diseases [Wellcome Trust Case Control Consortium, 2007]. These contributions significantly extended our knowledge of human biology in health and disease.

The human genome trilogy is propelled by the continuing advancement of high-throughput genome technology platforms. Capillary sequencers have played an essential role for the base-by-base sequencing in the human genome project. Array-based technology (i.e., microarrays) was first widely used for gene-expression analysis and then for genome-wide genotyping applications. The array-based technology has enabled investigators to leverage genetic variation across the entire genome, enabling the search of genetic underpinnings of complex disease. Using high-density SNP microarrays, and based on the previous genome-wide mapping of human SNPs in the HapMap project, large-scale population studies have allowed geneticists to uncover SNPs that are associated with common human traits or diseases. Mass-spectrometry instruments were widely used for characterizing protein-level gene functions. Faster, better sequencing and genotyping platforms have constantly been developed, making the highly-complex genomic information more accessible.

Array-Based Technology

Hundreds of thousands of SNPs can now be readily achieved in an efficient and highly accurate manner by array-based genome-wide genotyping to capture (tag) the bulk of the diversity in the genome [Steemers, 2006]. This strategy has great potential in tagging a disease-causing mutation if the study is sufficiently powered. Currently, these high-throughput genome-wide SNP genotyping technologies are led by two major vendors, Affymetrix (www.affymetrix.com) and Illumina (www.illumina.com). Assays from both companies can scale to as many SNPs as can be represented on the array and are readily adaptable to automation.

With the Affymetrix technology, input DNA is digested with the restriction enzymes NspI and StyI and ligated to adaptors that recognize the 4-bp overhangs. All fragments resulting from restriction enzyme digestion, regardless of size, represent substrates for adaptor ligation; a generic primer that recognizes the adaptor sequence is used to amplify adaptor-ligated DNA fragments. Polymerase chain reaction (PCR) is then used to preferentially amplify fragments in the 200- to 1100-bp size range. Amplification products for each restriction enzyme digest are combined and purified using polystyrene beads, and the amplified DNA is fragmented, labeled, and hybridized to a GeneChip. The samples are scanned, and the genotypes are scored using software developed by the company [Grant, 2008].

The Illumina technology represents a random assembly of oligonucleotide-containing beads in microwells located at the end of optical fiber bundles [Steemers, 2006]. The Infinium technology uses single-tube whole-genome amplification followed by primer extension without PCR or ligation. The three-day protocol begins with fragments approximately 300 to 600 bp. This genome representation is precipitated, resuspended, and hybridized onto a BeadChip. Single-base extension (SBE) uses a single probe sequence.

The cost per SNP chip has been reduced substantially over the past two years, making genome-wide association studies affordable to most research groups. It is now also feasible to genotype more than one individual on the same chip, The relatively recent availability of the Illumina's duo and quad chips, where two and four individuals, respectively, can be genotyped parallel on the same chip, has further reduced processing costs and increased throughput. Comparing the earlier version of both vendors' chips, Illumina beadchips are superior, since they are designed based on tag-SNP selection. However, both vendors now offer in excess of one million SNPs on each of their chip products, so the difference in information content is negligible. Furthermore, the SNP density is now also sufficient to detect copy number variations (CNVs) in the genome, which are known to play a role in the pathogenesis of complex disease. The Genome-Wide Human SNP Array 6.0 chip is currently the most updated chip product from Affymetrix, featuring 1.8 million genetic markers, including more than 906,600 single nucleotide polymorphisms (SNPs) and more than 946,000 probes for the detection of copy number variations. As for Illumina, the Infinium Human 1M-Duo BeadChip interrogates more than 1.1 million evenly distributed genomic markers per sample. Powered by Illumina's Infinium HD Assay, the chip provides the most comprehensive genome-wide coverage of SNPs and unrivaled ability to detect known and novel CNV regions. Both companies provide user-friendly software for downstream appraisal of the data generated, namely BeadStudio from Illumina and Birdseed from Affymetrix. Certified lab services are also available from these two companies.

Next-Generation Sequencing

Next-generation sequencing is the term for the faster and cheaper sequencing technologies that make it possible to obtain genomic sequence information on a previously unimaginable scale. The technology is now advanced enough to decode individual human genomes. In 2008, the first personalized genome sequencing projects were reported of James Watson's genome using 454 technology [Wheeler, 2008], and of the genomes of a Yoruban and a Han Chinese individual using Illumina technology. Next-generation sequencing is therefore poised to change the way the genetic basis for human complex traits, including disease risks, are studied.

The 454 FLX Pyrosequencer from Roch Applied Sciences (http://www.454.com/enabling-technology/the-system.asp) was the first commercially available next generation sequencer in 2004 [Margulies, 2005]. The platforms that followed were the Solexa 1G Genetic Analyzer from Illumina (http://www.illumina.com/pages.ilmn?ID=203) in 2006 [Bently, 2006], the SOLiD (Supported Oligonucleotide Ligation and Detection) system from Applied Biosystems (http://marketing.appliedbiosystems.com/images/Product/Solid_Knowledge/flash/102207/solid.html) in 2007, and the HeliScope from Helicos BioSciences (www.helicosbio.com) in 2008.

With Roche/454's Pyrosequencing technology, templates are prepared by emPCR, with 1–2 million beads deposited into PTP wells. Smaller beads with attached sulphurylase and luciferase surround the template beads. Individual dNTPs flow sequentially across the wells, dispensed in a predetermined order. On incorporation of the complement dNTP, released PPi is converted to ATP, producing light from the oxidation of luciferin to oxyluciferin. Reads averaging 400 bases are recorded as flowgrams. For homopolymer repeats up to six nucleotides, the number of dNTPs added is directly proportional to the light signal. Insertions are the most common error type, followed by deletions [Metzker, 2008].

With Illumina's reversible terminators methodology, bridge amplification of DNA fragments is randomly distributed across eight channels of a glass slide, to which high-density forward and reverse primers are covalently attached. The solid-phase amplification produces ~80 million MCs from individual ssDNA templates. A primer is annealed to the free ends of templates in each MC. The polymerase extends and then terminates DNA synthesis from a set of four RTs, each labeled with a different dye. Unincorporated RTs are washed away, base identification is performed by four-color imaging, and blocking and dye groups are removed by chemical cleavage to permit the next cycle. Color images for a given MC provide reads of ~45 bases. Substitutions are the most common error type [Metzker, 2008].

With Applied Biosystems's Sequencing by ligation approach, around 100 million emPCR-prepared template beads are deposited onto a glass slide. On annealing of a universal primer, a library of 1,2-probes is added.Appropriate conditions enable selective hybridization and ligation of probes to complementary positions. The first (Y) and second (Z) positions of the 1,2-probes are designed as interrogation bases, such that the 16 dinucleotides are encoded by four dyes. Following four-colour imaging, the ligated 1,2-probes are chemically cleaved to generate a 5′-PO4 group (P). The cycle of hybridization, ligation, imaging, and cleavage is repeated six more times. The extended primer is then stripped from the templates, and a second ligation round is performed with an n–1 primer, which resets the interrogation bases one position to the left. Interrogating each base twice improves the accuracy of the color call. Seven ligation cycles ensue, followed by three more ligation rounds. A string of 35 data bits, encoded in color space, is then aligned to a reference genome to decode the DNA sequence. Substitutions are the most common error type [Metzker, 2008].

With Helicos's single molecule sequencing with RTs, billions of unamplified ssDNA templates are prepared with poly(dA) tails that hybridize to poly(dT) primers covalently attached to a glass slide. For one-pass sequencing, this primer–template complex is sufficient. Two-pass sequencing involves copying the template strand, removing the original template, and annealing a primer directed towards the surface. Unlike Illumina's RTs, the four Helicos RTs are labeled with the same dye and dispensed individually in a predetermined order. An incorporation event results in a fluorescent signal. The problem of dephasing, in which thousands of copied templates within a given MC do not extend their primers efficiently, is eliminated using single molecules. Deletions, the most common error type, can be greatly reduced by two-pass sequencing providing ~25 base consensus reads [Metzker, 2008].

Third Generation Sequencing

Next-generation sequencing will help identify elusive genomic variants with different degrees of success, depending on the variant properties. Currently, the cost of DNA sequencing has dropped by a factor of 10 every year for the last four years, a faster rate of decline than even for computers (Dr. George M. Church, Harvard University). As costs decrease—with the maturing of the third generation of sequencing technology, for instance—it will become increasingly feasible to apply whole-genome sequencing directly to large population studies. In the meantime, current efforts will catalogue human genomic variation at finer resolution and help identify optimal methods for studies at population scale.

Looking to the future, the development of platforms from companies such as Pacific Biosciences SMRT instruments (www.pacificbiosciences.com), Dover Systems (Polonator G.007), Visigen Biotechnologies, LaserGen, Inc., Intelligent Bio-Systems, and Oxford Nanopore Technology is expected to further improve throughput and accuracy while lowering the cost.

Systems Biology

Fast instruments can generate relatively large volumes of data. The imminent challenge is to continuously utilize and interpret the vast amount of information in an effective, delicate and integrative manner. This would not be possible without the assistance of adequate software. Bioinformatics, the use of information technology for solving biomedical challenges, is an evolving and exciting multidisciplinary area featuring the integration and presentation of a complex dimension of data, making them more accessible and comprehensible. A genome browser is a typical bioinformatics tool with advanced graphical presentations, which can facilitate a user to browse or search crucial gene information such as exon and intron positions, splicing patterns, functional annotations and personal variants, chromosome by chromosome.

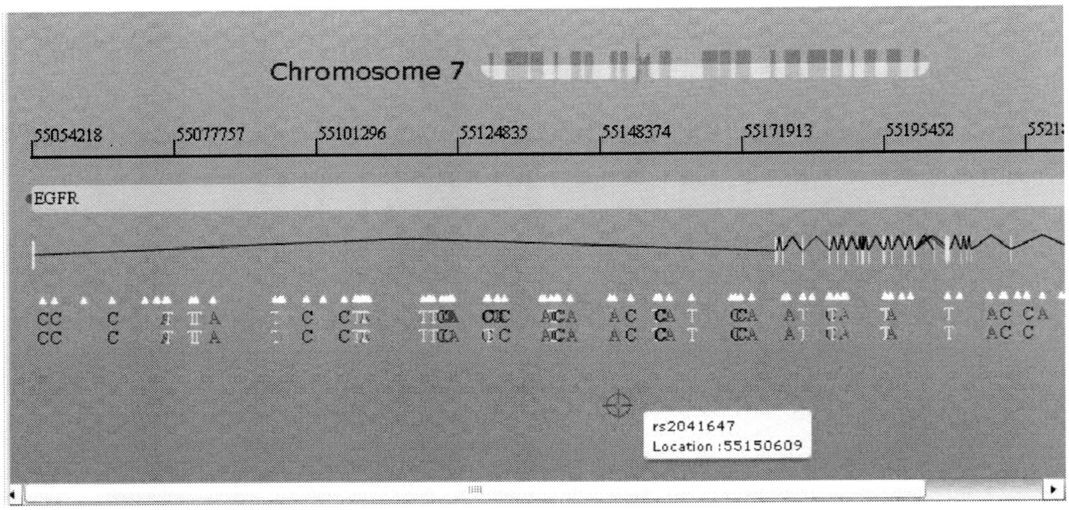

Figure 1. A typical genome browser can exhibit the gene and variant positions in the genome.

The vast volume of genomic data has granted us a holistic view of human biology. Thus, a paradigm shift is taking place that we no longer view the human biology from a single gene-based, locus-specific aspect. Instead, we will have a genome-wide, systematic view from the gene level, the protein level and the cell level alogether. This is called systems biology.

Structural Variants and CNVs

Recent high-density SNP platforms, such as Affymetrix 6.0 or Illumina HH 1M [Cooper, 2008], and array CGH platforms, had accidentally granted us a revolutionary insight on the structural variants (segmental duplications and deletions) of the genome occurred in normal people, particularly the copy number variations (CNV) which are structural variants longer than one kilo base in length. [Freeman, 2006]. In most part of the genome, we have two sets of DNA, one is from the father and one from the mother. However, recent studies show that segmental duplications and deletions occurred in the genome and passed down from generation to generation. Hence, the number of copies of DNA in these variant regions are either larger (for duplication regions) or smaller than 2 (for deletion regions). CNV regions are estimated to span a substantial portion (2%) of the entire genome. The CNV is now accessible in a genome-wide fashion illustrated in Figure 2.

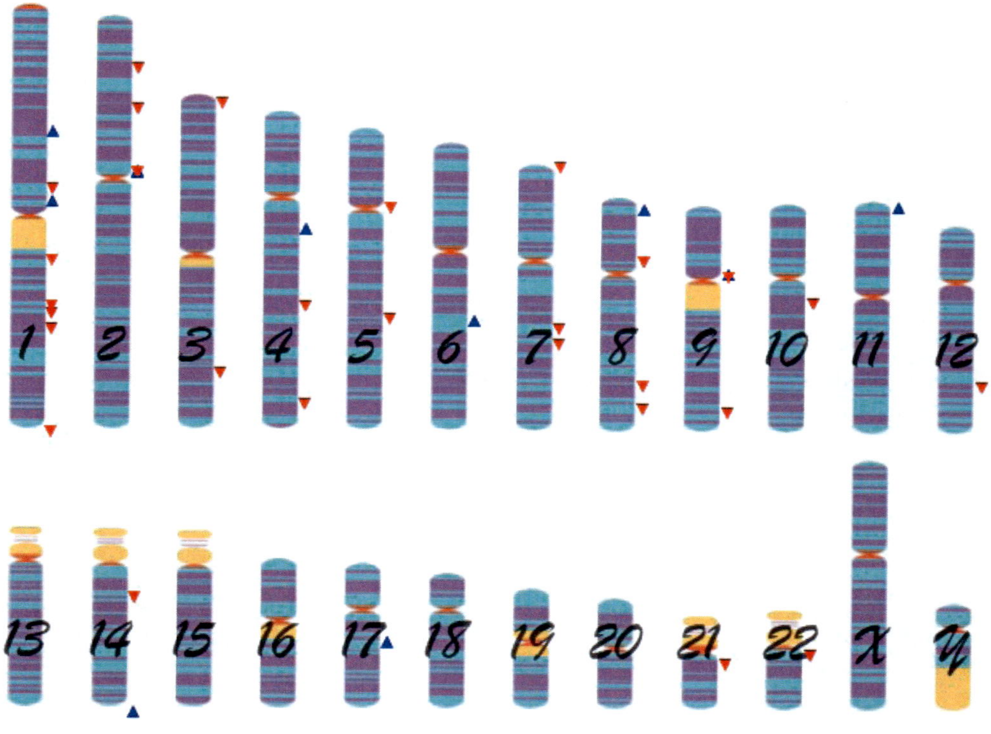

Figure 2. CNVs of a typical person. The blue triangles at the side of the chromosome represent spots of increased numbers of DNA segments (i.e., larger than 2) and the red triangles represent spots of decreased numbers of DNA segments (i.e., smaller than 2).

Structural variations used to be studied either on the mutated cancer tissues, or because they are responsible for several rare diseases such as Duchenne muscular dystrophy, Spinal muscular atrophy, Prader-Willi syndrome and Charcot Marie-Tooth disease [Freeman, 2006; Sebat, 2007]. The discovery of CNVs on normal people stimulates a multitude of insights into the human genome. The existence of CNV may help to explain the personal disease risks that cannot be explained by other forms of variants such as SNPs. Recent findings suggest that they are linked to neuro-developmental diseases such as autism, schizophrenia, mental retardation or other diseases [O'Donovan, 2008; Sebat, 2007]. One recent work identified three regions 1q21, 15q11, 15q13 to be associated to Schizophremia based on CNV data.

[Stefansson, 2008]. In addition, it is hypothesized that CNVs may to play important roles in rapid evolution [Guryev, 2007].

The International 1000 Genomes Project

Large-scale collaborative sequencing projects, aiming to decode individual genomes, continue after the human genome trilogy. Two projects, the international 1,000 genomes project, and a project aiming to sequence the genomes of at least 100 Chinese individuals at the Beijing Genomic Institute at Shenzen, were both initiated in 2008. Roche, Illumina, and AB platforms are being utilized in the 1,000 Genomes Project to produce a detailed map of human genetic variations. The goal of these efforts is to sequence many people to build a detailed resource of genomic variation, including single-nucleotide polymorphisms and structural variations. The 1,000 genomes project has just announced its initial release of SNP data from four of the individuals sequenced to high depth-of-coverage as part of the second pilot project (trios) in last December.

Applications

Modern medicine in both the diagnostic and therapeutic areas is anticipated to improve with the knowledge of human genome. Yet, this requires the involvement of private sectors to bring values to ordinary people. Many genetic tests have long been developed for detecting diseases such as cystic fibrosis. Companies were also established to provide various genetic tests for clinical suggestions, utilizing the information of multiple genes. For example, the Oncotype test developed by the Genomic Health company examine the breast tumor tissues of patients to check the aggressiveness of cancer, therefore making suggestions that whether a chemotherapy is necessary after surgical treatments of the cancer.

A new class of commercial personal genome services, including DecodeMe, 23andMe and Navigenics, emerged in 2006. They provide individual customers with access to their personal genome-wide information. The services usually comprise personal ancestry analysis and a series of disease-risk estimates, in addition to the raw genome-wide variants. These services heavily rely on the achievements of the human genome trilogy, particularly the population-specific variant frequencies derived from HapMap; and the disease-variant relationships derived from genome-wide association studies.

Personal Ancestry

Personal ancestry has been an interest of many people. The term “ancestry” has been used interchangeably orally with races or ethnic groups, yet the latter are more of the social and economical construct. Each ancestry group is characterized by a particular set of variant frequencies, the signature of that ancestry group. The HapMap project has achieved the signatures for European, African, Chinese and Japanese. Allele frequencies are critical pieces of information for an accurate estimation of personal lifetime disease risks discussed below, hence it is an important task to obtain the signatures of other ancestry groups in the years to come.

Personal genome documents the traces of ancestors passed down from generation to generation. It serves as the ultimate approach for showing personal ancestry. Skin colors, on the other hand, are not very accurate in showing ancestry [Parra, 2004]. Traditionally, ancestry analyses were mainly based on either the mitochondrial DNA [Behar, 2007] or the Y-chromosome [Underhill, 2000]. Although these methods can access the maternal and paternal ancestry respectively, their result may be of limited resolution. The availability of personal genomes now enables the population genetics / ancestry analysis to be grounded on the genome-wide scale using millions of genetic variants. The joint effects of differences of allele frequencies at multiple variants can effectively cluster people with the same continental ancestry [Mountain, 2004].

Personal Disease Risk Estimation

Disease-risk estimates were provided by the new-wave personal genome service providers. The estimations were mostly based on established statistical associations between genomic markers and diseases, which were found by genome-wide association studies involving tens of thousands of study subjects. The true functional causality of diseases may still be unclear, yet they should be in the proximity of the genetic markers identified. Using the odds ratios or relative risks found on the association studies, we would be able to calculate the personal lifetime risks based on personal genotypes.

The service providers need to combine expertise of two originally distinct disciplines so as to provide new values to the individual customer. To derive the estimates of personal lifetime disease risks, two separately estimated values need to be put together: the personal combined genetic risk, and the average lifetime risk from public health statistics. Lifetime-risk statistics play an important role in public health to quantify the risk of health-related conditions. An average lifetime risk can be estimated by the cumulative incidence in the field of epidemiology [Pencina, 2006]. On the other hand, personal genetic risks are derived from their genotypes which are confidently associated to a common complex disease based on established research findings.

A simplistic way to combine the two separately obtained numbers is to multiply them together, i.e.

Personal lifetime risk = Average lifetime risk * Personal combined genetic risk (1)

The underlying concept of this calculation is that the personal lifetime risk is linearly correlated to the personal combined genetic risks, i.e., a linear model. Although the linear model is taken for granted by most service providers, it is problematic because the estimated value may escalate to an astonishing high value, sometimes even higher then 100%. This occurs particularly when the average lifetime risk is already high. For example, the average lifetime risk of heart attack for male people at 50 years of age with the European ancestry is 51.7% [Lloyd-Jones, 2006]. If the estimated combined genetic risk is also high, say, two folds higher than average people, then the personal lifetime risk become 103.4%. The calculated value, if not mitigated, seems to suggest that this man is almost doomed to have a heart attack sometime in his life, and little can be done to prevent the onset of this disease. As a consequence, an upper limit (e.g., 80% or 95%) may be required to truncate the values above the limit. Acknowledging the fact that most common complex diseases are due to multiple factors such as the genetics and the environmental / behavioral factors, a high personal lifetime risk should only be observed from those with a family history of highly penetrant familial diseases. We therefore proposed a spring-force model here that can be used to calculate the personal lifetime risk on various common complex diseases.

We envision the value of personal lifetime risk as the consequence of balance of multiple influential forces such as the genetic, environmental and behavioral factors. Hence, we model the personal lifetime risk as a balanced state of two spring forces, where the first spring S1 represent the force of genetic odds and S2 all the other influential forces such as environmental and behavioral factors. Denote the spring constants of S1 and S2 as K1 and K2 respectively. The original length of both the two springs is 1. They are joined together in one end and each connected to a wall in the other end (Figure 3). The distance between the two parallel walls is 1, therefore, S1 and S2 are both compressed. The forces of springs S1 (which simulate the genetic factors) and S2 (which simulates all the other factors) balanced out with each other and the length of S1 represent the calculated lifetime risk. Denote the average lifetime risk and personal lifetime risk as P0 and P respectively. They are the consequence of balance of forces of S1 and S2. The length of S1 in the balanced state is 1- P0. The length of S2 is P0. We can write an equation to describe the balance of force

$$K1\ (1 - P0) = K2\ P0 \tag{2}$$

Assuming the personal genetic odds is r times in comparison with average people. To reflect this, the spring constant K1 for S1 becomes rK1. Hence, the personal risk P is deviated from the average risk P0 until a new balance is reached:

$$rK1\ (1 - P) = K2\ P \tag{3}$$

Divide (2) with (3), we obtain

$$(1 - P0)/r(1 - P) = P0\ /\ P \tag{4}$$
$$P = r\ P0/(1+(r-1)\ P0) \tag{5}$$

Equation (5) is the proposed equation for the calculation of personal lifetime risks. Note that the linear model described in Eq (1) can be revised using the mathematical notations as

$$P = r\ P0 \tag{6}$$

Comparing (5) and (6), the numerators are the same. Hence, the spring force model still remains the positive correlation structure of the linear model. The difference is on the denominator which adequately reflects the resistance of change exerted by all the other factors.

We calculated personal lifetime risks by setting the average lifetime risks as 0.1, 0.3, 0.5, 0.7 and 0.9 respectively. The genetic odds in the calculation are set between 1/10 and 10, which had covered most of the possible values of real genetic odds. Figure 3 illustrates the plots of personal lifetime risks for the linear model (see Methods, equation (6)). The calculated lifetime risk cannot be bounded between 0 and 1. Figure 4 shows the plots of personal lifetime risks calculated by the proposed spring force model (see Methods, equation (5)). No matter what the average lifetime risk is, the calculated personal genetic risk is bounded between 0 and 1.

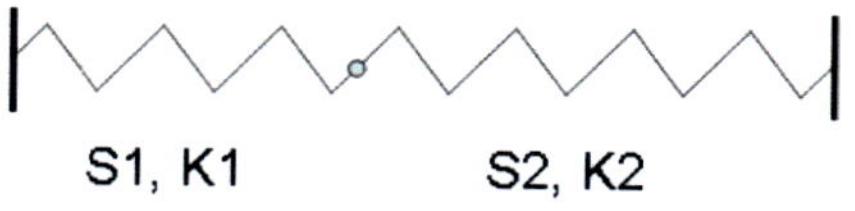

Figure 3. A schematic diagram of the spring model for calculating the personal lifetime risk estimates. S1 and S2 denote the two springs. The first spring S1 represent the force of genetic odds and S2 all the other influential forces such as environmental and behavioral factors. The spring constants were denoted as K1 and K2, respectively.

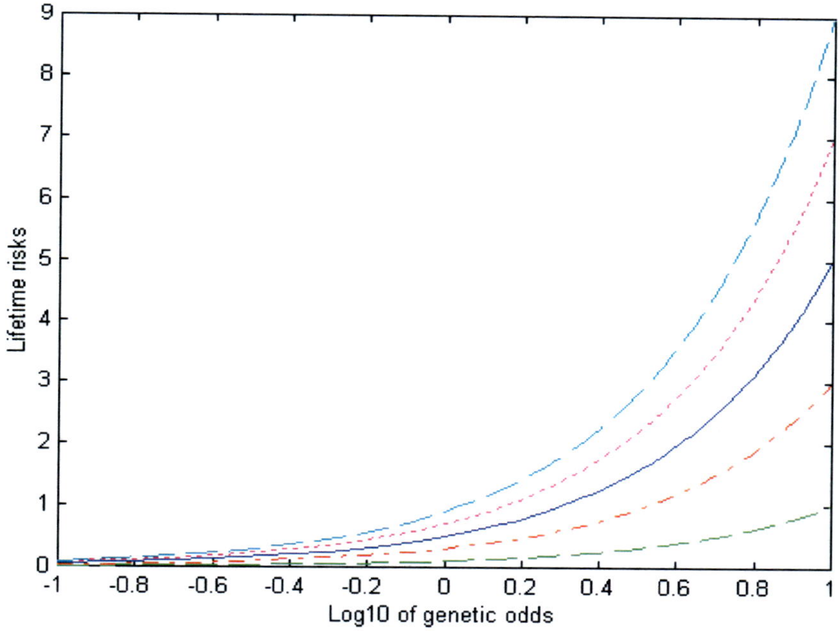

Figure 4. Curves of the estimated lifetime risk for the linear model. The horizontal axis is the log 10 number of the genetic odds. The vertical axis is the calculated personal lifetime risk. The average lifetime risks are 0.1, 0.3, 0.5, 0.7 and 0.9 which are illustrated by the green, red, blue, magenta and cyan colors, respectively.

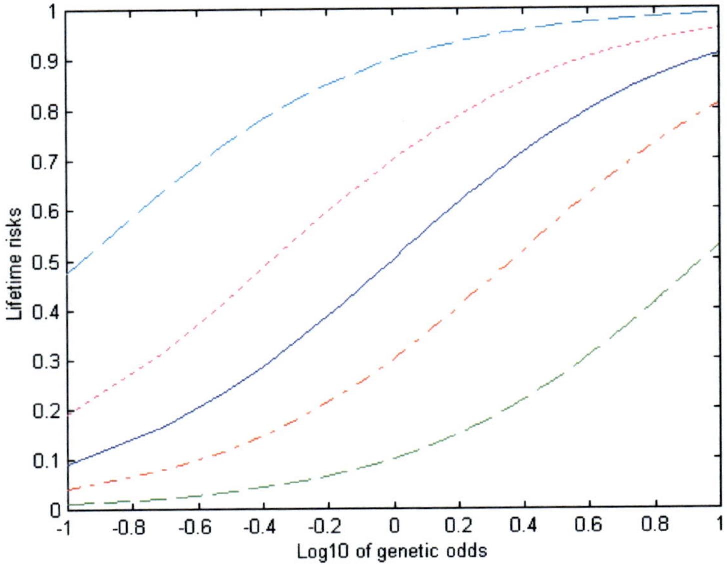

Figure 5. Curves of the estimated lifetime risk based on the spring force model. The horizontal axis is the log 10 number of the genetic odds. The vertical axis is the calculated personal lifetime risk. The average lifetime risks are 0.1, 0.3, 0.5, 0.7 and 0.9, which are illustrated by the green, red, blue, magenta and cyan colors, respectively. All values range between 0 and 1.

Disease risk estimations are intended to guide each person to an adequate lifestyle so as to avoid the onset of disease, i.e., a preventive medicine. Lists of high-risk items vary from people to people. With such information on hand, one can aggressively prevent the onset of diseases by following a healthful lifestyle. Evidence abounds that lifestyle and nutrition are important factors for one's long-term health status. A healthful diet with a balanced amount of nutrients can effectively reduce the risk for many diseases, if consumed consistently.

Obviously, this kind of service will trigger lots of attention, due to the immediate benefit and potential hazards due to the misuse of this powerful information. Hence, a responsible approach by the personal genome business is welcomed by leaders of large-scale GWA studies [Donelly, 2008].

Translational Preventive Medicine

Current evidence of genetic associations, the basis for lifetime risk estimations, has come mainly from projects funded by government and private sectors in the United States and Europe. As a result, the vast majority of genetic markers were derived from study subjects of European descent. It is a big challenge to translate this knowledge to other ancestry groups such as the Asian population. In theory, genetic markers (such as SNPs) are associated with disease because their genotypes can alter the underlying molecular mechanisms for the onset of disease. The underlying mechanisms of a common-complex disease should be universal, i.e., the same mechanisms are observed in different ancestry groups. This justifies the use of genetic markers derived from one ancestry group to predict the personal lifetime risks in another ancestry group.

A complex disease may have multiple subtypes that may involve mechanisms with different occurrence frequencies for different ancestry groups. Hence, multiple genetic markers may be required to characterize a complex disease. Yet, the allele frequencies of the genetic markers may vary from population to population. If the allele frequency of a marker is low, then this marker would be more difficult to find. This is why the genome-wide association studies conducted in different ancestry groups may result in different genetic markers [McCarthy, 2008]. Therefore, it would be ideal to combine the credible results of genetic association studies in different populations to form a panel of genetic markers, representing the superset of subtypes of a disease. One key element is to introduce the allele frequencies of the population (which are associated with the occurrence frequencies of the disease) into the calculation. The allele frequencies could come from the HapMap.

Pharmacogenomics

Personalized medicine, advocated by both the government and the pharmaceutical industry, is envisioned as a future trend of modern medicine. Pharmacogenomics studies over the last decade have accumulated sufficient information linking various drugs' effects to patients' genotypes. Given time, people will routinely have their genomes checked to predict their individual risk for disease and response to drugs.

Concerns

Personal genome services have sparked wide debate on various ethical issues such as the privacy and ownership of the genomic data. Until now, hundreds of persons have already had their genome sequenced or their genome-wide variant scanned for various projects. A few celebrities, exemplified by James Watson [Wheeler, 2008] and Craig Venter [Levy, 2007], chose to make their personal genomic information available to the public for research purposes. It seems that the disclosure of their genome has not caused any harm to them so far. Since the availability of the personal genome has been a fairly recent phenomenon, the general public still has great concerns about the privacy infringement potential of genomic information. This is why all of the personal genome service providers take the issue of the privacy of personal data very seriously.

Data Storage

Genome scan services usually detect more than one million SNPs for a customer, which poses a challenge in the data storage and organization of such an amount of complex data when the number of customers escalates. The information of personal variants has been successfully compressed into a very compact file with merely a few megabytes that could be delivered as an email attachment [Christley, 2009]. Of course, in practice we also need to keep the annotations and data indexes so as to facilitate the efficient search of different pieces of information. The resulting storage space will be much larger.

Toward Personal Genome Sequencing

According to Forbes, personal genomics will be one of the five technologies that will shape the next decade. The long-sought goal of the "$1,000 genome" now is close to becoming a reality. Personal genome sequencing will become widely available, paving the way toward personalized medicine that focuses on diseases to which patients are genetically predisposed or drugs to which patients are responsive. Knome (Cambridge, Mass.), which sequences and analyzes one's genome for between USD $100,000 and $350,000, is the first company to provide whole genome sequencing for private individuals. But Knome is already exploring farming out its sequencing to Complete Genomics (Mountain View, Calif.) for faster, less expensive sequencing. Complete Genomics aims to sell genome sequencing services wholesale for just $5,000 beginning in the second quarter of 2009 as their technology is 10 times more cost-effective than the current industry standard. Complete Genomics has already raised $46 million in venture capital, which they are putting toward building the world's largest commercial human genome sequencing center. They plan to perform 1,000 human genome sequences in 2009 and 20,000 in 2010, with goal of completing one million by 2013. Perhaps the frontrunner in the race to the $1,000 genome is California-based Pacific BioSciences, which claims it will be able to sequence an entire genome for $1,000 in a remarkable 30 minutes by 2013. The company's sequencing platform works by attaching different colored fluorescent tags to each of the four letters of the DNA, then reading their signature flashes each time the tags are incorporated into the genetic strand. The company has raised $80 million in funding and plans to start selling its system in the second half of 2010. Progress is moving even faster now, thanks to the Archon X Prize for genomics, which is offering $10 million to the first team who can sequence 100 human genomes in 10 days for under $10,000 per genome.

Conclusion

The era of the personal genome has arrived with the gradual maturation of science, technology and business models. People now have access to their genome at an affordable price. Unknown secrets still remain behind these long strings of codes, particularly regarding how they affect our personality and health status. These secrets will be unveiled gradually. Let us embrace our genetic heritage and learn what message our genome will deliver to us.

References

Behar, D. M., et al. (2007). The genographic project public participation mitochondrial DNA database. *PLos Genet., 3*, e104.

Bently, D. R. (2006). Whole-genome resequencing. *Curr. Opin. Genet.* Dev., *16*,545-52.

Christley, S., et al. (2009). Human genomes as email attachments. *Bioinformatics*, *25*, 274-275.

Cooper, G. M., Zerr, T., Eicher, E. E. & Nickerson, D. A. (2008). Systematic assessment of copy number variant detection via genome-wide SNP genotyping. *Nat. Genet.*, *40*, 1199-1203.

Donnelly, P. (2008). Progress and challenges in genome-wide association studies in humans. *Nature*, *456*, 728-731.

Freeman, et al. (2006). Copy number variation: new insights in genome diversity. *Genome Research*; *16*, 949-961

Grant, S. F. A. & Hakonarson, H. (2008). Microarray technology andaApplications in the arena of genome-wide association. *Clinical Chemistry*; *54*, 7 1116-1124

Kidd, et al. (2008). Mapping and sequencing of structural variation from eight human genomes. *Nature, 453*, 56-64

Komura, et al. (2006). Genome-wide detection of human copy number variants using high-density DNA oligonucleotide arrays. *Genome Research, 16,* 1575-1584

Korbel, J.O., et al. (2007). Paired-end mapping reveals extensive structural variation in the human genome. *Science, 318*, 420-426.

Levy, S., et al. (2007). The diploid genome sequence of an individual person. *PLoS Biology*, 2113-2144.

Lloyd-Jones, D.M., (2006). Prediction of lifetime risk for cardiovascular disease by risk factor burden at 50 years of age. *Circulation. 113*, 791-798.

Mardis, E.R. (2008). Next-generation DNA sequencing methods. *Annu.Rev. Genome. Human Genet.* 9, 387-402.

Margulies, M., Egholm, M., Altman, W. E., Attiya, S., Bader, J. S., et al. (2005). Genome sequencing in microfabricated high-density picolitre reactors. *Nature, 437,* 376-80.

McCaroll, S.A., et al. Integrated detection and population-genetic analysis of SNPs and copy number variation. *Nat. Genet., 40*, 1166-1174.

McCarthy, M. I. (2008). Casting a wider net for diabetes susceptibility genes. *Nat. Genet, 40*, 1039-1040.

Metzker, M. L. (2008). Sequencing technologies- the next generation. http://www.nature.com/reviews/posters/sequencing.

Mountain, J. L. Risch N. (2004). Assessing genetic contributions to phenotypic differences among 'racial' and 'ethnic' groups. *Nat. Genet.*, *36*, s48-s53.

O'Donovan, M. C. Kirov, G. & Owen, M. J. Phenotypic variations on the theme of CNVs. *Nat. Genet., 40*, 1392-1393.

Parra, E. J., Kittles, R.A. & Shriver, M. D. (2004). Implications of correlations between skin color and genetic ancestry for biomedical research. *Nat. Genet. 36*, s54-s60.

Pencina, M. J., Agostino, R. B., Beiser, A. S., Cobain, M. R. and Vasan, R. S. (2006). Estimating lifetime risk of developing high serum total cholesterol: adjustment for baseline prevalence and single-occasion measurements. *Am J. Epidemiology, 165(4)*, 464-472.

Redon, et al. (2006). Global variation in copy number in the human genome. *Nature, 444(23),* 444-454

Samani, N. J., Erdmann, J., Hall, A. S. (2007). Genomewide association analysis of coronary artery disease. *New England J Medicine*, *357,* 443-453.

Sebat, J. (2007). Major changes in our DNA lead to major changes in our thinking. *Nat. Genet., 39*, 53-55

Sladek, R. (2007). A genome-wide association study identifies novel risk loci for type 2 diabetes *Nature, 445*, 881 - 885.

SLEGEN. (2008). Genome-wide association scan in women with systemic lupus erythematosus identifies susceptibility variants in ITGAM, PXK, KIAA1542 and other loci. *Nat. Genet. 40(2)*, 204-210.

Steemers, F. J., Chang, W., Lee G., Barker, D. L., Shen, R. & Gunderson, K. L. (2006). Whole-genome genotyping with the single-base extension assays. *Nat. Methods. 3,* 31-33.

Stefansonn, H., et al. Large recurrent microdeletions associated with schizophremia. *Nature 455*, 232-236.

The International HapMap Consortium. (2007) A second generation human haplotype map of over 3.1 million SNPS. *Nature*, *449*, 851-862.

Thomas, G. (2008). Multiple loci identified in a genome-wide association study of prostate cancer. *Nat. Genet. 40,* 310 – 315.

Underhill P. A., et al. (2000). Y chromosome sequence variation and the history of human populations. *Nature Genetics*, *vol.26*, .358-361.

Waterston, R. H., Lander E. S., & Sulston J. E. (2002). On the sequencing of the human genome. *PNAS, 99*, 3712-3716.

Wellcome Trust. (2007). Genome-wide association study of 14000 cases of seven common diseases and 3000 shared controls. *Nature, 447,* 661-678.

Wheeler, D. A., et al. (2008). The complete genome of an individual by massively parallel DNA sequencing. *Nature, 452*, 872-876.

In: The Human Genome: Features, Variations...
Editor: Akio Matsumoto and Mai Nakano

ISBN: 978-1-60741-695-1

Chapter 10

Unstable Repeat Expansion and Human Disease

Miguel A. Varela
Fundación Pública Galega de Medicina Xenómica (Grupo de Medicina Xenómica), CIBERER, Hospital Clínico Universitario, Santiago de Compostela, A Coruña, Spain

Abstract

Microsatellites are abundant repetitive sequences accounting for 3% of the human genome. The list of diseases that are triggered by the unstable expansion of some of these sequences continues to increase, and includes disorders such as Huntington's Disease or Fragile X Syndrome. Diseases of unstable repeat expansion share peculiar genetic features. The size of the repetitive array correlates with the severity and the age of onset of the disease. Moreover, the microsatellite has a strong tendency to expand promoting earlier and more severe expression of the disease in successive generations. The most important factors determining this repeat instability seem to be related to structural properties. After DNA slippage, the more stable non-B DNA conformations serve as substrates for DNA repair and might therefore be excised. In contrast, some non-B DNA conformations could avoid the DNA repair systems. Furthermore, a recent study suggests the recruitment of microsatellites by genes that encode transcription factors and other regulatory genes, particularly in the nervous system. Therefore, some of these repeat polymorphisms may be associated with phenotypic traits that have the potential to increase fitness, but also susceptibility to unstable expansion diseases. The general pathogenic mechanisms involve altered protein function or aberrant RNA–protein interactions. Therapeutic strategies target the protein or counteract cellular defects reversing metabolic abnormalities. Additionally, gene therapy holds great promise hampering allele expansion or reducing the expression of the expanded allele using small interfering RNAs or viral-mediated approaches. Although much effort has been devoted to understanding the full disease process and the development of an effective therapy, many aspects of these disorders still remain to be fully understood and are addressed in this chapter.

Introduction

Microsatellites are repetitive sequences that comprise units from 1 to 6 nucleotides which are tandemly repeated. These sequences typically have a high level of instability and length polymorphism making them ideal for use as genetic markers (Goldstein and Schlötterer 1999). The mutational dynamics of microsatellite DNA has been well-studied and, as a consequence, we know that replication slippage seems to play a protagonist role in microsatellite variability, and most mutations involve the gain or loss of single repeat units (Ellegren 2000), but that larger jumps and point mutations that interrupt the repeat structure are also important (Di Rienzo et al. 1994; Kruglyak et al. 1998). Nevertheless, despite the effort to remove uncertainty around the complex mutational dynamics of microsatellites, some mutational processes occurring at microsatellite loci remain unclear. Thus, the malign expansion of some of these repetitive sequences is involved in a number of human neurodegenerative diseases (Gatchel and Zoghbi 2005) and understanding the processes by which these sequences expand remains a challenge.

The main factors that intervene in microsatellite expansion genome-wide are length, repetitive motif and purity of the repetitive array. Furthermore, there are other factors that can also contribute to variation in instability such as the occurrence of mutations at enzymes associated DNA replication and repair (Rubinsztein et al. 1995, Vigouroux et al. 2003) or interactions with flanking sequences (Glenn et al. 1996) including other microsatellites (Varela and Amos 2009). Nevertheless, there are a number of evidences that suggest that additional factors, probably related to the structural properties of some microsatellites, and factors at local level intervene in the expansion of microsatellites involved in human diseases and that either replication or a replication-associated process, such as DNA repair, contributes to microsatellite instability in diseases of unstable repeat expansion.

The importance of the structural properties of microsatellites in the mutational dynamics of these sequences in unstable repeat diseases can be easily observed as they determine the abundance and length distribution of microsatellites in vertebrate genomes. Thus, some non-B DNA conformations seem to avoid the DNA repair systems more often than others. Interestingly, the enrichment of such sequences in genes that encode transcription factors and regulatory genes, mostly related to nervous system activity, suggest that they could contribute to gene function but also confer susceptibility to disorders of unstable repeat expansion and, so far, there are no completely effective treatments for these diseases. In this context, gene therapy could provide the framework for the development of an effective treatment of disorders of unstable repeat expansion.

Mechanisms of Repeat Expansion

It is generally accepted that polymerase slippage plays a protagonist role in microsatellite instability. Microsatellites can misalign during replication causing extrahelical DNA loops that result in gain or loss of repeat units if slippage occurs on the daughter or the template strand respectively.

Microsatellite instability genome-wide depends mainly on the sequence of the repetitive motif and the length of the repeat tract (Xu et al. 2000; Ellegren 2000). Moreover, there are other factors that can contribute to variation in mutation rate such as the existence of differences at the enzymes associated DNA replication and repair (Rubinsztein et al. 1995, Vigouroux et al. 2003), or the occurrence of point mutations in repetitive sequences that stabilize the repetitive array and that could, at least impart, be responsible of the decay of microsatellites by the accumulation of base substitutions that decreases the rate of slippage and erodes the repetitive motif (Jin et al. 1996; Kruylag 1998; Varela et al. 2008). For example, disease alleles in Spinocerebellar Ataxia 1 (SCA1) contain uninterrupted tracts of CAG repeats while normal alleles have one to three interruptions in the middle that destabilize the intermolecular duplex structure hampering expansion (Balkwill et al. 2007).

Nevertheless, several lines of evidence point out that additional mechanisms must be operating in the expansion of microsatellites associated with neurodegenerative diseases. Namely, microsatellites in disease loci must be above a particular length threshold to be prone to expansion, DNA repair enzymes are intact and this unstable expansion is localized at one locus and it is tissue-specific (Kovtun and McMurray 2008).

The main common feature of microsatellites in disease loci is their ability to form secondary structures such as hairpins and triplexes that are thought to influence the expansion (McMurray et al. 1999; Kovtun and McMurray 2008). The structural properties of the microsatellites depend mainly on the repetitive motif. Thus, Z-DNA is easily formed by sequences altering purines and pyrimidines such as CG and CA tracts (Bichara et al. 1995). The importance of the structural properties on the expansion of these repeats can be easily observed as they determine the abundance and length distribution of microsatellites in vertebrate genomes (Bacolla et al. 2008). The structural properties of sequence and chromatin context might also be important in determining microsatellite stability (Kelkar et al. 2008; Varela and Amos 2009). Thus, the most expandable loci are those within CpG islands (Brock et al. 1999). Accordingly, methylation of CpG alter the stability of the CGG triplets at the Fragile X loci (Nichol and Pearson 2002; Kovtun and McMurray 2008). Furthermore, transgenic mice that had large pieces of human of human genomic sequence flanking the repeat showed more instability (Mangiarini et al. 1997)

In relation to the structural properties of these sequences, two classes of models for expansion in unstable repeat disorders have been proposed, one is replication-dependent, and the other is repair-dependent. The replication-dependent models explain that unstable repeats cause arrest of the replication fork in a length-dependent manner bringing about the replication stalling and the collapse of the replication fork, and expansion might occur through mechanisms needed to re-start replication (Kang et al. 1995; Mirkin 2007). The repair-dependent models propose that long unstable repeats are more susceptible to breakage and oxidative DNA damage and somatic instability could occur in the process of removing oxidized base lesions due to an error-prone repair of single-strand breaks (Freudenreich et al. 1998; Jankowski et al. 2000; Callahan et al. 2003; Kovtun et al. 2007). In relation to both replication slippage and DNA repair, the sequences that manifest the highest microsatellite instability also displayed the strongest base-stacking interactions that promote replication slippage and at the same time protect secondary structures from repair (Bacolla et al. 2008).

Therefore, after DNA slippage, some non-B DNA conformations could avoid the DNA repair systems giving rise to extended length distributions.

Interestingly, a recent study found a strong enrichment of these unstable sequences within coding regions of genes that encode transcription factors and other regulatory genes, mostly related to nervous system activity, suggesting that selection could preserve them in a large network of regulatory genes. The ability of gene-associated microsatellites to expand and contract within relatively short evolutionary time may have served to modulate gene or protein function, thereby contributing to gene and protein structure/function in vertebrate genomes increasing fitness, but also susceptibility to repeat expansion disorders in which they hamper cellular regulatory functions (Legendre et al. 2007).

On the other hand, analysis of human-chimpanzee-gorilla orthologs revealed that loci with large expansions are species-specific and that expansions have occurred after divergence from the common ancestor (Clark et al. 2006). These observations indicate that the nature of the factors governing microsatellite instability are still poorly understood and more complex than the presence of a particular sequence and its ability to adopt non-B DNA structures. The fact that expansion depends on both repair and replication proteins and that the frequency of expansion is influenced by the direction of DNA replication suggests that either replication or a replication-associated process, such as DNA repair, contributes to microsatellite instability but other locus-specific peculiarities and tissue-specific factors may be important components in the expansion process.

Human Diseases Linked to Microsatellite Expansion

Most microsatellites neither have any known function nor are linked to any known disease. However, the list of neurological disorders that are caused by unstable repeats continues to increase to include not only trinucleotide repeats, but also tetranucleotide, pentanucleotide and dodecamer repeats, leading to designation of this class of diseases as disorders of unstable repeat expansion.

Although there is much heterogeneity in these diseases regarding the pathogenic mechanism and the location of the microsatellite in coding or non coding sequences, these diseases have peculiar common features. In unstable repeat diseases there is a positive correlation between repeat size and the rate of symptom progression and the severity of the disease and the age of onset decreases in successive generations. The number of inherited repeats increases from one generation to another, usually via paternal transmission, thus causing earlier onset and faster progression. This phenomenon is known as anticipation, and it is a characteristic of most unstable repeat diseases (Gatchel and Zoghbi 2005). For example, in a patient with Huntington Disease (HD) the disease has no symptoms for many years until a sudden onset at an age that is inversely correlated with repeat number. Disease onset occurs when the microsatellite has expanded beyond a certain threshold in a sufficient number of cells, and progresses in severity, as more cells enter the pathological state. For example, in HD, the median onset age usually varies from 67 (for patients with 39 repeats) to 27 (for patients with 50 repeats) whereas a patient with more than 70 CAG repeats could manifest the disease during childhood, and with a much more aggressive course (Gatchel and Zoghbi 2005; Kaplan et al. 2007; Kovtun and McMurray 2008).

It is assumed that the level of toxicity depends on repeat number, such that longer repeats are more toxic and lead to a faster damage and earlier cell death (Kaplan et al. 2007). These large alleles that give rise to unstable repeat disorders seem to originate from the upper end of the normal allele length distribution as there is a relationship between the proportion of long normal alleles in a population and its frequency of Huntington Disease (HD), Myotonic Dystrophy (MD) and Dentatorubral Pallidoluysian Atrophy (DRPLA) (Rubinsztein and Amos 1998).

The unstable repeat disorders can be sorted into two classes depending on the location of the repeat. Diseases in the first class are caused by expansions of non-coding repeats in either 5′ and 3′ untranslated regions or introns that result in loss of protein function or altered RNA function. For example, the Fragile X Syndrome (FXS) and the Myotonic Dystrophy (MD) are associated with large expansions of untranslated repeats that can grow to thousands of units (Wong et al. 1995). These alleles many times larger than the parental allele are characteristically seen in non coding regions and occur almost exclusively on maternal transmission.

On the other hand, diseases in the second class only show small changes in repeat number. These diseases are caused by smaller expansions of CAG repeats that code for repeated stretches of the amino acid glutamine within the protein coding portion of the affected gene, which result in altered protein function. For example, in the HD gene, the CAG tract is within the protein-coding region and rarely exceeds 130 units, which is a small expansion in comparison with the FXS, presumably, due to selective pressure (Chong et al. 1997). Besides, these expansions show a paternal transmission bias.

The pathogenic mechanisms in unstable repeat disorders can involve loss or altered protein function. In polyglutamine diseases altered conformation of the mutant protein triggers aberrant interactions that lead to aggregation in the cytoplasm and nucleus. Because expanded polyglutamine proteins adopt energetically stable structures they resist unfolding and therefore avoid clearance by the proteasome (Shao and Diammond 2007). These polyglutamine proteins interact with numerous regulators that affect many downstream processes, for example, they sequester essential nuclear factors required for transcription which gives rise to cumulative damage in the affected cells leading to progressive neuronal dysfunction and neurodegeneration. Intriguingly, despite the ubiquitous expression of the mutant genes, it is common that only a specific subset of neurons is vulnerable to neurodegeneration in each disease (Shao and Diammond 2007). Probably related to this observation, cells that form inclusions earlier survive longer, whereas other cells, such as purkinje cells, are among the last to form inclusions and also the most vulnerable (Koyano et al. 2002; Haass and Selkoe 2007).

Dentatorubropallidoluysian Atrophy (DRPLA) is an autosomal dominant progressive disorder characterized by ataxia and dementia in adults, whereas children with the disease may also be affected by involuntary movements, epilepsy and progressive intellectual decline. The mean age of onset is 30 years. DRPLA is caused by an expanded trinucleotide CAG repeat in the DRPLA gene that lies on chromosome 12 (Koide et al. 1994). The protein product of the DRPLA gene is called atrophin-1. Typically, normal alleles have between 6 and 35 copies of the repeat and disease-causing alleles range from 48 to 93. This disorder is very uncommon except in Japan (Katsuno et al. 2008).

Progressive Myoclonus Epilepsy of Unverricht-Lundborg type 1 (EPM1) is an autosomal recessively inherited disorder characterized by brief contractions (known as myoclonus) of muscles that occurs spontaneously or after a sensory stimulus (Berkovic et al. 1986), ataxia and epileptic seizures. The onset of symptoms occurs between 6-15 years of age and the course of the disease is slow and progressive although it is not a severe life-threatening disease like other unstable repeat disorders (Koskiniemi et al. 1974; Alakurtti 2006).

The most common underlying mutation is the repeat expansion of an unstable dodecamer (5′-CCCCGCCCCGCG-3′) in the promoter region of the gene cystatin B (CSTB) that is a cysteine protease inhibitor (Lalioti et al. 1997; Virtaneva et al. 1997). This repeat unit occurs in 2 to 22 copies in normal chromosomes, whereas the repeat copy number in expanded alleles varies from 30 to 125 (Lalioti et al. 1997; Virtaneva et al. 1997; Lalioti et al. 1998).

Fragile X Syndrome (FXS) is the most common cause of inherited mental impairment. Some of the symptoms seen in this disorder include mental retardation, deformed features and hyperactivity. The disease has an X linked recessive inheritance and it is caused by a repeat in the 5′ untranslated region of the Fragile X Mental Retardation gene (FMR1). Normally there are 6 to 53 repeats in this region. In the disease-causing allele there are more than 230 repeats (Mathews et al. 2001).

The expansion of the CGG repeat results in methylation that causes the transcriptional silencing of the gene FMR1, leading to loss of normal gene function and hampering the expression of the Fragile X Mental Retardation Protein (FMRP). This methylation also results in constriction of the X chromosome given a light appearance at that point of the chromosome, what gave the disease its name (Debacker and Kooy 2007).

Friedreich ataxia (FA) is an autosomal-recessive disease primarily characterized by a progressive neurodegeneration usually with an age at onset younger than 25 years that is strongly correlated with the size of the shorter allele (Zühlke et al. 2007). Patients usually develop deficiencies in the central nervous system and in the heart, mainly ataxia and hypertrophic cardiomyopathy, respectively. Normal alleles contain less than 30 GAA repeats in an Alu sequence localized at the first intron the of the FXN locus that lies on chromosome 9. Patients with FA have expanded alleles ranged from 200 repeats up to more than 1,000 repeats on both arms of the chromosome. This expansion is present in homozygosity in more than 95% of the cases, whereas the remaining patients are compound heterozygotes and carry a GAA expansion on one allele, and a point mutation in the other pathogenic allele (Zühlke et al. 2007).

The intronic expansion interferes with the transcription of frataxin, which functions as a chaperone and is involved in the mitochondrial iron metabolism (Li et al. 2008). Nowadays, therapeutic efforts focus on antioxidant therapies, iron removal from mitochondria and a treatment that maximizes residual frataxin expression increasing cytosolic iron levels (Hebert and Whittom 2007; Li et al. 2008). This disease is only developed by Indo-Europeans (Labuda et al. 2000).

Huntington's Disease (HD) is an autosomal dominant neurodegenerative disease caused by a CAG trinucleotide repeat expansion in the first exon of the human HD gene (HTT) that encodes a protein called huntingtin and lies on chromosome 4. The age of onset is between 35 and 45 years. There is an intermediate range of rare alleles between 27 and 35 CAG repeats with a higher risk for further expansion in subsequent generations. Patients typically show symptoms of involuntary movements and dementia (Imarisio et al. 2008).

Despite ubiquitous expression of mutant and normal huntingtin, HD causes selective neurodegeneration in some regions of the brain. Neurons of the basal ganglia that show the highest levels of expression of adenosine A2A receptors, are the most vulnerable in HD suggesting that these receptors might be important in the pathogenetic mechanism of HD (Popoli et al. 2007). The mutant protein may cause toxicity after a series of cleavage events leading to the production of N-terminal fragments that may perturb specific transcriptional pathways in the nucleus or inhibit mitochondrial function and proteasome activity (Gatchel and Zoghbi 2005; Imarisio et al. 2008).

Myotonic Dystrophy (DM) is an autosomal dominant disease and the most common form of muscular dystrophy in adults. DM can be caused by a mutation on either chromosome 19 (DM1) or chromosome 3 (DM2). The disease has a highly variable phenotype including muscular dystrophy, myotonia, cardiac defects, cataracts, endocrine disorders, and, in some cases, mental handicap (Harper 2001).

Myotonic Dystrophy type 1 (DM1) is caused by an expanded CTG triplet repeat in the 3′ untranslated region of the Dystrophia Myotonica-Protein Kinase (DMPK) gene (Mahadevan et al. 1992). The CTG expansion causes loss of function in this gene and in two flanking genes. Additionally, the CTG expanded transcript could also gain toxic function by interfering with processing of RNAs. Myotonic Dystrophy type 2 (DM2) is caused by a CTG expansion in intron 1 of the zinc finger protein 9 (ZNF9) gene. The DM2 repeat tract is a complex repeat motif that is conserved in mammals suggesting a conserved biological function. In normal chromosomes, the repeat tract is interrupted (Liquori et al. 2001), whereas expanded alleles do not show these interruptions but a high somatic instability (Day et al. 2003). This loss of interruptions predisposes alleles to expansion presumably resulted from a common founder in Northern Europe where this disease is more common (Liquori et al. 2003).

Spinal Bulbar Muscular Atrophy (SBMA) also known as Kennedy's disease is caused by the expansion of a CAG tract in the first exon of the gene that codes for the androgen receptor protein on the X-chromosome (La Spada et al. 1991). SBMA is a neurodegenerative disorder characterized by muscular atrophy of facial, limb and bulbar muscles, hand tremor and gynaecomastia and, in some cases, reduced fertility (Sobue et al. 1989). The normal gene contains 9 to 36 repeats and the repetitive array of the mutant gene usually ranges from 38 to 62 repeats.

The pathogenic mechanism involves nuclear accumulation of the mutant protein with expanded polyglutamines in motoneurons (Li et al. 1998). The neurological symptoms of SBMA occur mainly in adult males. Related to this, testosterone, the ligand of the androgen receptor, probably plays an important role in the pathogenesis of SBMA (Katsuno et al. 2006; Chahin et al. 2008).

Spinocerebellar Ataxias (SCAs) are autosomal dominant neurodegenerative disorders characterized by a late-onset and slowly progressive ataxia, difficulty in articulating speech and eventually sensory loss. Most of these diseases result from expanded CAG tracts in the encoded protein while two forms, SCA8 and SCA12, are associated with repeat expansions in non-translated regions (Mutsuddi and Rebay 2005). Despite their clinical and genetic heterogeneity even within a subtype, the main symptoms relate to neurodegeneration in the cerebellum, spinal cord and brain stem (Bowater et al. 2001; Cummings et al. 2000).

Interestingly, interruptions seem to very important conferring stability and preventing pathological expansion of repeat tracts in some forms of SCAs. Thus, in the gene involved in SCA1, asymptomatic individuals have between 6 and 44 copies of the repeat CAG. In non disease causing alleles with more than 20 copies and less than 44, there are between 1 and 4 interruptions of the codon CAT in the tract of CAG repeats, whereas the expanded alleles do not harbour these interruptions and the allele size ranges from 39 to 81 repeats. In the SCA2 gene, the CAG repeat in normal individuals varies in length from 14 to 31 repeats and it is frequently interrupted by one or more CAA triplets. However, in a person with the disease, expanded alleles contain a pure uninterrupted stretch of 34 to 59 CAG repeats (Choudhry et al. 2001).

Microsatellite Instability as a Hallmark of Cancer Other important aspect of repeat expansion related to human disease is their instability in some kinds of cancer. Microsatellite instability associated with cancer is observed mainly in Hereditary Non-Polyposis Colorectal Cancer (HNPCC), also known as Lynch syndrome, where it is a prognostic factor and a screening marker for identifying patients with this disease. In general, it is detected in approximately 15% of colorectal, gastric and endometrial cancers, and in lower frequencies in other tumors (Miyaki et al. 1997; Wijnen et al. 1999; Imai and Yamamoto 2008; Kovtun and McMurray 2008).

The underlying cause of microsatellite instability in HNPCC is a mutation in at least one component of the mismatch repair (MMR) system (Liu et al. 1996). Promoter methylation gives rise to silencing of gene expression and enzyme deficiency in Mlh1, Msh2, Msh3, Msh6 and Pms2 (Imai and Yamamoto 2008; Kovtun and McMurray 2008). This DNA mismatch repair (MMR) deficiency results in a genome-wide increase in spontaneous mutation rate and high microsatellite instability due to the inability of MMR to correct post-replicative errors (Kovtun and McMurray 2008).

In constrast to the unstable repeat expansions of polyglutamine disorders, the analysis of the magnitude and direction of mutations at microsatellites in HNPCC has revealed that short sequences can be affected by small insertions, but most of mutations are deletions, including the majority of the mutations at the HD and the SCA1 locus (Goellner et al. 1997). Thus, in HNPCC, unstable repeats in disease loci behave as typical microsatellites when post-replicative repair is defective, and do not expand as observed in human neurodegenerative diseases such as HD suggesting that the mechanisms involved in microsatellite instability in some kinds of cancer and in unstable repeat disorders are different (Imai and Yamamoto 2008; Kovtun and McMurray 2008).

Potential Therapies in Diseases of Unstable Repeat Expansion

The potential therapeutic strategies fall into two categories. The first category includes methods that try to reverse transcriptional and metabolic abnormalities induced by the toxic protein. In this category pharmacological intervention tries to counteract cellular perturbations, for example, using compounds that target the energy metabolism such as creatine, or antioxidants such as the coenzyme Q10 (Beal and Ferrante 2004; Shao and Diamond 2007).

The second category of therapies specifically targets the toxic protein. Some of them use small interfering RNAs to halt the expression of the mutant protein. This procedure has already been carried out in mouse models of polyglutamine disease successfully (Xia et al. 2004). Other treatments inhibit or inactivate protein-splitting enzymes such as certain caspases to block the release of the toxic N-terminal fragments (Wellington and Hayden 2000). Other methods try to stimulate cellular degradation pathways that target disease proteins preferentially. For example, rapamycin, which stimulates autophagy to clear toxin proteins from cells (Rubinsztein et al. 2006). Alternatively, some molecules could directly interfere with polyglutamine protein aggregation (Heiser et al. 2000). For example, pharmacological induction of molecular chaperones could aid in protein refolding and degradation although even without aggregation protein monomers could retain their capacity for pathogenesis (Shao and Diamond 2007). Finally, other methods use molecules that influence the equilibrium between a toxic and non-toxic conformation trying to stabilize the conformation of the polyglutamine protein in a non-toxic form (Shao and Diamond 2007).

Although nowadays there are no completely effective treatments, some of these procedures hold great promise. In the future, the development of gene therapy could provide the framework for a mutant gene-specific method that successfully interferes with microsatellite expansion.

Conclusion

Much effort has already been devoted to remove uncertainty around some mutational processes occurring at microsatellite loci. However, many aspects of unstable repeat disorders still remain unclear. We already know that either replication or a replication-associated process, such as DNA repair, contributes to microsatellite instability but other locus-specific peculiarities and tissue-specific factors that are poorly understood may be important components in the expansion process. The most important factors determining this repeat instability seem to be related to structural properties of the DNA. Thus, after DNA slippage, some DNA conformations could avoid the DNA repair systems in certain locations of the genome leading to altered protein function or aberrant RNA–protein interactions and, therefore, increase the susceptibility to diseases of unstable repeat expansion. There are a number of treatments available for these diseases. Some therapeutic strategies target the mutant protein, whereas other therapeutic strategies try to counteract cellular defects reversing metabolic abnormalities but, so far, none of them has been proved to be completely effective. In this context, the development of new gene therapy methods could provide the tools for an improvement of the treatments. Developing a general procedure for unstable repeat diseases that only affects the allele with the expanding microsatellite remains a significant challenge.

Acknowledgements

M.V. is funded by an *Angeles Alvariño* fellowship from *Xunta de Galicia* (IN840D and IN809A).

References

Alakurtti, K. (2006). *Molecular biology of progressive myoclonus epilepsy of Unverricht-Lundborg type (EPM1).* Helsinki: University of Helsinki.

Bacolla, A., Larson, J. E., Collins, J. R., Li, J., Milosavljevic, A., Stenson, P. D., Cooper, D. N. & Wells, R. D. (2008). Abundance and length of simple repeats in vertebrate genomes are determined by their structural properties. *Genome Res., 18,* 1545-1553.

Balkwill, G. D., Williams, H. E. & Searle, M. S. (2007). Structure and folding dynamics of a DNA hairpin with a stabilising d(GNA) trinucleotide loop, influence of base pair mismatches and point mutations on conformational equilibria. *Org. Biomol. Chem., 5,* 832-839.

Beal, M. F. & Ferrante, R. J. (2004). Experimental therapeutics in transgenic mouse models of Huntington's disease. *Nat. Rev. Neurosci., 5,* 373-384.

Berkovic, S. F., Carpenter, S. & Andermann, F. (1986). Atypical inclusion body progressive myoclonus epilepsy, a fifth case? *Neurology, 36,* 1275-1276.

Bichara, M., Schumacher, S. & Fuchs, R. P. (1995). Genetic instability within monotonous runs of CpG sequences in *Escherichia coli. Genetics, 140,* 897-907.

Bowater, R. P. & Wells, R. D. (2001). The intrinsically unstable life of DNA triplet repeats associated with human hereditary disorders. *Prog. Nucleic Acid. Res. Mol. Biol., 66,* 159-202.

Brock, G. J., Anderson, N. H. & Monckton, D. G. (1999). Cis-acting modifiers of expanded CAG/CTG triplet repeat expandability, associations with flanking GC content and proximity to CpG islands. *Hum. Mol. Genet., 8,* 1061-1067.

Callahan, J. L., Andrews, K. J., Zakian, V. A. & Freudenreich, C. H. (2003). Mutations in yeast replication proteins that increase CAG/CTG expansions also increase repeat fragility. *Mol. Cell. Biol., 23,* 7849-7860.

Chahin, N., Klein, C., Mandrekar, J. & Sorenson, E. (2008). Natural history of spinal-bulbar muscular atrophy. *Neurology, 70,* 1967-1971.

Chong, S. S., Almqvist, E., Telenius, H., et al. (1997). Contribution of DNA sequence and CAG size to mutation frequencies of intermediate alleles for Huntington disease, evidence from single sperm analysis. *Hum. Mol. Genet., 6,* 301-309.

Choudhry, S., Mukerji, M., Srivastava, A. K., Jain, S., Brahmachari, S. K. (2001). CAG repeat instability at SCA2 locus, anchoring CAA interruptions and linked single nucleotide polymorphisms. *Hum. Mol. Genet., 10,* 2437-2446.

Clark, R. M., Bhaskar, S. S., Miyahara, M., *Dalgliesh, G. L.* & Bidichandani, S. I. (2006). Expansion of GAA trinucleotide repeats in mammals. *Genomics, 87,* 57-67.

Cummings, C. J. & Zoghbi, H. Y. (2000). Trinucleotide repeats, mechanisms and pathophysiology. *Annu. Rev. Genomics Hum. Genet., 1,* 281-328.

Day, J. W., Ricker, K., Jacobsen, J. F., Rasmussen, L. J., Dick, K. A., Kress, W., Schneider, C., Koch, M. C., Beilman, G. J., Harrison, A. R., Dalton, J. C. & Ranum, L. P. (2003). Myotonic dystrophy type 2: molecular, diagnostic and clinical spectrum. *Neurology, 60,* 657-664.

Debacker, K. & Kooy, R. F. (2007). Fragile sites and human disease. *Hum. Mol. Genet., 16,* 150-158.

Di Rienzo, A., Peterson, A. C., Garza, J. C., Valdes, A. M. & Slatkin, M. (1994). Mutational processes of simple sequence repeat loci in human populations. *Proc. Natl. Acad. Sci. USA, 91,* 3166-3170.

Ellegren, H. (2000). Microsatellite mutations in the germline, implications for evolutionary inference. *Trends Genet., 16,* 551-558.

Freudenreich, C. H., Kantro, S. M. & Zakian, V. A. (1998). Expansion and length-dependent fragility of CTG repeats in yeast. *Science, 270,* 853-856.

Gatchel, J. R. & Zoghbi, H. Y. (2005). Diseases of unstable repeat expansion, mechanisms and common principles. *Nat. Rev. Genet., 6,* 743-755.

Gentles, A. J. & Karlin, S. (2001). Genome-scale compositional comparisons in eukaryotes. *Genome Res., 11,* 540-546.

Glenn, T. C., Stephan, W., Dessauer, H. C. & Braun, M. J. (1996). Allelic diversity in alligator microsatellite loci is negatively correlated with GC content of flanking sequences and evolutionary conservation of PCR amplifiability. *Mol. Biol. Evol., 13,* 1151-1154.

Goellner, G. M., Tester, D., Thibodeau, S., Almqvist, E., Goldberg, Y. P., Hayden, M. R. & McMurray, C. T. (1997). Different mechanisms underlie DNA instability in Huntington disease and colorectal cancer. *Am. J. Hum. Genet., 60,* 879-890.

Goldstein, D. B. & Schlötterer, C. (1999). *Microsatellites, evolution and applications.* Oxford: Oxford University Press.

Haass, C. & Selkoe, D. J. (2007). Soluble protein oligomers in neurodegeneration, lessons from the Alzheimer's amyloid beta-peptide. *Nat. Rev. Mol. Cell Biol., 8,* 101-112.

Harper, P. S. (2001). *Myotonic dystrophy. Major Problems in Neurology Series.* (3rd Ed). London: Harcourt.

Hebert, M. D. & Whittom, A. A. (2007). Gene-based approaches toward Friedreich ataxia therapeutics. *Cell Mol. Life Sci., 64,* 3034-3043.

Heiser, V., Scherzinger, E., Boeddrich, A., Nordhoff, E., Lurz, R., Schugardt, N., Lehrach, H. & Wanker, E. E. (2000). Inhibition of huntingtin fibrillogenesis by specific antibodies and small molecules, implications for Huntington's disease therapy. *Proc. Natl. Acad. Sci. USA, 97,* 6739-6744.

Imai, K. & Yamamoto, H. (2008). Carcinogenesis and microsatellite instability, the interrelationship between genetics and epigenetics. *Carcinogenesis, 29,* 673-680.

Imarisio, S., Carmichael, J., Korolchuk, V., Chen, C. W., Saiki, S., Rose, C., Krishna, G., Davies, J. E., Ttofi, E., Underwood, B. R. & Rubinsztein, D. C. (2008). Huntington's disease, from pathology and genetics to potential therapies. *Biochem. J., 412,* 191-209.

Jankowski, C., Nasar, F. & Nag, D. K. (2000). Meiotic instability of CAG repeat tracts occurs by double-strand break repair in yeast. *Proc. Natl. Acad. Sci. USA, 97,* 2134-2139.

Jin, L., Macaubas, C., Hallmayer, J., Kimura, A. & Mignot, E. (1996). Mutation rate varies among alleles at a microsatellite locus, phylogenetic evidence. *Proc. Natl. Acad. Sci. USA, 93,* 15285-15288.

Kang, S., Ohshima, K., Shimizu, M., Amirhaeri, S. & Wells, R. D. (1995). Pausing of DNA synthesis in vitro at specific loci in CTG and CGG triplet repeats from human hereditary disease genes. *J. Biol. Chem., 270,* 27014-27021.

Kaplan, S., Shalev, I. & Ehud, S. (2007). A universal mechanism ties genotype to phenotype in trinucleotide diseases. *PLoS Comput Biol., 3,* e235.

Katsuno, M., Adachi, H., Waza, M., Banno, H., Suzuki, K., Tanaka, F., Doyu, M. & Sobue, G. (2006). Pathogenesis, animal models and therapeutics in spinal and bulbar muscular atrophy (SBMA). *Exp. Neurol., 1,* 8-18.

Katsuno, M., Banno, H., Suzuki, K., Takeuchi, Y., Kawashima, M., Tanaka, F., Adachi, H. & Sobue, G. (2008). Molecular genetics and biomarkers of polyglutamine diseases. *Curr. Mol. Med., 8,* 221-234.

Kelkar, Y. D., Tyekucheva, S., Chiaromonte, F. & Makova, K. D. (2008). The genome-wide determinants of human and chimpanzee microsatellite evolution. *Genome Res., 18,* 30-38.

Koide, R., Ikeuchi, T., Onodera, O., Tanaka, H., Igarashi, S., Endo, K., Takahashi, H., Kondo, R., Ishikawa, A., Hayashi, T., et al. (1994). Unstable expansion of CAG repeat in hereditary dentatorubral-pallidoluysian atrophy (DRPLA). *Nat. Genet., 6,* 9-13.

Koskiniemi, M. (1974). Psychological findings in progressive myoclonus epilepsy without Lafora bodies. *Epilepsia, 15,* 537-545.

Kovtun, I. V., Liu, Y., Bjoras, M., Klungland, A., Wilson, S. H. & McMurray, C. T. (2007). OGG1 initiates age-dependent CAG trinucleotide expansion in somatic cells. *Nature, 24,* 447-452.

Kovtun, I. V. & McMurray, C. T. (2008). Features of trinucleotide repeat instability in vivo. *Cell. Res., 18,* 198-213.

Koyano, S., Iwabuchi, K., Yagishita, S., Kuroiwa, Y. & Uchihara, T. (2002). Paradoxical absence of nuclear inclusion in cerebellar Purkinje cells of hereditary ataxias linked to CAG expansion. *JNNP, 73,* 450-452.

Kruglyak, S., Durrett, R. T., Schug, M. D. & Aquadro, C. F. (1998). Equilibrium distributions of microsatellite repeat length resulting from a balance between slippage events and point mutations. *Proc. Natl. Acad. Sci. USA, 95,* 10774-10778.

Labuda, M., Labuda, D., Miranda, C., Poirier, J., Soong, B.-W., Barucha, N. E. & Pandolfo, M. (2000). Unique origin and specific ethnic distribution of the Friedreich ataxia GAA expansion. *Neurology, 54,* 2322-2324.

Lalioti, M. D., Mirotsou, M., Buresi, C., Peitsch, M. C., Rossier, C., Ouazzani, R., Baldy-Moulinier, M., Bottani, A., Malafosse, A. & Antonarakis, S. E. (1997). Identification of mutations in cystatin B, the gene responsible for the Unverricht-Lundborg type of progressive myoclonus epilepsy (EPM1). *Am. J. Hum. Genet., 60,* 342-351.

Lalioti, M. D., Scott, H. S., Genton, P., Grid, D., Ouazzani, R., M'Rabet, A., Ibrahim, S., Gouider, R., Dravet, C., Chkili, T., Bottani, A., Buresi, C., Malafosse, A. & Antonarakis, S. E. (1998). A PCR amplification method reveals instability of the dodecamer repeat in progressive myoclonus epilepsy (EPM1) and no correlation between the size of the repeat and age at onset. *Am. J. Hum. Genet., 62,* 842-847.

La Spada, A. R., Wilson, E. M, Lubahn, D. B., Harding, A. E. & Fischbeck, K. H. (1991). Androgen receptor gene mutations in X-linked spinal and bulbar muscular atrophy. *Nature, 352,* 77-79.

Legendre, M., Pochet, N., Pak, T. & Verstrepen, K. J. (2007). Sequence-based estimation of minisatellite and microsatellite repeat variability. *Genome Res., 17,* 1787-1796.

Levinson, G. & Gutman, G. A. (1987). Slipped-strand mispairing: a major mechanism for DNA sequence evolution. *Mol. Biol. Evol., 4,* 203-221.

Li, K., Besse, E. K., Ha, D., Kovtunovych, G. & Rouault, T. A. (2008). Iron-dependent regulation of frataxin expression, implications for treatment of Friedreich ataxia. *Hum. Mol. Genet., 17,* 2265-2273.

Li, M., Miwa, S., Kobayashi, Y., Merry, D. E, Yamamoto, M., Tanaka, F., Doyu, M., Hashizume, Y., Fischbeck, K. H. & Sobue, G. (1998). Nuclear inclusions of the androgen receptor protein in spinal and bulbar muscular atrophy. *Ann. Neurol., 44,* 249-254.

Liquori, C. L., Ikeda, Y., Weatherspoon, M., Ricker, K., Benedikt, G., Schoser, H., Dalton, J. C., Day, J. W. & Ranum, L. P. W. (2003). Myotonic dystrophy type 2, human founder haplotype and evolutionary conservation of the repeat tract. *Am. J. Hum. Genet., 73,* 849-862.

Liquori, C. L., Ricker, K., Moseley, M. L., Jacobsen, J. F., Kress, W., Naylor, S. L., Day, J. W. & Ranum, L. P. (2001). Myotonic dystrophy type 2 caused by a CCTG expansion in intron 1 of ZNF9. *Science, 293,* 864-867.

Liu, B., Parsons, R., Papadopoulos, N., et al. (1996). Analysis of mismatch repair genes in hereditary non-polyposis colorectal cancer patients. *Nat. Med., 2,* 169-174.

Mahadevan, M., Tsilfidis, C., Sabourin, L., Shutler, G., Amemiya, C., Jansen, G., Neville, C., Narang, M., Barcelo, J., O'Hoy, K., Leblond, S., Earle-Macdonald, J., Jong, P., Wieringa, B. & Komeluk, R. G. (1992). Myotonic dystrophy mutation, an unstable CTG repeat in the 3' untranslated region of the gene. *Science, 255,* 1253-1255.

Mangiarini, L., Sathasivam, K., Mahal, A., Mott, R., Seller, M. & Bates, G. P. (1997). Instability of highly expanded CAG repeats in mice transgenic for the Huntington's disease mutation. *Nat. Genet., 15,* 197-200.

Mathews, D. J., Kashuk, C., Brightwell, G., Eichler, E. E. & Chakravarti, A. (2001). Sequence variation within the fragile X locus. *Genome Res., 11,* 1382-1391.

McMurray, C. T. (1999). DNA secondary structure, a common and causative factor for expansion in human disease. *Proc. Natl. Acad. Sci. USA, 96,* 1823-1825.

Mirkin, S. M. (2007). Expandable DNA repeats and human disease. *Nature, 447,* 932-940.

Miyaki, M., Konishi, M., Tanaka, K. et al. (1997). Germline mutation of MSH6 as the cause of hereditary nonpolyposis colorectal cancer. *Nat. Genet., 17,* 271-272.

Mutsuddi, M. & Rebay, I. (2005). Molecular genetics of spinocerebellar ataxia type 8 (SCA8). *RNA Biol., 2,* 49-52.

Nichol, K. & Pearson, C. E. (2002). CpG methylation modifies the genetic stability of cloned repeat. *Genome Res.,* 12, 1246-1256.

Pearson, C. E., Edamura K. & Cleary, J. D. (2005). Repeat instability, mechanisms of dynamic mutations. *Nat. Rev. Genet., 6,* 729-742

Popoli, P., Blum, D., Martire, A., Ledent, C., Ceruti, S. & Abbracchio, M. P. (2007). Functions, dysfunctions and possible therapeutic relevance of adenosine A2A receptors in Huntington's disease. *Prog. Neurobiol., 81,* 331-348.

Rubinsztein, D. C. (2006). The roles of intracellular protein-degradation pathways in neurodegeneration. *Nature, 443,* 780-786.

Rubinsztein, D. C. & Amos, W. (1998). Trinucleotide repeat mutation processes. In: Rubinsztein DC, Hayden MR editors. *Analysis of triplet repeat disorders*. Oxford: BIOS; pp. 257-268.

Rubinsztein, D. C., Leggo, J. & Amos, W. (1995). Microsatellites evolve more rapidly in humans than in chimpanzees. *Genomics, 30,* 610-612.

Shao, J. & Diamond, M. I. (2007). Polyglutamine diseases, emerging concepts in pathogenesis and therapy. *Hum. Mol. Genet., 16,* 115-123.

Sobue, G., Hashizume, Y., Mukai, E., Hirayama, M., Mitsuma, T. & Takahash, A. (1989). X-linked recessive bulbospinal neuronopathy. A clinico pathological study. *Brain, 112,* 209-232.

Varela, M. A., Sanmiguel, R., Gonzalez-Tizon, A. & Martinez-Lage, A. (2008). Heterogeneous nature and distribution of interruptions in dinucleotides may indicate the existence of biased substitutions underlying microsatellite evolution. *J. Mol. Evol., 66,* 575-580.

Varela, M. A. & Amos, W. (2009). Evidence for non-independent evolution of adjacent microsatellites in the human genome. *J. Mol. Evol., 68,* 160-170.

Vigouroux, Y., Matsuoka, Y. & Doebley, J. (2003). Directional evolution for microsatellite size in maize. *Mol. Biol. Evol., 20,* 1480-1483.

Virtaneva, K., D'Amato, E., Miao, J., Koskiniemi, M., Norio, R., Avanzini, G., Franceschetti, S., Michelucci, R., Tassinari, C. A., Omer, S., Pennacchio, L. A., Myers, R. M., Dieguez-Lucena, J. L., Krahe, R., de la Chapelle, A. & Lehesjoki, A.-E. (1997). Unstable minisatellite expansion causing recessively inherited myoclonus epilepsy, EPM1. *Nature Genet., 15,* 393-396.

Wellington, C. L. & Hayden, M. R. (2000). Caspases and neurodegeneration, on the cutting edge of new therapeutic approaches. *Clin. Genet., 57,* 1-10.

Wijnen, J., de Leeuw W., Vasen, H., et al. (1999). Familial endometrial cancer in female carriers of MSH6 germline mutations. *Nat. Genet., 23,* 142-144.

Wilder, J. & Hollocher, H. (2001). Mobile elements and the genesis of microsatellites in dipterans. *Mol. Biol. Evol., 18,* 384-392.

Wong, L. J., Ashizawa., T., Monckton, D. G., Caskey, C. T. & Richards, C. S. (1995). Somatic heterogeneity of the CTG repeat in myotonic dystrophy is age and size dependent. *Am. J. Hum. Genet., 56,* 114-122.

Xia, H., Mao, Q., Eliason, S. L., Harper, S. Q., Martins, I. H., Orr, H. T., Paulson, H. L., Yang, L., Kotin, R. M. & Davidson, B. L. (2004). RNAi suppresses polyglutamine-induced neurodegeneration in a model of spinocerebellar ataxia. *Nat. Med., 10,* 816-820.

Xu, X., Peng, M., Fang, Z. & Xu, X. (2000). The direction of microsatellite mutations is dependent upon allele length. *Nat. Genet., 24,* 396-399.

Zühlke, C., Bernard, V. & Gillessen-Kaesbach, G. (2007). Investigation of recessive ataxia loci in patients with young age of onset. *Neuropediatrics, 38,* 207-209.

In: The Human Genome: Features, Variations...
Editor: Akio Matsumoto and Mai Nakano
ISBN: 978-1-60741-695-1

Chapter 11

SNPs and CNVs in Human Disorders

Barkur S. Shastry
Department of Biological Sciences, Oakland University, Rochester, MI, USA

Abstract

The genetic make up of an individual, at least in part, determines disease susceptibility and response to drug treatment. It is because of this reason a tremendous progress has been made in cataloging human sequence variations. It is thought that a high-density map of variations will provide necessary tools to develop genetic based diagnostic and therapeutic options. The most common type of variation is called single nucleotide polymorphism (SNP). These are highly abundant, stable and distributed throughout the genome. They are also associated with diversity in the population, individuality, susceptibility to diseases and response to medicine. It has been suggested that SNPs can be used for heterogeneity testing, pharmacogenetic studies and to identify and map complex, common diseases such as high blood pressure, diabetes and heart diseases. Consistent with this proposal is the identification of the patterns of SNPs in conditions such as schizophrenia, blood pressure homeostasis and diabetes. Recently, a new form of genetic variation known as copy number variation (CNV) has also been identified. By using different types of genome wide scanning procedures, CNVs have been shown to be associated with several complex and common disorders including nervous system disorders. One of the common features of the regions associated with the complex and common disorders identified so far is the presence of CNVs and segmental duplications. Segmental duplications lead to genome instability. Because of their location and nature (several of them contain genes) many CNVs have functional consequences such as gene dosage alteration, disruption of genes and modulation of activities of other genes. Therefore, these genetic variations will have influence on phenotypes, susceptibility of an individual to disease, drug response and human genome evolution. These types of variants (gain and loss of DNA) are not restricted to humans but they have also been identified in other organisms. Because most common, complex disorders are caused by the combined effects of multiple genes and non-genetic environmental factors, it is likely that sequence variation alone is not sufficient to predict the risk of disease susceptibility, particularly in homeostatic organisms like humans. Nevertheless, these variations (SNPs) may provide a starting point for future inquiry. Our current knowledge on CNVs and their heritability is still rudimentary because of their location in regions of complex genomic structure. Future advances in the technology will help in constructing a new CNV map that can be used to (a) find genes underlying common diseases (b) understand the familial genetic conditions (c) uncover the severe developmental defects in humans and other organisms and (d) genome evolution.

Introduction

It is generally believed that the genomes between two randomly selected individuals contain approximately 0.1% difference or variation. This variation is called polymorphism and it arises because of mutations. The simplest form of DNA variation among individuals is the substitution of one single nucleotide for another. This type of change is called single nucleotide polymorphism (SNP) and they are found to be more common than other types of polymorphisms. It is estimated that they occur at a frequency of approximately 1 in 1000 base pairs (bp) throughout the genome (Brooks, 1999) and there are more than 3 million SNPs that had been charted so far. These simple changes are believed to be stable and not deleterious to organisms. According to a published report, 50% of SNPs occur in the non-coding regions, 25% lead to missense mutations and the remaining 25% are silent mutations (Halushka et al. 1999). These silent SNPs are called synonymous SNPs because they do not change the encoded amino acids. On the other hand, non-synonymous SNPs alter amino acids, may produce pathology and may be subjected to natural selection. SNPs can be observed between individuals in a population, may influence promoter activity (gene expression), pre-mRNA conformation (stability) and translational efficiency (Lohrer and Tangen, 2000). Therefore, they may be responsible for the susceptibility of an individual to many common diseases, medicinal drug metabolism and genome evolution. They may also play a direct role with or without other factors in the phenotypic expression of diseases or traits such as tallness, curly hair and individuality (Martin et al. 1997; Krawezak et al. 1992; Lohrer and Tangen, 2000; LeVan et al. 2001).

SNPs in Gene Discovery

SNPS, the most common type of segregating DNA sequence variations have a variety of applications. They can be used to identify genes responsible for disease predisposition by an association study (Riley et al. 2000; Kruglyak, 1999). This is because when alleles are closely located to each other, it is likely that they are inherited together. Therefore, markers that are closer to the gene are likely to be inherited along with that gene. When a large number of disease-carrying patients and unrelated groups of individuals are analyzed for allelic association, a particular allele may occur more frequently in patients than the normal control. It is possible that this allele could be linked to the disease susceptibility gene (reviewed in Shastry, 2002; 2003). SNP markers can also be used to identify drug response, ecology, evolutionary biology, diagnostic test, heterogeneity testing and to identify quantitative trait loci that contribute to polygenic variations (Gary et al. 2000; Schork et al. 2000). However, one major problem with them is that many SNPs may not display any variability in the family that is being studied and hence the causative role of identified genes must be independently confirmed. In addition, population structure, different levels of linkage disequilibrium (LD), allelic and non-allelic heterogeneity of phenotypes and epistatic interactions of alleles may pose problems (Schork et al. 2000; Chakravarti, 1999; Weiss and Terwilliger, 2000). Many variants may also be common to all populations and others may have a very restricted distribution (Salisbury et al. 2003). Despite these limitations, there has been some success in identifying disease susceptibility genes and a number of SNPs are now shown to be associated with various disorders. A partial list of diseases associated with SNPs is presented in Table1.

Table 1. A partial list of human disorders associated with SNPs

Disease	Gene*	Disease	Gene*
Arrhythmias	KCNQ1	Lung cancer	MMP-1
Asthma	IL-10	Migraine	Insulin receptor
Autism	Tryptophan 2,3 dioxygenase	Myocardial infarction	lymphotoxin alpha
Blood pressure	TAFI	Obesity	PAI-1
Biliary cirrhosis	Mannose binding lectin	POAG	TIGR
Calcium stone	E-cadherin	Prostate cancer	COX-2
Type II diabetes	GFPT-2	Panic disorder	Adenosine 2A receptor
Dyslipidemia	Lipase	Rheumatoid arthritis	PADI-4 SLC22A4
Hyperandrogenism	SHBG	Systematic lupus erythematosis	PDCD1
Knee and hip osteoarthritis	COL 2A and COL 9A3	Systemic sclerosis	Fibrillin-1
Late onset Parkinson	Tau	Urinary bladder cancer	Cyclin D1

* individual references are found in the reference, Shastry, 2004.

KCNQ1 = potassium channel; TAFI = thrombin-activable fibrinolysis inhibitor; IL = interleukin; GFPT2 = glutamine fructose- 6-phosphate amidotransferase –2; SHBG = sex hormone binding globulin; COL = collagen; MMP1 = matrix metalloproteinase 1; PAI-1 = plasminogen activator inhibitor; TIGR = trebacualr mesh-work inducible glucocorticoid response; COX = cyclooxygenase; PADI = peptidylargininedeiminase; SLC = an organic cation transporter; PDCD1 = programmed cell death –1 gene.

Genomic Variation and Drug Metabolizing Genes

SNPs in genes encoding drug metabolizing enzymes, drug transporters and receptors contribute, at least in part, to the inter-individual variability in drug response (reviewed in Shastry 2003; 2004; 2007; Evans and Johnson, 2001). These factors affect drug absorption, distribution, metabolism and excretion. As a result, some drugs work better in some patients than others and some drugs may be highly toxic to certain patients (Ansari and Kajinovic, 2007). This type of anti-drug reaction has been observed in several diseases such as pulmonary hypertension, epilepsy, cardiac arrhythmia, renal cell carcinoma, leukemia and liver cancer (reviewed in Shastry, 2006; Roses, 2000). There are now several high-density SNP maps of genes encoding proteins of medical importance (Iida et al. 2003) and there is strong evidence that links SNPs to inter-individual differences in drug response (Nothen and Cichon, 2002; Ansari and Krajinovic, 2007). Patients with more active drug metabolizing enzymes may require higher doses of drug and those who do not have active enzyme may exhibit toxicity. In the future, a high-density SNP map may provide the necessary tools to develop genetically based diagnostic and therapeutic tests.

Properties of CNVs

Chromosome structural abnormalities have been known for a long time at the cytogenetic and molecular level and are associated with what is now called genome disorders. Their importance however, has been realized only after the completion of the human genome project with the use of several scanning technologies including array based comparative genomic hybridization (CGH). Recently it has been found that some individuals can have deletion or multiple copies of the same gene (Figure 1). These kinds of changes (DNA fragment of approximately 1 kilobases or larger) are called copy number variation (CNV) or copy number polymorphism (Hind et al. 2006; McCarroll et al. 2006; Conard et al. 2006). They also appear to be widespread in normal individuals. Although a true number of CNVs and their frequencies in human populations are not known they may result in altered level of gene expression. As a result, they may account for a significant normal phenotypic variation within species (because they often encompass genes), susceptibility or resistance to disease, drug response, complex and common disorders and evolution of genome itself (Cappuzzo et al. 2005; Gonzalez et al. 2005; Aitman et al. 2006; Sebat et al. 2007; Reddon et al. 2006; McCarroll et al. 2007; Freeman et al. 2006; Marshall et al. 2008).

The size of DNA variation in CNV could range from a few kilobases to megabases (Sebat et al. 2004; Iafrate et al. 2004; Buckland 2003). It is not necessary that all CNVs are related to genome disorders and some of them are in fact present in healthy individuals with no obvious genetic disorders (de Stahl et al. 2008). At present there are approximately 1447 CNVs that have been identified (http://projects.tcag.ca/variation) and they cover 12 % of the estimated 15% of the human genome. This number could be slightly smaller for common human CNVs (Perry et al. 2008). Additionally, somatic mosaicism has been reported for CNV in differentiated human tissues (Piotrowski et al. 2008). This type of high degree of variability in the human genome and somatic mosaicism challenges the definition of normality (Kehrer-Sawatzki, 2007) because it is generally believed that normal cells are genetically identical. Large CNVs are often found to occur in regions containing large homologous repeats or segmental duplication (Iafrate et al. 2004; Sharp et al. 2005; Tuzun et al. 2005) while smaller CNVs may occur because of non-homology driven mutational mechanisms. A DNA replication based mechanism has been suggested to explain the formation of CNVs (Lee et al. 2007). According to this mechanism, the presence of many repeats of nucleotides (they form an unusual structure) stalls the replication fork and then it switches to a different template. That results in copy number changes. This kind of stalling and template switching could occur in any region of DNA and may play a role in evolution thereby helping the organism to survive in a stressed environment. Widespread CNVs are not unique for humans; they have also been identified in inbred strains of laboratory mice (Li et al. 2004; Adams et al. 2005). In these also, non-allelic homologous recombination mechanism may play a role in the genesis of CNV (Lupski 1998). They also contribute to the phenotypic variation among mouse strains along with SNPs. In addition, different mouse strains show comparable CNVs when compared to human but such a variation is more locally restricted (Perry et al. 2006; She et al. 2008).

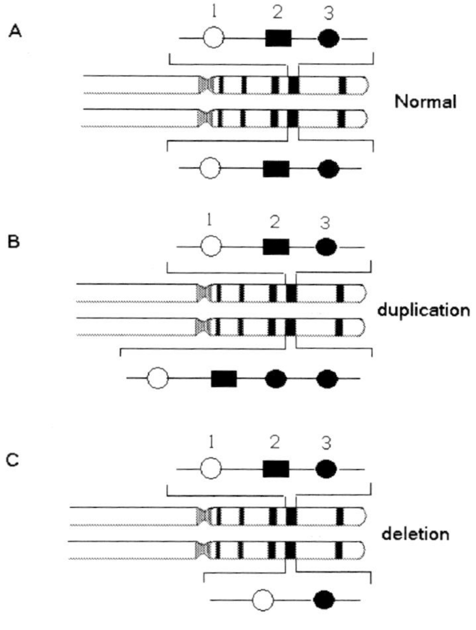

Figure 1. An example of CNV is shown. Panel A shows a normal pattern whereas panels B (duplication) and C (deletion) show CNVs. The numbers 1, 2 and 3 represent genes in a chromosomal segment.

CNVs may alter the gene dosage without abolishing the gene function or they may affect the gene structure and regulation (Stranger et al. 2007). Because of their mild effect on gene function, it is possible that some of the common CNVs alter phenotypes in complex and sporadic diseases (Inoue et al. 2002) including nervous system disorders (Lee and Lupski 2006; Esch et al. 2005). They also underlie inherited disorders (Lupski 2007). A partial list of some of the disease-related chromosomal regions containing CNVs is presented in Table 2. However, it is not clear to what extent genetic disease is caused by CNVs. According to the literature, their contribution to genetic disease appears to be smaller than that of SNPs (Stranger et al. 2007) and their contribution to gene expression is also independent of SNPs.

Table 2. Examples of CNVs* identified in the genomic regions that are associated with diseases

Disorders	Chromosomal regions	References
Bipolar	3q13.3	Lachman et al. 2007
Spinal muscular atrophy	5q13.2	Sebat et al. 2004 Iafrate et al. 2004 Sharp et al. 2005
Crohn disease	8p23.1	Fellermann et al. 2006
Prader-Willi and Angelman syndrome	15q11.3	Conard et al. 2006 McCarroll et al. 2006 Iafrate et al. 2004
Autism	16p11.2	Marshall et al. 2008
Charcot-Marie-tooth disease type I	17p12	McCarroll et al. 2006 Sharp et al. 2005
Autism	20p13	Sebat et al. 2007
DiGeorge/Velocardiofacial syndrome	22q11.2	Sharp et al. 2005 Conard et al. 2006 McCarroll et al. 2006 Babcock et al. 2007
Schizophrenia	?	Vrijenhoek et al. 2008

* In some cases, CNVs are also present in normal individuals

At present, it is difficult to correlate the relationship between CNVs and phenotypes. This is because of insufficient data collected and existing techniques are not accurate enough in measuring the association between CNVs and phenotype. Many CNVs are also located in regions of complex genomic structure. An effective genome wide CNVs genotyping method is needed. One important technical development is the hybrid oligonucleotide arrays containing both SNPs and copy number probes. In this method both SNPs and CNVs are used together for association studies (Zhao et al. 2004; Franke et al. 2008). Recently, microarray comparative genome hybridization (CGH) has also been used. This method detects differences in CNVs across diverse species (Perry et al. 2006) and also assesses CNVs at multiple loci. Additionally, CNVs can be determined by using oligonucleotide expression microarrays (Auer et al. 2007), BAC (bacterial artificial chromosome) arrays (de Stahl et al. 2008), SNP arrays and genotyping data. A review about the different genotyping platforms has been recently published (Carter 2007). Then there are variety of fine mapping techniques researchers use to test for disease association (Gonzalez et al. 2005; Aitman et al. 2006; Yang et al. 2007; Fanciulli et al. 2007) either with family based (traditional) or population based designs. These methods may work if CNVs are large because such variations are usually taken as functional and hence causative. However, smaller CNVs that are mostly benign and hence cannot be taken as functional, require statistical challenges to assess their association with diseases. Similarly, there are other limitations such as insufficient recombination events, requirement of a large pedigree, low penetrance and population stratification that have to be considered in determining the association. Some of the methods may not be able to detect low penetrant variants that are associated with common diseases. Family based studies are the easier way to prevent stratification and hence preferred by many investigators (Thomas et al. 2005). Additionally, quality of DNA and sensitivity of methods such as CGH may introduce technical artifacts in association studies (Fiegler et al. 2006). Therefore, it is necessary to apply a highest standard throughout the process of research.

Conclusion

The use of SNPs and CNVs in disease gene isolation and genetic testing is truly exciting. It is clear that some SNPs do change the amino acid sequence or regulatory site but it is not clear how useful they are in diagnostic testing and how genotype alone predicts disease susceptibility. It is not necessary that all DNA sequence variations (SNPs and CNVs) will result in clinical manifestation of a disease. Hence functional significance must be evaluated before their use (Jordan et al. 2002). This is because of variation in frequency of polymorphism among different populations (Wakeley et al. 2000; Nielsen et al. 2000; Nielsen and Slatkin 2000), complexity of human disease and the involvement of environmental factors that determine the phenotype. For instance, identical twins have identical genotypes. But they exhibit 50% concordance for more common diseases (Weiss and Terwilliger, 2000; Plomin et al. 1994). Moreover, we must also consider the importance of the developmental programs such as DNA methylation, X-inactivation and environmental conditions during postnatal development. Additionally, in multigenic diseases the contribution of susceptible individual genes to the disorder is very weak (Martin et al. 1997). On the other hand, some SNPs and CNVs are certainly important for susceptibility to various diseases and drug metabolism. Research in the CNV field also suggests that large-scale variation in the human genome could be due to segmental duplications. Segmental duplications lead to genome instability and hence are associated with genome disorders as well as evolution of the genome. They are the hotspots of chromosomal rearrangement (de Stahl et al. 2008). However, our understanding of their organization and heritability is still in its infancy. The long-term goal of CNV research is to prepare a comprehensive CNV map of the human genome. This includes correlation of variations to phenotypes, evolutionary and mutational aspects. In addition, high priority should be given to validation. A suitable scanning technology must be developed. When new technologies become available, primary results must be verified. In the coming years it is hoped that CNV research will not only provide insight into the human genetic variation, but also contribute to the better understanding of the mechanism of human genetic disease and evolution. When an increasing number of functional SNPs are identified, it is also possible to develop useful genetic markers.

Acknowledgement

My apologies to those whose work or original publications could not be cited in this short article because of limitations to the number of references.

References

Adams, D. J., Dermitzakis, E. T., Cox, T., Smith, J., Davies, R., Banerjee, R., Bonfield, J., Mullikin, J. C., Chung, Y. J., Rogers, J. & Bradely, A. (2005). Complex haplotypes, copy number polymorphisms and coding variation in two recently divergent mouse strains. *Nat Genet, 37,* 532-536.

Aitman, T. J., Dong, R., Vyse, T. J., Norsworthy, P. J., Johnson, M. D., Smith, J., Mangion, J., Roberton-Lowe, C., Marshall, A. J., Petretto, E., Hodges, M. D., Bhangal, G., Patel, S. G., Sheehan-Rooney, K., Duda, M., Cook, P. R., Evans, D. J., Domin, J., Flint, J., Boyle, J. J., Pusey, C. D. & Cook, H. T. (2006). Copy number polymorphism in Fcgr 3 predisposes to glomerulonephritis in rats and humans. *Nature, 439,* 851-855.

Ansari, M. & Krajinovic, M. (2007). Pharmacogenomics in cancer treatment defining genetic basis for inter-individual differences in response to chemotherapy. *Curr Opin Pediatr, 19,* 15-22.

Auer, H., Newsom, D. L., Nowak, N. J., McHugh, K. M., Sunita, S., Yu, C-Y, Yang, Y., Wenger, G., D., Gastier-Foster, J. M. & Kornacker, K. L. (2007). Gene resolution analysis of DNA copy number variation using oligonucleotide expression microarrays. *BMC Genomics, 8,* 111-117.

Babcock, M., Yatsenko, S., Hopkins, J., Brenton, M., Cao, Q., deJong, P., Stankiewicz, P., Lupski, J. R., Sikela, J. M. & Morrow, B. E. (2007). Hominoid lineage specific amplification of low copy repeats on 22q11.2 (LCR22s) associated with velo-cardo-facial/digeorge syndrome. *Hum Mol Genet, 16,* 2560-2571.

Brookes, A. J. (1999). The essence of SNPs. *Gene, 234,* 177-186.

Buckland, P. R. (2003). Polymorphically duplicated genes: their relevance to phenotypic variation in humans. *Ann Med, 35,* 308-315.

Cappuzzo, F., Hirsch, F. R., Rossi, E., Bartolini, S., Ceresoli, G. L., Bemis, L., Haney, J., Witta, S., Danenberg, K., Domenichini, I., Ludovini, V., Magrini, E., Gregorc, V., Doglioni, C, Sidoni, A., Tanato, M., Franklin, W. A., Crino, L., Bunn, P. A. Jr & Varelia-Garcia, M. (2005). Epidermal growth factor receptor gene and protein and gefitinib sensitivity in non-small-cell lung cancer. *J Natl Can Inst, 97,* 643-655.

Carter, N. P. (2007). Methods and strategies for analyzing copy number variation using DNA microarrays. *Nat Genet, 39,* S16-S21

Chakravarti, A. (1999). Population genetics: making sense out of sequence. *Nat Genet, 21,* 56-60.

Conard, D. F., Andrews, T. D., Carter, N. P., Hurles, M. E. & Pritchard, J. K. (2006). A high resolution survey of deletion polymorphism in the human genome. *Nat Genet, 38,* 75-81.

de Stahl, T. D., Sandgren, J., Piotrowski, A., Nord, H., Andersson, R., Menzel, U., Bogdan, A., Thuresson, A. C., Poplawski, A., von Tell, D., Hansson, C. M., Elshafie, A. I., Elghazali,

G., Imreh, S., Nordenskjold, M., Upadhyaya, M., Komorowski, J., Bruder, C. E. & Dumanski, J. P. (2008). Profiling of copy number variations (CNVs) in healthy individuals from three ethnic groups using a human genome 32 KBAC-cloned-base array. *Hum Mutat, 29,* 398-408.

Esch, H. V., Bauters, M., Ignatius, J., Jansen, M., Raynaud, M., Hollanders, K., Lugtenberg, D., Bienvenu, T., Jensen, L. R., Gecz, J., Moraine, C., Marynen, P., Fryns, J-P. & Froyen, G. (2005). Duplication of the MECP2 region is a frequent cause of severe mental retardation and progressive neurological symptoms in males. *Am J Hum Genet, 77,* 442-453.

Evans, W. E. & Johnson, J. A. (2001). Pharmacogenomics: the inherited basis for inter-individual differences in drug response. *Ann Rev Genomics Hum genet, 2,* 9-39.

Fanciulli, M., Norsworthy, P. J., Petretto, E., Dong, R., Harper, L., Kamesh, L., Heward, J. M., Gough, S. C., de Smith, A., Blakemore, A. I., Froguel, P., Owen, C. J., Pearce, S. H., Teixeira, L., Guillevin, L., Graham, D. S., Pusey, C. D., Cook, H. T., Vyse, T. J. & Aitman, T. J. (2007). FCGR3B copy number variation is associated with susceptibility to systemic but not organ-specific autoimmunity. *Nat Genet, 39,* 721-723.

Fellermann, K., Stange, D. E., Shaeffeler, E., Schmalzl, H., Wehkamp, J., Bevins, C. L., Reinisch, W., Teml, A., Schwab, M., Lichter, P., Radlwimmer, B. & Stange, E. F. (2006). A chromosome 8 gene cluster polymorphism with low human beta-defensin 2 gene copy number predisposes to Crohn disease of the colon. *Am J Hum Genet, 79,* 439-448.

Fiegler, H., Reddon, R., Andrews, D., Scott, C., Andrews, R., Carder, C., Clark, R., Dovey, O., Ellis, P., Feuk, L., French, L., Hunt, P., Kalaitzopoulos, D., Larkin, J., Montgomery, L., Perry, G. H., Plumb, B. W., Porter, K., Rigby, R. E., Rigler, D., Valsesia, A., Langford, C., Humphray, S. J., Scherer, S. W., Lee, C., Hurles, M. E. & Carter, N. P. (2006). Accurate and reliable high-throughput detection of copy number variation in the human genome. *Genome Res, 16,* 1566-1574.

Franke, L, de Kovel, C. G., Aulchenko, Y. S., Trynka, G., Zhernakova, A., Hunt, K. A., Blauw, H. M., van den Berg, L. H., Ophoff, R., Deloukas, P., van Heel, D. A. & Wijmenga, C. (2008). Deletion, imputation and association analysis of small deletions and null alleles on oligonucleotide arrays. *Am J Hum Genet, 82,* 1316-1333.

Freeman, J. L., Perry, G. H., Feuk, L., Reddon, R., McCarroll, S. A., Altshuler, D. M., Aburatani, H., Jones, K. W., Tyler-Smith, C., Hurles, M. E., Carter, N. P., Scherer, S. W. & Lee, C. (2006). Copy number variation: new insight in genome diversity. *Genome Res, 16,* 949-961.

Gary, I. C., Campbell, D. A. & Spurr, N. K. (2000). Single nucleotide polymorphisms as tools in human genetics. *Hum Mol Genet, 9,* 2403-2408.

Gonzalez, E., Kulkarni, H., Bolivar, H., Mangano, A., Sanchez, R., Catano, G., Nibbs, R. J., Freedman, B. I., Quinones, M. P., Bamshad, M. J., Murthy, K. K., Rovin, B. H., Bradley, W., Clark, R. A., Anderson, S. A., O'Connell, R. J., Agan, B. K., Ahuja, S. S., Bologna, R., Sen, L., Dolan, M. J. & Ahuja, S. K. (2005). The influence of CCL3L1gene-containing segmental duplication on HIV-1/AID susceptibility. *Science, 307,* 1434-1440.

Halushka, M. K., Fan, J. B., Bentley, K., Hsie, L., Shen, N. P., Weder, A., Cooper, R., Lipshutz, R. & Chakravarti, A. (1999). Patterns of single nucleotide polymorphisms in candidate genes for blood pressure homeostasis. *Nat Genet, 22,* 239-247.

Hind, D. A., Kloek, A. P., Jen, M., Chen, X. & Frazer, K. A. (2006). Common deletions and SNPs are in linkage disequilibrium in the human genome. *Nat Genet, 38,* 82-85.

Iafrate AJ, Feuk L, Rivera MN, Listewnik ML, Donahoe PK, Qi Y, Scherer, S. W. & Lee, C. (2004). Detection of large scale variation in the human genome. *Nat Genet, 36,* 941-951.

Iida, A., Saito, S., Sekine, A., Mishima, C., Kitamura, Y., Kondo, K., Harigae, S., Osawa, S. & Nakamura, Y. (2003). Catalog of 668 SNPs detected among 31 genes encoding potential drug targets on the cell surface. *J Hum Genet, 48,* 23-46.

Inoue, K. & Lupski, J. R. (2002). Molecular mechanisms for genomic disorders. Ann Rev Genomics *Hum Genet, 3,* 199-242.

Jordan, B., Charest, A., Dowd, J. F., Blumenstiel, J. P., Yeh, R. F., Osman, A., Housman, D. E. & Lander, J. E. (2002). Genome complexity reduction for SNP genotyping analysis. *Proc Natl Acad Sci USA, 99,* 2942-2947.

Kehrer-Sawatzki, H. (2007). What a difference copy number variation makes. *BioEssays, 29,* 311-313.

Krawezak, M., Reiss, J. & Cooper, D. N. (1992). The mutational spectrum of single base – pair substitutions in mRNA splice junctions of human genes: causes and consequences. *Hum Genet, 90,* 41-54.

Kruglyak, L. (1999). Prospects for whole genome linkage disequilibrium mapping of common disease gene. *Nat Genet, 22,* 139-144.

Lachman, H. M., Perdosa, E., Petruolo, O. A., Cocker ham, M., Papolos, A., Novak, T., Papolos, D. F. & Stopkova, P. (2007). Increase in GSK3 beta gene copy number variation in bipolar disorder. *Am J Med Genet B Neuropsychiatric Genet, 144B,* 259-265.

Lee, J. A., Carvalho, C. M. & Lupski, J. R. (2007). A DNA replication mechanism for generating nonrecurrent rearrangements associated with genomic disorders. *Cell, 131,* 1235-1247.

Lee, J. A. & Lupski, J. R. (2006). Genomic rearrangements and copy number alterations as a cause of nervous system disorders. *Neuron, 52,* 103-121.

LeVan, T. D., Bloom, J. W., Bailey, T. J., Karp, C. L., Halonen, M., Martinez, F. D. & Vercelli, D. (2001). A common single nucleotide polymorphism in the CD 14 promoter decreases the affinity of Sp protein binding and enhances transcriptional activity. *J Immunol, 167,* 5838-5844.

Li, J., Jiang, T., Mao, J. H., Balmain, A., Peterson, L., Harris, C., Rao, P. H., Havlak, P., Gibbs, R. & Cai, W. W. (2004). Genomic segmental polymorphisms in inbred mouse strains. *Nat Genet, 36,* 952-954.

Lohrer, H. D. & Tangen, U. (2000). Investigations into the molecular effects of single nucleotide polymorphism. *Pathobiology, 68,* 283-290.

Lupski, J. R. (1998). Genomic disorders: structural features of the genome can lead to DNA rearrangements and human disease traits. *Trends Genet, 14,* 417-422.

Lupski, J. R. (2007). Genomic rearrangements and sporadic disease. *Nat Genet, 39,* S43-S47.

Marshall, C. R., Noor, A., Vincent, J. B., Lionel, A. C., Feuk, L., Skaug, J., Shago, M., Moessner, R., Pinto, D., Ren, Y., Thiruvahindrapduram, B., Fiebig, A., Schrelber, S., Friedman, J., Ketelaars, C. E., Vos, Y. J., Ficicioglu, C., Kirkpatrick, S., Nicolson, R., Sloman, L., Summers, A., Gibbons, C. A., Teebi, A., Chitayat, D., Weksberg, R., Thompson, A., Vardy, C., Crosbie, V., Luscombe, S., Baatjes, R., Zwaigenbaum, L.,

Roberts, W., Fernandez, B., Szatmari, P. & Scherer, S. W. (2008). Structural variation of chromosomes in autism spectrum disorder. *Am J. Hum Genet, 82,* 477-488.

Martin, N., Boomsma, D. & Machin, G. (1997). A twin-pronged attacks on complex traits. *Nat Genet, 17,* 387-392.

McCarroll, S. A. & Altshuler, D. M. (2007). Copy number variation and association studies of human disease. *Nat Genet, 39,* S37-S42.

McCarroll, S. A., Hadnott, T. N., Perry, G. H., Sabeti, P. C., Zody, M. C., Barrett, J. C., Dallaire, S., Gabriel, S. B., Lee, C., Daly, M. J, & Altshuler, D. M. (2006). Common deletion polymorphisms in the human genome. *Nat Genet, 38,* 86-92.

Nielsen, R. & Slatkin, M. (2000). Likely-hood analysis of ongoing gene flow and historical association. *Evolution, 54,* 44-50.

Nielsen, R. (2000). Estimation of population parameters and recombination rates from single nucleotide polymorphisms. *Genetics, 154,* 931-942.

Nothen, M. M. & Cichon, S. (2002). Linking single nucleotide polymorphisms. *Pharmacogenetics, 12,* 89-90.

Perry, G. H., Ben-Dor, A., Tsalenko, A., Sampas, N., Rodriguez-Revenga, L., Tran, C. W., Scheffer, A., Steinfeld, I., Tsang, P., Yamada, N. A., Park, H. S., Kim, J. I., Seo, J. S., Yakhini, Z., Laderman, S., Bruhn, L. & Lee, C. (2008). The fine scale and complex architecture of human copy number variation. *Am J Hum Genet, 82,* 685-695.

Perry, G. H., Tchinda, J., McGrath, S. D., Zhang, J., Picker, S. R., Caceres, A. M., Iafrate, A. J., Tyler-Smith, C., Scherer, S. W., Eichler, E. E., Stone, A. C. & Lee, C. (2006). Hotspots for copy number variation in chimpanzees and humans. *Proc Nat Acad Sci USA, 103,* 8006-8011.

Piotrowski, A., Bruder, C. E., Andersson, R., de Stahl, T. D., Menzel, U., Sandgren, J., Poplawski, A., von Tell, D., Crasto, C., Bogdan, A., Bartoszewski, R., Bebok, Z., Krzyzanowski, M., Jankowski, Z., Partridge, E. C., Komorowski, J. & Dumanski, J. P. (2008). Somatic mosaicism for copy number variation in differentiated human tissues. *Hum Mutat, 29,* 1118-1124.

Plomin, R., Owen, M. V. J. & McGuffin, P. (1994). The genetic basis of complex human behaviors. *Science, 264,* 1733-1739.

Reddon, R., Ishikawa, S., Fitch, K. R., Feuk, L., Perry, G. H., Andrews, T. D., Fiegler, H., Shapero, M. H., Carson, A. R., Chen, W., Cho, E. K., Dallaire, S., Freeman, J. L., Gonzalez, J. R., Gratacos, M., Huang, J., Kalaitzopoulos, D., Komura, D., MacDonald, J. R., Marshall, C. R., Mei, R., Montgomery, L., Nishimura, K., Okamura, K., Shen, F., Somerville, M. J., Tchinda, J., Valsesia, A., Woodwark, C., Yang, F., Zhang, J., Zerjal, T., Zhang, J., Armengol, L., Conard, D. F., Estivill, X., Tyler-Smith, C., Carter, N. P., Aburatani, H., Lee, C., Jones, K. W., Schere, S. W. & Hurles, M. E. (2006). Global variation in copy number in the human genome. *Nature, 444,* 444-454.

Riley, J-H., Allan, C. J., Lai, E. & Roses, A. (2000). The use of single nucleotide polymorphism in the isolation of common disease gene. *Pharmacogenomics, 1,* 39-47.

Roses, A. D. (2000). Pharmacogenetics and the practice of medicine. *Nature, 405,* 857-865.

Salisbury, B. A., Pungliya, M., Choi, J. Y., Jiang, R. H., Sun, X. J. & Stephens, J. C. (2003). SNP and haplotype variation in the human genome. *Mutat Res-Fund Mol Mech Mutagenesis, 526,* 53-61.

Schork, N. J., Fallin, D. & Lanchbury, J. S. (2000). Single nucleotide polymorphisms and the future of genetic epidemiology. *Clin Genet, 58,* 250-264.

Sebat, J., Lakshmi, B., Malhotra, D., Troge, J., Lese-Martin, C., Walsh, T., Yamrom, B., Yoon, S., Krasnitz, A., Kendall, J., Leotta, A., Pai, D., Zhang, R., Lee, Y-H., Hicks, J., Spence, S. J., Lee, A. T., Puura, K., Lehtimaki, T., Ledbetter, D., Gregersen, P. K., Bregman, J., Sutcliffe, J. S., Jobanputra, V., Chung, W., Warburton, D., Kingm, M-C., Skuse, D., Geschwind, D. H., Gilliam, T. C., Ye, K. & Wigler, M. (2007). Strong association of de novo copy number mutations with autism. *Science, 316,* 445-449.

Sebat, J., Lakshmi, B., Troge, J., Alexander, J., Young, J., Lundin, P., Maner, S., Massa, H., Walker, M., Chi, M., Navin, N., Lucito, R., Healy, J., Hicks, J., Ye, K., Reiner, A., Gilliam, T. C., Trask, B., Patterson, N., Zetterberg, A. & Wigler, M. (2004). Large scale copy number polymorphism in the human genome. *Science, 305,* 525-528.

Sharp, A. J., Locke, D. P., McGrath, S. D., Cheng, Z., Bailey, J. A., Vallente, R. U., Pertz, L. M., Clark, R. A., Schwartz, S., Segraves, R., Oseroff, V. V., Albertson, D. G., Pinkel, D. & Eichler, E. E. (2005). Segmental duplication and copy number variation in the human genome. *Am J Hum Genet, 77,* 78-88.

Shastry, B. S. (2002). SNP alleles in human disease and evolution. *J Hum Genet, 47,* 561-566.

Shastry, B. S. (2003). SNPs and haplotypes: markers for disease and drug response. *Int J Mol Med, 11,* 379-382.

Shastry, B. S. (2004). Role of SNP/haplotype map in gene discovery and drug development: an over view. *Drug Dev Res, 62,* 143-150.

Shastry, B. S. (2006). Pharmacogenetics and the concept of individualized medicine. *Pharmacogenomic J, 6,* 16-21.

Shastry, B. S. (2007). SNPs in disease gene mapping, medicinal drug development and evolution. *J Hum Genet, 52,* 871-880.

She, X., Cheng, Z., Zollner, S., Church, D. M. & Eichler, E. E. (2008). Mouse segment duplication and copy number variation *Nat Genet, 40,* 909-914.

Stranger, B. E., Forrest, M. S., Dunning, M., Ingle, C. E., Beazley, C., Thome, N., Reddon, R., Bird, C. P., de Grassi, A., Lee, C., Tyler-Smith, C., Scherer, S. W., Tavare, S., Deloukas, P., Hurles, M. E. & Dermitzakis, E. T. (2007). Relative impact of nucleotide and copy number variation on gene expression phenotypes. *Science, 315,* 848-853.

Thomas, D. C., Haile, R. W. & Duggan, D. (2005). Recent developments in genomewide association scan: a workshop summary and review. *Am J Hum Genet, 77,* 337-345.

Tuzun, E., Sharp, A. J., Bailey, J. A., Kaul, R., Morrison, V. A., Pertz, L. M., Haugen, E., Hayden, H., Albertson, D., Pinkel, D., Olson, M. V. & Eichler, E. E. (2005). Fine-scale structural variation of the human genome. *Nat Genet, 37,* 727-732.

Vrijenhoek, T., Buizer-Voskamp, J. E., van der Stelt, I. & Strengman, E. Genetic Risk and Outcome in Psychosis (group) Consortium, Sabatti C, van Kessel AG, Brunner HG, Ophoff, R. A. & Veltman, J. A. (2008). Recurrent CNVs disrupt three candidate genes in schizophrenia patients. *Am J Hum Genet, 83,* 504-510.

Wakeley, J., Nielsen, R., Liu-Cordero, S. N. & Ardile, K. (2000). The discovery of single nucleotide polymorphisms and inference about human demographic history. *Am J Hum Genet, 69,* 1332-1347.

Weiss, K. M. & Terwilliger, J. D. (2000). How many diseases does it take to map a gene with SNPs. *Nat Genet, 26,* 151-157.

Yang, Y., Chung, E. K., Wu, Y. L., Savelli, S. L., Nagaraja, H. N., Zhou, B., Hebert, M., Jones, K. N., Shu, Y., Kitzmiller, K., Bianchong, C. A., McBride, K. L., Higgins, G. C., Rennebohm, R. M., Rice, R. R., Hackshaw, K. V., Roubey, R. A., Grossman, J. M., Tsao, B. P., Birmingham, D. J., Rovin, B. H, Hebert, L. A. & Yu, C. Y. (2007). Gene copy number variation and associated polymorphisms of complement component C4 in human systemic lupus erythematosus (SLE): low copy number is a risk factor for and high copy number is a protective factor against SLE susceptibility in European Americans. *Am J Hum Genet, 80,* 1037-1054.

Zhao, X., Li, C., Paez, J. G., Chin, K., Janne, P. A., Chen, T. H., Girard, L., Minna, J., Christiani, D., Leo, C., Gray, J. W., Sellers, W. R. & Meyerson, M. (2004). An integrated view of copy number and allelic alterations in the cancer genome using single oligonucleotide polymorphism arrays. *Cancer Res, 64,* 3060-3071.

In: The Human Genome: Features, Variations...
Editor: Akio Matsumoto and Mai Nakano

ISBN: 978-1-60741-695-1

Chapter 12

An Outlook on Uterine Neoplasms: From Hormonal and DNA Damaging to Cervical and Endometrial Cancer Development and Minimally Invasive Management

[1]Andrea Tinelli*, [2]Antonio Malvasi, [3]Giuseppe Leo, [4]Ughetta Vergari, [5]Roberta Martignago, [6]Daniele Vergara, [7]Vito Lorusso, [8]Marcello Guido, [9]Antonella Zizza and [10]Maurizio Pisanò

[1] Department of Gynecology and Obstetric, Vito Fazzi Hospital, Lecce, Italy,
[2] Department of Gynecology and Obstetric, Vito Fazzi Hospital, Lecce, Italy,
[3] Molecular Biology and Experimental Oncology Lab, Oncological Hospital, Lecce, Italy,
[4] International Center of Bioethics and Human Rights, University of Salento, Lecce, Italy,
[5] Department of Biological and Environmental Sciences and Technologies, Laboratory of Human Anatomy, University of Salento, Lecce, Italy,
[6] National Nanotechnology Laboratory (NNL), CNR-INFM, University of Salento, Lecce, Italy,
[7] Department of Oncology, Oncological Hospital, Lecce, Italy,
[8] Laboratory of Hygiene, Department of Biological and Environmental Sciences and Technologies, Di.S.Te.B.A., Faculty of Sciences, University of Salento, Lecce, Italy,
[10] Molecular Biology and Experimental Oncology Lab, Oncological Hospital, Lecce, Italy

* Corresponding Author: Dr. Tinelli Andrea, MD. Vito Fazzi Hospital, Piazza F. Muratore, 73100, Lecce, Italy, Cellular Number 0039/339/2074078, Fax Number 0039/0832/661511, *E-mail: andreatinelli@gmail.com*

Abstract

Uterine neoplasms are common tumors, formed by endometrial and cervical cancers; endometrial cancer is the fourth most frequently diagnosed cancer in developed countries and the eighth leading cause of cancer death in women, and cervical cancer is the second most common cancer in women worldwide and is a leading cause of cancer-related death in women in underdeveloped countries.

Cervical cancer arises by HPV DNA damaging; in fact cervical cancer starts in the cells on the surface of the cervix, exposed to viral infective agents, as HPV, founded in 80% of patients affected by cervical cancer. Thus, more than 99% of cervical uterine cancer cases show HPV presence.

Nevertheless, Endometrial cancer involves cancerous growth of the endometrium, and increasing evidence indicates that different biological and genetic factors play relevant roles its onset so as carcinogenesis generally develops by hormonal modifications.

Both tumors can be safely and feasibly managed from minimally invasive surgical techniques till to endoscopic radical operations, such as hysterectomy, bilateral salpingo-oophorectomy, pelvic and para-aortic lymphadenectomy for surgical treatment.

The authors reviewed several excellent reviews and studies in the area of hormonal, viral and genetical risk factors associated with endometrial and cervical cancer risk and development, analyzing the area of biologic markers, all papers dealing with serum and plasma markers involved in uterine cancer detection, development, progression and minimally invasive treatment.

Keywords: Endometrial cancer, cervical cancer, DNA, HPV, cancerogenesis, minimally invasive treatment, laparoscopy, endoscopy

Introduction

Endometrial Cancer involves cancerous growth of the endometrium, the lining of the uterus. EC is one of the most common invasive gynaecologic malignancy and it represents the fourth most frequent diagnosed cancer among women (1); Europe and North America are the countries with the major incidence of this tumor, while is lowest in Africa, Central and South America and Asia (2).

The cancer is typically one of post-menopausal age, with a main age-related incidence between 50 and 70 years. Increasing evidence indicates that different biological and genetic factors play relevant roles its onset and are parameters that can be used as indicators of pathological progression.

The endometrial carcinomas was classified into two pathogenetic groups: the Type I, that occur most commonly in pre- and peri-menopausal women, often with a history of unopposed estrogen exposure and/or endometrial hyperplasia, of the low-grade endometrioid type, and carry a good prognosis; the Type II, that occur in older, post-menopausal women, more common in African-Americans, not associated with increased exposure to estrogen, typically of the high-grade endometrioid, papillary serous or clear cell types, and carry a generally poor prognosis, not associated with hyper estrogenic factors and accounts for 20% of the cases (3).

General risk factors for endometrial cancer include: age at menarche, age at menopause, history of infertility, obesity, diabetes, oestrogen therapy, polycystic ovarian syndrome, prior pelvic radiation therapy, hereditary non-polyposis colon cancer (HPNCC) and westernisation of lifestyle. (2,3)

Worldwide, cervical cancer is the second most common cancer in women worldwide and is a leading cause of cancer-related death in women in underdeveloped countries. (4)

Worldwide, approximately 500,000 cases of cervical cancer are diagnosed each year: approximately 13,000 cases of invasive cervical cancer and 50,000 cases of cervical carcinoma in situ (i.e., localized cancer) are diagnosed yearly in USA (5); over the next 40 years, the death rate from cervical cancer decreased by more than 70% because preinvasive lesions and cervical cancers were detected at an earlier stage.

Cervical cancer starts in the cells on the surface of the cervix: it develops in the lining of the cervix, and is always associated to HPV infection, since carcinogenic human papillomavirus (HPV) infection is necessary for the development of cervical cancer (6).

More than 99% of cervical uterine cancer cases show HPV presence, with more than 100 HPV types identified, and types 16 and 18 being found in approximately 70 percent of cases.

But infection alone is not sufficient to cause cervical cancer, since infections become undetectable within 1–2 years (6).

Cervical cancer risk seems to be influenced by other variables too, like smoking, while other factors, like alcohol consumption and diet, don't seem to have any influence. Infection with other sexually transmitted viruses seems to act as a cofactor in the development of cervical cancer (7)

In this article, the authors will focus on the area of hormonal, viral and genetical risk factors associated with endometrial and cervical cancer risk and development; they analyze the area of hormonal cancerogenesis, particularly on molecular mechanisms involved in endometrial cancer of type I, and the area of the molecular mechanisms involved in the onset of the cervical carcinoma.

Finally, it will focus the possible minimally invasive surgery in cervical and endometrial tumors, as an appropriate, if not preferred, alternative in many wide ranging surgical procedures, till to radical endoscopic treatments, such as hysterectomy, bilateral salpingo-oophorectomy, pelvic and para-aortic lymphadenectomy for surgical treatment.

Endometrial Cancer and Hormonal Carcinogenesis

Approximately 95% of uterine cancers are of endometrial origin, while the remaining 5% include carcinosarcoma, leiomyosarcoma, stromal sarcoma, and adenosarcoma.

As said, the endometrial adenocarcinomas were divided into two pathogenetic groups, and the more common Type I, is associated with a history of estrogen exposure and this is now considered the most important etiological factor.

Type 1 endometrial cancer mimics the endometrium and is usually the endometrioid cell type; it is most common in obese peri-menopausal and post-menopausal women, or in women taking unopposed estrogen, and arises on the basis of endometrial hyperplasia.

Estrogens exert their effects by binding to specific receptors (ERα and ERβ); this action that can be blocked by some drugs called selective estrogens receptor modulators (SERMs), such as tamoxifen, which is associated with a strong anti-estrogenic activity in breast cancer, but an increased risk of endometrial cancer. Tamoxifen and estrogens have a distinctive but overlapping target gene profile, PAX2 (paired box 2 gene), which can promote the onset of endometrial cancer (8).

Exogenous estrogen sources, substances without progestins, increase the risk of endometrial cancer, while combined therapy actually decreases the risk of cancer (9).

Genetic and allelic polymorphisms of genes involved in estrogen metabolism may contribute to inter-individual differences that can lead to an increased or decreased endometrial cancer risk (10).

The HSD17B1 (17β-hydroxysteroid dehydrogenase type 1) gene produces the enzyme that catalyzes the final step of estradiol biosynthesis; single nucleotide polymorphisms (SNPs) within the gene have nor been shown to be associated with endometrial cancer risk, although a significant increase in the level of estradiol was present in lean postmenopausal women carriers of the +1954A/A genotype (Ser312Gly) (10).

Studying the correlation between the use of estrogen replacement therapy and the risk of endometrial cancer, other authors found a higher risk for women carriers of the CYP17 (17α-hydroxylase/17,20-lyase) alleles A1/A1 (11).

Berstein et al compared CYP17 and CYP19 aromatase gene polymorphisms in patients with endometrial cancer and controls, reporting that genotypes with the longest alleles of CYP19 (A6 or A7) were over-represented and the A2/A2 CYP17 genotype under-represented in endometrial cancer patients (12); they demonstrated that the intra-tumoral aromatase activity of patients with the A2/A2 genotype was significantly lower than in carriers of the A1/A1 genotype, while in carriers with the longest CYP19 alleles aromatase activity was higher than in carriers of all other CYP19 allele variants, suggesting a more rapid breakdown metabolism in these patients compared to others. This results, on the contrary, has not be confirmed in other studies (13-15) and the precise mechanisms leading to the association between CYP SNPs and endometrial cancer are not clear, even if all these findings could indicate that certain polymorphisms of the CYP17 and CYP19 genes may increase the risk of endometrial cancer.

Another evidence is that Cytochrome P450 1B1 (CYP1B1) catalyzes the conversion of 17β-estradiol to 4-hydroxyestradiol (4-OH-E2) and 2-hydroxyestradiol (2-OH-E2); in the Syrian hamster, 4-OH-E2 has been shown to induce tumours (16,17); the glucuronidation of 2-OH-E2, by preventing the formation of 2-MeO-E2, that exhibits anti-tumorigenic and anti-angiogenic effects (18), results in an increased cancer risk and specific polymorphisms in UDP-glucurosyltransferase gene (UGT) may lead to an altered risk of endometrial cancer (19,20).

Several alleles of CYP1B1 have been described by the Human Cytochrome P450 (CYP) Allele Nomenclature Committee, but only certain combinations have a significant effect on cancer susceptibility.

Progesterone, a hormone involved in endometrial control, acts in reverse to the tumorigenic effect of estrogens and its effects are dependent on the progesterone receptor (PR).

A single gene encodes for two proteins, termed progesterone receptor A (PRA) and progesterone receptor B (PRB) (21), with a different sub-cellular localization in endometrial cancer cells (22,23).

Many studies have been focused on elucidating the precise role of these two isoforms in endometrial carcinogenesis and it is clear that differential expression of the two isoforms can represent an important prognostic factor (24).

An author reported reduced PR expression levels in tumours compared with normal glands and areas of complex atypical hyperplasia within the same specimen: a predominance of PR isoforms in hyperplasia and in higher grade tumours and an inverse correlation between PR expression and clinical grade (25).

The expression of the PR isoforms was investigated by immunohistochemical analysis in 141 tissue samples of endometrial carcinoma of endometrioid type, since the expression of both progesterone receptors (A and B) decreases with the increasing of the tumor growth (26).

The PRB appeared to have a crucial role as a prognostic indicator of endometrial adenocarcinoma of type I and an inverse correlation with clinical pathological prognostic factors, including myometrial invasion and lymph-vascular space involvement and the FIGO (International Federation of Gynecology and Obstetrics) stage, while the PRA was only inversely correlated with myometrial invasion and, in addition, the expression of both isoforms was significantly lower in hyperplasia and in poorly differentiated adenocarcinoma than in endometrial carcinomas (26).

It has been suggested that genetic polymorphisms in PR genes could affect their expression.

A promoter region polymorphism +331G/A appears to result in a unique transcriptional start site and the +331A form seemed to favour the production of PRB expression in an endometrial cancer cell, suggesting an association between the +331G/A SNP and endometrial cancer risk (27).

Another PR polymorphism, known as PROGINS, has recently been identified (28,29): an author showed that in the presence of PR expression, PROGINS was significantly predictive of the risk of recurrence (30) and that this polymorphism seemed to be associated with an increased risk of endometrial cancer (31).

The PRB is inactivated by methylation of the CpG dinucleotides, whereas the PRA was un-methylated in all of the cell lines studied (32); low levels of DNA methylation in the promoter region are correlated with active expression, whereas high levels of methylation are associated with gene silencing.

A recent study showed that sixty-two of 83 endometrial cancer samples had only methylated bands of PRB and were all negative on immuno-histochemical analysis of PRB (33) and an aberrant methylation status of the PR gene in endometrial cancer is supported by others reports (34, 35).

It was demonstrated a hypermethylation of the first exon of the gene in two cancer cell lines; the gene expression was restored by a combination of 5-aza-20-deoxycytidine and trichostatin A (34); ADC is an inhibitor of DNA methyltransferase, the enzyme responsible for the methylation of CpG residues; trichostatin acts by blocking the enzyme histone deacetylase and the use of these two agents has been proposed as a powerful therapy for the treatment of cancer by restoring the normal functions of many genes (35) and by promoting cell-cycle and apoptosis-related protein expression (36).

Finally, in hormonal carcinogenesis, also androgens have a crucial role in endometrial cancer risk and polymorphisms in androgen receptor gene are found in the cancer (37).

The first exon of the androgen receptor gene contains two polymorphic short tandem repeats of GGC and CAG, highly polymorphic in length, that influence its activity (38,39).

Endometrial cancer patients had longer CAG and GGC alleles than healthy subjects (40), while an association between short alleles and a more favourable prognosis was reported in another research (41).

In contrast, an author demonstrated an inverse association with increased androgen receptor CAG repeat length and endometrial cancer risk (42) and no correlation with age at the time of diagnosis of endometrial cancer was found in a study involving 43 patients (43).

Endometrial Hyperplasia: A Possible Pre-Cancer

Endometrial hyperplasia is generally considered a precursor to endometrial cancer.

The presence of unopposed estrogen, for example, may result from exogenous estrogen therapy, anovulatory cycles, polycystic ovary syndrome, or obesity may lead to endometrial hyperplasia, and, successively, to endometrial cancer (44).

As said, by histopathology, endometrial cancer is usually an endometrioid adenocarcinoma: it appears on a background of endometrial hyperplasia.

In endometrial hyperplasia, the tumor cells are atypical and form irregular glands, with multiple lumens, pluristratification, the stroma is reduced, producing the "*back to back*" aspect and, with evolution of the disease, the myometrium is infiltrated.

Endometrial hyperplasia must be ruled out particularly in patients older than 35 years of age who present with these conditions; the most common method used to diagnose hyperplasia and cancer is endometrial aspiration with a pippelle, also known as an endometrial biopsy (45).

The important consideration when dealing specifically with the treatment of endometrial hyperplasia is the classification of the type of hyperplasia that is present; endometrial hyperplasia is defined as a proliferation of glands of irregular size and shape with an increase in the glands/stroma ratio (46).

By pathologists, there are 4 types of hyperplasia: simple, complex, simple with atypia, and complex with atypia (47); carcinoma developed in 25 percent of patients with atypia as opposed to less than 2 percent of those without atypia (48).

In fact, it is well established that obesity, weight change and body size are associated with an increased risk of endometrial cancer, as more than the 40% of its incidence can be attributed to excess body weight, in affluent societies.

This can be explained by the fact that excess weight results in increased estrogens concentrations from peripheral conversion of androgens (mainly A) to estrogens (mainly E1) in adipose tissue by aromatase enzyme (49).

Another important risk factor for the development of endometrial cancer by endometrial hyperplasia, among pre- and post-menopausal women, seems to be chronic hyperinsulinemia, as this condition appears to be associated with the development of hyperandrogenism, and so progesterone deficiency and with a reduction of IGFBP-1 (Insulin Growth factor Binding Protein-1) levels also an increasing of IGF-I (Insulin Growth Factor-I) levels, mediated by insulin, that then acts as a growth factor for the tumour mass (50).

IGF-1 and IGF-2 are also involved in the regulation of endometrial growth as well as the interactions between the epithelial and stromal compartments of the endometrium (51).

Plasma levels of IGFBP-3 were significantly lower in women with endometrial cancer than in control subjects, demonstrating an inverse relationship between the plasma level of IGFBP-3 and the risk of developing endometrial cancer. In contrast to other solid tumor sites, they have found lower levels of IGF-1 in the endometrial cancer cases as compared to the controls (52).

Although there is a very high rate of spontaneous regression (80% in cases without atypia and over 50% in complex with atypia), therapy should be instituted since some patients will progress to cancer nonetheless; for patients who present with cellular atypia, the general recommendation is to strictly follow up the patients, for the high risk to develop an endometrial cancer (53).

Several recent studies have suggested that the distinguishing feature of endometrial hyperplasia and cancer is the presence or absence of cytologic atypia, since endometrial hyperplasia and endometrial cancer are two different entities (54).

A series of studies has provided a correlation between genomics, endogenous and exogenous estrogens and development of endometrial cancer from endometrial hyperplasia.

Mutations in DNA-mismatch repair (MMR) genes and microsatellite instability (MSI) are strongly associated with the onset of hyperplasia and many studies have proved their involvement in the endometrial cancer process.

An author examined the association between MMR genes and the increased risk of endometrial cancer: they showed that the two common variant alleles of the *MLH1* and *MSH2* genes give a significant contribution to endometrial cancer incidence (55).

Genetic and allelic polymorphisms of the genes involved in the oestrogen metabolism may contribute to predisposition to EC: the *HSD17B1* produces the enzyme that catalyzes the final step of estradiol biosynthesis and it has been observed that three common SNPs (single nucleotide polymorphisms) within the gene were not associated with endometrial cancer risk (56).

Beiner et al. compared *CYP17* (17α-hydroxylase/17,20-lyase) and *CYP19* (aromatase) gene polymorphisms in patients with EC respect to controls; they reported that genotypes with longest alleles of *CYP19* (A6 or A7) were over-represented and A2/A2 *CYP17* genotype was under-represented in patients as compared to controls (57).

Minimally Invasive Surgical Treatment of Endometrial Hyperplasia

In women affected by excess body weight, with a Body Mass Index (BMI) >29 kg/m^2, particularly during the perimenopausal period (age 40-50 years), with chronic hyperinsulinemia, hyperandrogenism, progesterone deficiency, with a reduction of IGFBP-1 (Insulin Growth factor Binding Protein-1) and IGFBP-3 levels, an increasing of IGF-I (Insulin Growth Factor-I) levels, the endometrial hyperplasia development and the successive cancer development is at very high risk.

In add, the presence of DNA-mismatch repair (MMR) genes and micro satellite instability (MSI) strongly associated with the onset of EC, so as the two common variant alleles of the *MLH1* and *MSH2* genes and the inactivation of the progesterone receptors (PR) isoforms, PR-B, in patients with endometrial hyperplasia with cellular atypia, should lead to a minimally invasive surgical treatment to avoid endometrial cancer (58).

The minimally invasive treatment of high risk endometrial pathologies is hysteroscopy, by biopsy or by endometrial removing by ablation; endometrial ablation has traditionally been done using an operative hysteroscope, the resectoscope, inserted through the cervix, after a dilation of the cervix, always done in an operating room.

Minor procedures can be done under local anesthesia, but most women prefer general anesthesia; the resectoscope has a built in wire loop that uses high-frequency electrical energy to cut or coagulate tissue (59).

This instrument has the advantage of being able to remove polyps and some fibroids at the time of ablation; in results reported to the FDA where resectoscopic endometrial ablation was done by experts, the success rate was approximately 95%, with 40% of women having no bleeding whatsoever in 1 year (60).

The resectoscope is far more efficient at removing tissue than conventional instruments; although the resectoscope provides excellent results in experienced hands, the technique is difficult to master: resection of the endometrium is superior to destructive techniques because it provides tissue for pathologic evaluation (61) and endometrial carcinoma after hysteroscopic endometrial ablation is still a possibility even when strict selection criteria are applied (62,63).

An other method of ablation, FDA approved, was the Thermachoice balloon: this device uses a balloon placed in the uterine cavity through the cervix and hot water is circulated inside the balloon to destroy the endometrium.

Some experts are concerned about the balloon's ability to reach the cornual areas of the uterus, although the balloon's "success" rate in FDA studies was reasonable: it had a much lower rate of amenorrhea the other currently available device (only 13%), but it was seen some disadvantages connected to it's use (64).

The Hydrothermablator also uses hot water, but allows it to circulate freely in the endometrial cavity; it is done under direct vision through a hysteroscope [123]: once the proper temperature is reached, the hot water circulates for 10 minutes; there's the possibility of fluid leaking out the fallopian tubes and burning intestines and, although this did not happen in clinical studies, a case of an intestinal burn is being reviewed by the FDA (65).

After endometrial ablation, most women are able to go home within an hour after the an endometrial ablation and there may be mild cramping, which can usually be relieved by ibuprofen (66).

It is normal to be tired for a few days, but most women are able to return to most normal activities in a day or two; intercourse and very strenuous activity is usually restricted for 2 weeks; it is possible to have a increased discharge for 2 to 4 weeks afterward, as the lining is shedding (66).

After endometrial removing or biopsy, if the histological exam is positive for cancer, it is indicated to remove uterus with ovaries.

Minimally Invasive Surgical Treatment of Endometrial Cancer

Generally, the majority of patients with endometrial cancer are diagnosed with early stage disease with a favorable 5 year prognosis, therefore any new surgical approach must be carefully validated to assure that the technique is at least equivalent.

Minimally invasive hysterectomy can be performed in some ways: total laparoscopic hysterectomy (TLH), laparoscopic assisted vaginal hysterectomy (LAVH) and vaginal hysterectomy (VH).

A LAVH or LH is often less invasive than an abdominal hysterectomy, but more invasive than a vaginal hysterectomy, that is performed more quickly but without direct vision of the pelvis and the abdomen, and without the possibility to remove lymph nodes (67).

As we said, hysterectomy can be done vaginally, and there are no data showing that LAVH is superior to vaginal hysterectomy (68).

Whether a patient is well staged by pre-surgical assessment and via laparoscopy, the oncologic procedure should be performed by endoscopy.

Normally a surgical treatment of an endometrial cancer should consist of, at least, cytological sampling of the peritoneal fluid, pelvic-abdominal exploration, palpation and biopsy of suspicious lymph nodes, hysterectomy and bilateral salpingo-ophorectomy (BSO); these findings cannot be done by vaginal approach or by laparoscopic helping.

In case of advanced endometrial cancer, an hysterectomy, bilateral-salpingo oophorectomy, pelvic and para-aortic lymph node sampling vs lymphadenectomy is performed, also by endoscopy.

Lymphadenectomy, or removal of pelvic and para-aortic lymph nodes, is performed for tumors that have high risk features, such as pathologic grade 3 serous or clear-cell tumors, invasion of more than 1/2 the myometrium, or extension to the cervix or adnexa.

In advanced cases, removal of the omentum is also performed (69).

With the appropriate treatment, the 5-year survival rate for endometrial cancer is: 75% to 95% for stage 1, 50% for stage 2, 30% for stage 3 and less than 5% for stage 4 (70).

A prospective German study compared laparoscopy to laparotomy in a randomized trial of 70 patients: 37 patients were treated in the laparoscopic versus 33 patients in the laparotomy group, the blood loss and transfusion rates were significantly lower in the laparoscopic group, yield of pelvic and para-aortic lymph nodes, duration of surgery, and incidence of postoperative complications were similar for both groups and overall and recurrence-free survival did not differ significantly for both groups (71).

In another study, these same investigators reported their experience with 650 pelvic and para-aortic lymphadenectomies performed for gynecologic malignancies, 112 of which were for endometrial cancer; after a learning period of approximately 20 procedures, a constant number of pelvic lymph nodes (16.9-21.9) were removed, the number of removed para-aortic lymph nodes increased over time, from 5.5 to 18.5 and the number of removed lymph nodes was independent of the body mass index (BMI).

Duration of pelvic lymphadenectomy was independent of BMI, but right-sided paraaortic lymphadenectomy lasted significantly longer in obese women (35 vs. 41 min, P = 0.011), while the overall complication rate was 8.7% with 2.9% intraoperative (vessel or bowel injury) and 5.8% postoperative complications (72).

A case series of 203 patients demonstrated success with laparoscopic staging. the conversion rate was 8% for adhesions or poor exposure; the mean hospital stay was 2.8 days, one recent report showed a significant increase in the use of laparoscopy over a 12 year period at a single institution involving 1312 patients (73).

An Italian study showed that laparoscopy provides equivalent lymph node yield compared to laparotomy: this retrospective study identified 110 patients with apparent early-stage endometrial cancer, 55 (50%) were treated by laparoscopic-assisted vaginal hysterectomy (LAVH) and 55 (50%) by total abdominal hysterectomy (TAH).

All patients underwent pelvic lymphadenectomy: the mean number of lymph nodes removed was 17 for the LAVH group and 18.5 for the TAH group ($p = 0.294$).

Compared with TAH, LAVH required a significantly longer operating time (220 vs. 175 min; $p < 0.01$), but shorter hospital stay (4 vs. 8.5 days; $p < 0.001$) and less estimated blood loss (177 cm3 vs. 285 cm3; $p = 0.02$).

Overall, there were fewer post-operative complications in the LAVH group (6 vs. 11 cases; $p < 0.001$); the conclusion of this study is limited by the biases inherent in retrospective analysis, and lack of complete staging (74).

The Gynecologic Oncology Group (GOG) has completed a phase III randomized study (LAP 2) comparing laparoscopy vs. laparotomy in endometrial cancer: laparoscopic surgical staging could be performed in 76.3% of cases.

No difference in stage, positive cytology, or lymphatic metastasis could be attributed to the laparoscopic approach: quality of life and physical functioning are improved 6 weeks post-operatively following laparoscopy, but these differences were not significant by 6 months; while laparoscopic staging may be a technically feasible option for surgical management, long-term data regarding recurrence and survival have yet to be characterized (75,76).

The Laparoscopic Approach to Cancer of the Endometrium (LACE) is a randomized controlled trial comparing total abdominal hysterectomy to total laparoscopic hysterectomy in stage 1 endometrial cancer patients, the study is ongoing and expected to be completed in 2009 and will report on quality of life and disease free survival in these two groups; survival outcomes with minimally invasive surgery have also been reported (77).

A prospective randomized study demonstrated comparable survival at a median follow-up of 44 months between laparoscopy and laparotomy: among 122 patients, the overall survival for the laparoscopy vs. laparotomy group, respectively, was 82.7% and 86.5%, and these data are consistent with reported retrospective data demonstrating equivalent survival between these two groups (78,79).

The only question is the impact of laparoscopy on the incidence of positive cytology, that is controversial; some investigators report a significantly higher incidence of positive peritoneal cytology in patients undergoing laparoscopy for endometrial cancer.

This may be due to the retrograde dissemination of cancer cells into the peritoneal cavity during uterine manipulation, the clinical significance of these findings is not clear and should be individualized based upon uterine pathology and lymph node status (80,81).

Finally, there are potential economic advantages to minimally invasive surgery; a retrospective report suggested that for early-stage endometrial cancer, patients treated with laparoscopy had significantly shorter hospitalization and fewer complications, resulting in less overall hospital charges when compared to patients treated via laparotomy (82-85).

Cervical Tumors and HPV Correlation

Invasive cervical cancer is more common in women middle aged and older and in women of poor socioeconomic status, who are less likely to receive regular screening and early treatment.

There is also a higher rate of incidence among African American, Hispanic, and Native American women.

The etiological role of HPV in large part of cervical tumors is well consolidated, since more than 99% of cervical uterine cancer cases show HPV presence, with more than 100 HPV types identified and, of these, up to 40 different types are able to infect the ano-genital tract.

Low-risk types are frequently connected to genital warts (condyloma acuminata), while types with a medium and high oncogenic risk are occasionally or commonly found in patients with high degree lesions and cervical invasive cancer (86).

The HPV (Human Papilloma Virus) is a virus responsible for many female and male genital tract diseases, especially in uterine cervix; these clinical conditions, from simple feet and hands warts to genital cancer, can be classified in mucosal and cutaneous viruses: within each of these groups, HPV types are divided into low-risk, intermediate risk and high-risk types on the basis of their association with benign or malignant lesions (87).

The cervix, usually, has HPV lesions with a flat aspect that are not visible to the naked eye, often require colposcopy for detection and are often associated to dysplastic lesions; cervical dysplasia describes the presence of abnormal, precancerous cells on the surface of the cervix or its canal and, about the number of diagnosis for year, between 250,000 and 1 million women in the United States are diagnosed with cervical dysplasia (88).

While it can occur at any age, the peak incidence is in women between the ages of 25 to 35; most of dysplasia cases can be cured with proper treatment and follow-up: without treatment, from 30% to 50% may progress to invasive cancer, even if it can take 10 years or longer for cervical dysplasia to develop into cancer (6).

Generally, the literature recognize two types of dysplasia: low-grade squamous intraepithelial lesions (LGSIL) and high-grade squamous intraepithelial lesions (HGSIL).

The Bethesda 2001 system classifies squamous cell abnormalities in four categories: first type or ASC (atypical squamous cells), second type or LSIL (low-grade squamous intraepithelial lesions), third type or HSIL (high-grade squamous intraepithelial lesions), fourth type or cervical squamous cell carcinoma (89); the ASC category contains two subcategories: ASC-US (atypical squamous cells of undetermined significance) and ASC-H (atypical squamous cells cannot exclude HSIL) (90).

The abnormal cells present in LGSIL usually return to normal on their own within 18 to 24 months, but the HGSIL cells, if not treated, can progress to cancer of the cervix, and, to detect these changes early, it is essential to have regular Pap smears (91).

Genomic Detection of HPV by DNA Chips: A New Frontier

Since tumors develop as a result of accumulated molecular genetic or genomic alterations including amplification, deletion, point mutation, and translocation and the studies of these are critical to understand the molecular basis of cancer and provide potential diagnostic/outcome markers and therapeutic targets for cancer patients.

But gene expression is dynamic and this is the reason why it is difficult to discover new oncogenes and tumor suppressors.

The introduction of the DNA chips use have resulted in an era of genome-wide approaches to prognostication and outcome prediction in patients with cancer because the traditional clinicopathological parameters such as tumor size, involvement of axillary lymph nodes (LNs), histological grade, nuclear grade, have some limits, considering also the molecular-genetic heterogeneity and the large number of genes involved in controlling cell proliferation, apoptosis and differentiation (92).

DNA sensors and DNA chips evolve from the combination of the principle of nucleic acid hybridization with the sensitivity of optical, electrochemical or gravimetric transducers and have many applications including medical diagnostic and genetic screening.

A DNA microarray consists of a collection of thousands of microscopic spots of DNA to hybridize a cDNA or cRNA sample (a fluorophore-labeled targets) under high-stringency conditions: this hybridization is usually detected and quantified by fluorescence with the aim to determine changes in expression levels or to detect single nucleotide polymorphisms (SNPs) in the target.

DNA microarrays can be used to detect DNA (as in comparative genomic hybridization), or RNA (most commonly as cDNA after reverse transcription) that may or may not be translated into proteins.

The fluorescence, electrochemical, optical, electrical or microgravimetric signal analyzes DNA concentrations as small as 10−18 μ, typically required for medical diagnostic applications; in standard microarrays, the probes are attached to a solid surface by a covalent bond to a chemical matrix (via epoxy-silane, amino-silane, lysine, polyacrylamide or others).

So, it is possible to speak of *DNA chip* (also called *genome chip*, *gene chip* or *gene array*) when the solid surface is glass, plastic or silicon chip.

Finally, DNA chips consist of immobilized biomolecules, often PCR products or oligonucleotides, on planar surfaces, and, after viral DNA extraction, the amplification products obtained with a primer set designed for specific viral ORF are hybridized to microarrays prepared with specific probes to discriminate different genotypes.

These DNA chips are the new frontiers of HPV study, since so many researches are conducted in these field since the viral genome consists of a single circular DNA molecule, containing about 7900 basepairs (bp) associated with histonic proteins.

By the genomic side, the HPV genome is functionally distinct in three regions: the first of these is a noncoding region or URR (Upstream Regulatory Region), also known as LCR (Long Control Region), that it is about 400-1000 basepair long, and it contains the promoter p97 along with enhancer and silencer sequences, that regulate transcription of the ORFs (Open Reading Frame).

The second region is the early region or ER (ORF E1, E2, E4, E5, E6 and E7), is about 4000 bp long and encodes non-structural proteins involved in viral replication and oncogenesis.

In particular, the E1 and E2 proteins start viral replication by binding to particular motifs in the replication origin; furthermore, the products of these two ORFs insure a correct division of the viral genome during cell replication, with different strategies for each type (93).

The E2 protein encoded by HPV 16 show a stronger transcriptional activity and a DNA binding affinity that E2 encoded by other virus types.

The third region is the late region that encodes the two structural proteins of the viral capside, L1 and L2.

Vázquez-Ortíz et al. used, in 2005, a cDNA array (ULTRArray Advantage System array blots, Ambion Inc.) to evaluate expression of 8400 genes in cellular lines derived by human cervical cancer and control cell lines, with the aim to identify new genes involved in cervical cancer biology. Five genes were found to be consistently high in all malignant cells and tissues (94).

Currently in the diagnostic field, many microarray systems are already available for HPV genotyping and, at the present, the challenge for researchers is to compare these systems to standard genotyping methods for validation purposes (95).

Detection of HPV DNA is generally based on immunochemical or fluorescent methods, by using of biotinylated or fluorescent primers: the E6, E7 and L1 ORF have been extensively used to diagnose and to genotype HPVs (96), allowing the simultaneous discrimination of 53 different genotypes (97).

Mechanisms of Pathogenesis of Cervical Tumors by HPV

One of the key events in HPV induced carcinogenesis is viral integration into the host genome.

Molecular basis for the malignant potential of these viruses has been determined in the dysregulation of the cell cycle by the viral oncogenes E6 and E7.

After viral integration, expression of the E6 and E7 genes is maintained, whereas the interruption of the circular viral genome in the E2 region leads to a loss of its specific repressor function and to an increased stability of HPV 16 E6 and E7 mRNA (98).

The E6 gene is one of the most variable regions of the HPV-16 genome and some studies suggested a relationship between E6 variants and the clinical manifestation of viral infection.

Viral integration in the host genome occurs at preferred sites, causing changes in the expression of surrounding genes, likely contributing to tumor development. (99).

The E7 protein binds to the retinoblastoma protein (pRB) and inhibits its ability to modulate the function of E2F transcription factors, while the E6 protein forms a ternary complex with p53 activity and E6-AP, leading to degradation of p53 and, in addition, the E6 and E7 proteins play important roles in the episomale maintenance viral genome (100).

In normal cells, the function of the Rb protein is regulated by the cyclin/cyclin-dependent kinase (CDK) complexes.

Inhibitors, such as $p16^{INK4a}$, prevent the phosphorylation of Rb; in HPV positive cervical cancer a good correlation was found between cervical lesions and $p16^{INK4a}$ expression, as a result of the inactivation of pRb by the HPV-E7 protein (101); in addition, overexpression of $p16^{INK4a}$ seems to be connected to the viral type, being higher with genotypes associated with high degree lesions (102).

The expression of these proteins occurs with a different mechanism based on cervical lesion types, with a marked expression of both proteins in CIN3 and carcinoma cases (103).

In cervical cancer cells with mutations of *p53*, genomic instability is detected: mutations in the p53 gene are frequently associated with gynecological malignancies (104).

The molecular mechanisms involved in the onset of the cervical carcinoma are well defined and they are mainly associated to the ability of the proteins E6 and E7 to neutralize the activity of p53 and the pRB respectively.

Minimally Invasive Surgical Treatment of Benign Cervical Pathologies

Because of the HGSIL pathologies and cervical neoplasia "*in situ*", if not treated, can progress to cancer of the cervix; it is clear that HGSIL dysplasia and, specially if HPV positive, must be treat by minimally invasive surgical options.

Cryotherapy, or freezing, is done by placing a probe against the cervix which cools the cervix to sub-zero temperatures: the cells damaged by freezing are shed over the next month in a heavy watery discharge (105).

The main advantages of freezing are that it is simple to do and uses inexpensive equipment but one problem with freezing is that the depth cannot be precisely controlled, so abnormal cells may be left behind and this is less of a problem with small areas of mild to moderate dysplasia, and more of a problem with severe dysplasia and carcinoma-in-situ (106).

Cryotherapy has a high failure rate for treating large areas of dysplasia and dysplasia that extends into the cervical canal, so other methods are preferable when they are available (107).

It is possible to obtain the target zone cervix resection also by laser treatment; laser technology (LASER – Light Amplification by Stimulated Emission of Radiation) is one of the most rapidly medical areas and, since many years, laser technology is applied in endoscopical gynaecology (108).

In laser surgery basically are applied Er-YAG, Nd-YAG and CO_2 lasers; among them the special place occupies the CO_2 laser because of its unique properties (depth of penetration 20-50 microns., zone of damage – maximal 300 microns), of a universality and its versatility in gynaecology (107,108).

The CO_2 laser uses a tiny beam of light to vaporize the abnormal cells, in the medical office with no or very little discomfort, with laser directed through the colposcope so that the area and depth of treatment can be controlled precisely (109).

Healing after laser treatment is much faster than after freezing because dead tissue is not left behind; studies using the latest techniques of laser treatment are showing lower failure rates with the laser than with freezing (110).

An important advantage is that the cervix usually heals with the squamo-columnar junction visible, so that future evaluation is easily carried out, but the major disadvantage of the laser over the cryo is that it requires sophisticated equipment, and most gynecologists do not have a laser in their office, also if laser is much more expensive to do laser if it has to be done in the hospital (110).

Also known as "LLETZ" or "LEEP", loop excision uses a fine wire loop with electrical energy flowing through it to remove the abnormal area of the cervix: the tissue removed is sent to the laboratory for examination (111).

LEEP, therefore, can often treat and diagnose the problem at the same time, it is commonly done under local anesthesia and usually causes little discomfort and it can often be used as a substitute for cone biopsy; the advantage of LEEP is that the problem is treated at the time of diagnosis, so it is not necessary to wait for lab results before treatment: at other times, a tiny sample may be taken at the time of the initial evaluation (112).

A cone biopsy removes a cone-shaped or cylinder-shaped piece of the cervix, it is usually done in an operating room and it can be done with a laser or with conventional surgical instruments (cold-cone).

A cone biopsy may be done for diagnosis or for treatment, although a diagnostic cone may treat the problem at the same time; although laser vaporization and cryotherapy are effective treatments for dysplasia, they are not suitable for invasive cancer; a cone biopsy may also be selected as treatment of dysplasia or carcinoma-in-situ: this treatment has a high success rate, but a "cold-cone" has a higher complication rate than a laser cone, cryo, or loop.

In a small percentage of cases, a cone biopsy may interfere with childbearing; actually, many cases requiring cold cone biopsy in the past can be treated, with the laser or with the loop, with a lower chance of complications (113).

Cervical Cancer and Endoscopic Surgical Treatments

The development of cervical cancer is very slow, since one of the key events in HPV induced carcinogenesis is viral integration into the host genome; as said, normal cervical cells may gradually undergo changes to become precancerous and then cancerous, by cervical dysplasia.

The treatment options for early and advanced cervical cancer are surgical and include:

- LEEP
- Cryotherapy
- Laser therapy
- Radical Trachelectomy
- Radical Hysterectomy

About the first tree options we discussed before, but on the last two options we have to explain more in details, because traditionally radical hysterectomy has formed the mainstay of treatment for early stage cervical carcinoma.

Because of cervical cancer occurs frequently in young women who would like to preserve their childbearing potential, since 10 years radical trachelectomy and laparoscopic lymphadenectomy have been introduced to allow preservation of fertility in early stage invasive lesions (114): this is a new complicate approach to fertility-sparing surgery that preserves the functions of the uterus, the technique is similar to a standard radical hysterectomy and lymphadenectomy, combines laparoscopic (for pelvic lymphadenectomy) and transvaginal approaches, the ovarian vessels are not ligated and, following lymphadenectomy and skeletonisation of the uterine arteries, the cervix, parametrium and vaginal cuff are excised (115).

The residuum of the cervix is then sutured to the vagina and the uterine arteries re-anastomosed; the mean duration of the radical hysterectomy, for laparoscopic and vaginal steps, is more than two hours and half.

In young patients affected by early invasive cervical carcinoma, radical trachelectomy does not appear to increase the rate of cancer recurrence; it carries a relative risk of infertility and late miscarriage but makes it possible for some patients to become pregnant and give birth to normal newborns (115).

Thus, it seems reasonable to offer this procedure in selected cases, provided that each patient is fully informed and the surgeon properly trained.

However, in more advanced disease, a radical hysterectomy may be performed (4, 116).

Generally, radical hysterectomy is performed, in the large part of General Hospital, in laparotomy; the history of radical hysterectomy is very interesting and it began in the beginning of the last century.

In 1898, Wertheim, a Viennese physician, developed the radical total hysterectomy with removal of the pelvic lymph nodes and the parametrium; in 1905, Wertheim reported the outcomes of his first 270 patients: the operative mortality rate was 18%, and the major morbidity rate was 31%.

In 1901, Schauta described the radical vaginal hysterectomy and reported a lower operative mortality rate than the abdominal approach; than, in the late 20th century, radiation therapy became the favored approach because of the high mortality and morbidity of the surgical approach.

In 1944, Meigs repopularized the surgical approach when he developed a modified Wertheim operation with removal of all pelvic nodes and he reported a survival rate of 75% for patients with stage I disease and demonstrated an operative mortality rate of 1% when these procedures were performed by a specially trained gynecologist.

Throughout the remainder of the 20th century, various modifications have been made for this radical procedure, especially in light of improvements in the areas of anesthesia, intensive care, antibiotics, and blood product transfusion science (113-115).

This type of hysterectomy removes the uterus, ovaries and much of the surrounding tissues, including internal lymph nodes and upper part of the vagina and it was initially developed as a surgical treatment for cervical cancer due to the absence of other modalities for treatment (116).

The pelvic lymphadenectomy is performed in a systematic fashion; the anatomy of this procedure involves stripping all fatty tissue from the mid portion of the common iliac vessels and the internal and external iliac vessels to the level of the circumflex iliac vein distally, with preservation of the genitofemoral nerve on the psoas muscle.

The nodal tissue in the obturator fossa is removed from above the obturator nerve to the external iliac vein superiorly and laterally to the pelvic sidewall.

Care must be taken in the obturator fossa to avoid injury to the obturator nerve or to an accessory obturator vein, which is present in approximately 20% of patients (4,116)

It is very important to stage cervical cancer initially by International Federation of Obstetric & Gynaecology (FIGO) classification.

This classification of CC is as follows:

- Stage I: Cervical carcinoma is confined to the uterus.
- Stage IA1: Invasive carcinoma is diagnosable only by microscopy. Stromal invasion is 3 mm deep or less and 7 mm or less in horizontal dimension.
- Stage IA2: The microscopic depth of invasion is greater than 3 mm and less than 5 mm. Horizontal spread is 7 mm or less.
- Stage IB1: The lesion is grossly visible and less than 4 cm in diameter. The microscopic lesion has a depth of invasion greater than 5 mm or a horizontal spread greater than 7 mm.
- Stage IB2: The lesion is grossly visible and greater than 4 cm in diameter.
- Stage II: Cervical carcinoma invades beyond the uterus but not to the pelvic sidewall or to the lower third of the vagina.
- Stage IIA: Cervical carcinoma extends down the vagina but does not exceed two thirds of the vaginal length.
- Stage IIB: Cervical carcinoma extends out into the parametrium but does not extend all the way to the pelvic sidewall.
- Stage III: Cervical carcinoma extends out to the pelvic sidewall, and/or involves the distal third of the vagina, or causes hydronephrosis or a nonfunctioning kidney.
- Stage IIIA: Cervical carcinoma involves the lower third of the vagina without extension to the pelvic wall.
- Stage IIIB: Cervical carcinoma extends out to the pelvic sidewall or causes hydronephrosis or a nonfunctioning kidney.
- Stage IVA: Tumor invades the bladder or rectal mucosa and/or extends beyond the true pelvis.
- Stage IVB: Distant metastasis is present

For patients with stage IA1 lesions, an extrafascial hysterectomy or cold-knife cone with adequate negative margins may be performed if future fertility is an issue. In these patients, the risk of having tumor in the pelvic lymph nodes is 0.5-1.5%.

Radical hysterectomy is indicated for patients with FIGO stage IA2-IIA cervical cancer who are medically fit enough to tolerate an aggressive surgical approach and wish to avoid the long-term adverse effects of radiation therapy. Prospective randomized trials have validated equal curative rates from radical surgery and radiotherapy (overall survival similar at 83%).

Currently, with stage IB patients, approximately 54% of patients with tumors size 4 cm or less (stage IB1) and 84% with tumors greater than 4 cm (stage IB2) will require postoperative adjuvant radiotherapy (117).

Recent encouraging data for improved outcomes with combined chemo-radiation therapy and the increased morbidity noted with the combined surgical and adjuvant radiotherapy has brought into question the role of radical surgery with stage IB2 and stage IIA.

In the setting of recurrence, radical hysterectomy has been performed for very small, centrally recurrent or persistent cancers after radiation therapy.

Current evidence on the safety and efficacy of laparoscopic radical hysterectomy confirm that laparoscopic radical hysterectomy can be used to treat stage I (cervical cancer confined to the cervix) and stage IIA (cancer spread to the top of the vagina, but not into the uterus).

Laparoscopic radical hysterectomy involves surgical removal of the uterus, the supporting ligaments and the upper vagina, together with removal of the pelvic lymph nodes and sometimes the para-aortic lymph nodes (4,116).

Radical hysterectomy is also indicated for other disease processes that involve the cervix (eg, primary upper vaginal carcinoma, endometrial cancer with involvement of the lower uterine segment or cervix).

Intraoperative complications include damage to surrounding structures during the intended procedure. Injury may occur to the bladder, bowel, ureters, pelvic vessels, and nerves (118).

Large-volume blood loss and subsequent need for transfusion may occur; as with any abdominopelvic operation for cancer, these patients are at an extremely high risk for deep venous thrombosis and subsequent embolism.

Because the upper 2 cm of the vagina are removed, some patients may note vaginal shortening, particularly if even more vagina was removed because of stage IIA disease or in the event that postoperative adjuvant radiation therapy was administered (119).

Postoperative complications include wound complications that lead to wound skin separation, wound abscess, and wound dehiscence (eg, seroma, hematoma), while postoperative issues involving the ureter, which may be significantly devitalized during the dissection, include ureteral stricture and fistula. Vesicovaginal fistulae may occur in the postoperative period.

Patients with bulky tumors (>4 cm) are at a higher risk for both nodal metastasis and pelvic recurrence and patients with deep stromal invasion, positive vaginal margins, or positive parametrial margins are at increased risk for recurrence.

Cervical Cancer and Vaccine

The demonstrated effectiveness of HPV prophylactic vaccination opens a new era of hope for both health professionals and women.

In June of 2006, the Food and Drug Administration (FDA) approved a cervical cancer vaccine for girls and women between the ages of 9 and 26, which prevents infection against the two types of HPV responsible for the majority of cervical cancer cases.

These vaccine (Gardasil, Cervarix) has been shown to protect against the HPV (120).

Studies have shown that the vaccine appears to prevent early-stage cervical cancer and precancerous lesions (121).

The HPV vaccines are the first vaccines presented as an anti-cancer immunization; indeed, these prophylactic vaccines, to protect against precancerous and cancerous lesions associated with HPV, shall save lives, reduce costly treatment interventions, and have an individual and collective benefit that should not be neglected.

The clinical studies of vaccines against papillomavirus based on the use of viral like particles (VLPs), constituted of the major protein L1 of the capsid of the virus, without any viral genetic material — immunogenic while not infectious and non-transforming — demonstrated their remarkable efficacy in preventing cervical precancers and cancers, as proven for the quadrivalent [against HPV types 6,11,16,18] and the bivalent [against HPV types 16,18] vaccines (122).

Their level of clinical efficacy in the ''per-protocol'' analysis (consisting of women who were naive to vaccine targeted HPV types at baseline as determined by serology testing for the presence of HPV type-specific antibodies or polymerase chain reaction (PCR) testing of genital samples for the presence of HPV DNA) is unprecedented in the history of vaccination: close to 100% (123).

The highest efficacy is demonstrated in young women naive to the virus types associated to the vaccines, whom seem to have no therapeutic effect on existing lesions or on the course of viral infections already carried by healthy individuals (121-123).

Four large trials of either a HPV 16 monovalent vaccine or the quadrivalent HPV vaccine demonstrated a vaccine efficacy of 44% for preventing HPV 16/18 associated CIN 2,3 or AIS in the ''intent-to-treat'' population (consisting of all women who were enrolled into the trial) after a mean follow-up of 3 years (124).

Results with a limited benefit have been reported for the bivalent HPV 16 and 18 vaccine: the vaccines also have been shown to not accelerate clearance of infections in women already infected with HPV 16 and 18 (125).

In practice the effectiveness of HPV vaccines are limited by two factors: all genital cancers and precancerous lesions are not induced exclusively by HPV types 16 or 18, and the optimal benefit is demonstrated in adolescents and young women before they have encountered these viruses (126).

The question of vaccination before or after sexual debut is controversial, and depends on the concept of individual or collective benefits and arguments of effectiveness over efficacy.

Regardless, continued and regular screening with a Pap Test for all vaccinated and unvaccinated populations effectively lowers the risk for developing invasive cervical cancer, by detecting precancerous changes in cervical cells; women who do not receive regular Pap smears have a higher risk for the condition.

Conclusions

Endometrial cancer is one of the most common invasive gynaecologic malignancies, its incidence can be attributed to obesity, weight change and body size in affluent societies, because these conditions result in increased estrogen concentrations.

The endogenous and exogenous estrogen hyper-stimulation, along with cancer biomarkers, may play an important role in the early detection, progression and survival after the diagnosis of endometrial cancer.

In add, the interaction between polymorphisms in genes coding for metabolism and biosynthesis of estrogens, androgens and progesterone and endometrial cancer risk is of particular interest because different data support that the cancer development could be driven by different allelic variants.

Increasing our understanding of the role of DNA damaging in the aetiology and the course of endometrial cancer and combining the genomic data with known risk factors have a great potential to facilitate the development of new early detection and treatment modalities for this challenging disease.

On the other side, persistent cervical infections by approximately 15 carcinogenic genotypes of human papillomavirus (HPV) cause virtually all cases of cervical cancer and its immediate precancerous precursor, cervical intraepithelial neoplasia grade 3 or carcinoma in situ.

Two standard laboratory methods have been used in epidemiology studies to identify HPV infection: HPV DNA detection and serum antibody detection.

Type specific HPV DNA is identified in exfoliated cells sampled from the cervix or vagina by PCR consensus primers or occasionally performed after detection with a cocktail probe of multiple HPV types.

To HPV detection, new diagnostic systems, such as DNA microarrays, can detect viral genotypes in a sensitive and rapid way, since DNA-chips have the ability to determine rare genotypes, variants and multiple infections with different viral types of HPV, giving to this method excellent prospects both for research and diagnostic applications.

In fact, molecular biology and biomicrotechnology have promising tools for the early detection and disease monitoring of HPV and endometrial DNA modifications, but future studies in this area should concentrate on examining the longitudinal changes in endometrial and cervical cells and in serum concentrations of these biomarkers and investigating their associations with tumor treatment response, relapse, complications and survival.

After the tumor development, minimally invasive and laparoscopic treatment of cervical and endometrial tumors are an appropriate and reasonable therapeutic option for young women with low-stage disease who wish to preserve their childbearing potential.

Finally, minimally invasive gynecological surgery should only be considered if the benefits of removing pre-cancerous lesions from a patient outweigh the risks of the tumor advancing.

References

[1] Landis, S. H., Murray, T., Bolden, S. & Wingo, P, A. (1999). Cancer statistics. *Cancer J Clin*. 49(1):8-31.

[2] Max Parkin, D., Bray, F., Ferlay, J. & Pisani, P. (2005). Global Cancer Statistics. *Cancer J Clin.* 55:74-108.

[3] Tinelli, A., Vergara, D., Martignago, R., Leo, G., Malvasi, A. & Tinelli, R. (2008). Endometrial cancer: emerging roles of sociobiological and genetic factors: clinical review. *Acta Obstet Gynecol Scand* 87:1101-1113.

[4] Malzoni, M., Tinelli, R., Cosentino, F., Perone, C., Iuzzolino, D., Rasile, M. & Tinelli, A. (2008). Laparoscopic radical hysterectomy with lymphadenectomy in patients with early cervical cancer: our instruments and technique. *Surg Oncol*. Sep 18.

[5] Dobson, R. (2007). UK ranks among lowest in Europe on cervical cancer survival. *BMJ*. Apr 14;334(7597):764.

[6] Koshiol, J., Lindsay, L., Pimenta, J. M., Poole, C., Jenkins, D. & Smith, J. S. (2008). Persistent Human Papillomavirus Infection and Cervical Neoplasia: A Systematic Review and Meta-Analysis. *Am J Epidemiol* 168:123-137.

[7] Burd, E. M. (2003). Human Papillomavirus and Cervical Cancer. *Clin Microbiol Rev* 16(1):1-17.

[8] Wu, H., Chen, Y., Liang, J., Shi, B., Wu, G., Zhang, Y., et al. (2005). Hypomethylation-linked activation of PAX2 mediates tamoxifen-stimulated endometrial carcinogenesis. *Nature,* 438:981-987.

[9] Amant, F., Moeman, P., Neven, P., Timmerman, D., Limbergen, E. & Vergot, I. (2005). Endometrial cancer. *Lancet,* 366:491-505.

[10] Setiawan, V. W., Hankinson, S. E., Colditz, G. A., Hunter, D. J. & De Vivo, I. (2004). HSD17B1 Gene Polymorphisms and Risk of Endometrial and Breast Cancer. *Cancer Epidemiol Biomarkers Prev*. 13:213-219.

[11] McKean-Cowdin, R., Spencer Feigelson, H., Pike, M. C., Coetzee, G. A., Kolonel, L. N. & Henderson, B. E. (2001). Risk of endometrial cancer and estrogen replacement therapy history by cyp17 genotype. *Cancer Res*. 61(1):848-849.

[12] Berstein, L. M., Imyanitova, E. N., Kovalevskij, A. J., Maximov, S. J., Vasilyev, D. A. & Buslov, K. G. et al. (2004). CYP17 and CYP19 genetic polymorphisms in endometrial cancer: association with intratumoral aromatase activity. *Cancer Lett*. 207:191-196.

[13] Szyllo, K., Smolarz, B., Romanowicz-Makowska, H., Lewy, J. & Kulig, B. (2006). The T/C polymorphism of the CYP17 gene and G/A polymorphism of the CYP19 gene in endometrial cancer. *J Exp Clin Cancer Res*. 25(3):411-6.

[14] Liehr, J. G., Fang, W. F., Sirbasko, D. A. & Ari-Ulubelen, A. (1986). Carcinogenicity of catechol estrogens in Syrian hamsters. *J Steroid Biochem*. 24:353-6.

[15] Hayashi, N., Hasegawa, K., Komine, A., Tanaka, Y., McLachian, J. A., Barrett, J. C. & Tsutsui, T. (1996). Estrogen-induced cell transformation and DNA adduct formation in cultured Syrian hamster embryo cells. *Mol Carcinog*. 16(3):149-56.

[16] Fotsis, T., Zhang, Y., Pepper, M. S., Adlercreutz, H., Montesano, R., Nawroth, P. P. & Schweigerer, L. (1994). The endogenous estrogen metabolite 2-methoxyestradiol inhibits angiogenesis and suppresses tumor growth. *Nature* 368:237-9.

[17] Duguay, Y., McGrath, M., Lépine, J., Gagnè, J. F., Hankinson, S. E. & Colditz, G. A. et al. (2004). The Functional UGT1A1 Promoter Polymorphism Decreases Endometrial Cancer Risk. *Cancer Res*. 64:1202-1207.

[18] Thibaudeau, J., Lepine, J., Tojcic, J., Dugugay, Y., Pelletier, G., Plante, M. et al. (2006). Characterization of common UGT1A8, UGT1A9, and UGT2B7 variants with

different capacities to inactivate mutagenic 4-hydroxylated metabolites of estradiol and estrone. *Cancer Res.* 66(1):125-133.

[19] Sasaki, M., Tanaka, Y., Kaneuchi, M., Sakuragi, N. & Dahiya, R. (2003). CYP1B1 Gene polymorphisms have higher risk for endometrial cancer, and positive correlations with estrogen receptor α and estrogen receptor β expressions. *Cancer res.* 15(63):3913-3918.

[20] McGrath, M., Hankinson, S. E., Arbeitman, L., Colditz, G. A., Hunter, D. J. & De Vivo, I. (2004). Cytochrome P450 1B1 and cathecol-*O*-mehtyltransferase polymorphisms and endometrial cancer susceptibility. *Carcinogenesis.* 25(4):559-565.

[21] Kastner, P., Krust, A., Turcotte, B., Stropp, U., Tora, L., Gronemeyer, H. & Chambon, P. (1990). Two distinct estrogen-regulated promoters generate transcripts encoding the two functionally different human progesterone receptor forms A and B.*EMBO J.* 9:1603-14.

[22] Leslie, K. K., Stein, M. P., Kumar, N. S., Dai, D., Stephens, J., Wandinger-Ness, A., Glueck, D. H. (2005). Progesterone receptor isoform identification and subcellular localization in endometrial cancer. *Gynecol Oncol.* 96(1):32-41.

[23] Dai, D., Wolf, D. M., Litman, E. S., White, M. J. & Leslie, K. K. (2002). Progesterone Inhibits Human Endometrial Cancer Cell Growth and Invasiveness: Down-Regulation of Cellular Adhesion Molecules through Progesterone B Receptors. *Cancer Res.* 62:881-886.

[24] Saito, S., Ito, K., Nagase, S., Suzuki, T., Akahira, J-I., Okamura, K., Yaegashi, N. & Sasano, H. (2006). Progesterone receptor isoforms as a prognostic marker in human endometrial carcinoma. *Cancer Sci.* 97:1308-1314.

[25] Arnett-Mansfield, R. L., de Fazio, A., Wain, G. V., Jaworski, R. C., Byth, K., Mote, P. A. & Clarke, C. L. (2001). Relative Expression of Progesterone Receptors A and B in Endometrioid Cancers of the Endometrium. *Cancer Res.* 61:4576-4582.

[26] Miyamoto, T., Watanabe, J., Hata, H., Jobo, T., Kawaguchi, M., Hattori, M., Saito, M. & Kuramoto, H. (2004). Significance of progesterone receptor-A and B expressions in endometrial adenocarcinoma. *J Steroid Biochem Mol Biol.* 92:111-118.

[27] De Vivo, I., Huggins, G. S., Hankinson, S. E., Lescault, P. J., Boezen, M., Colditz, G. A. & Hunter, D. J. (2002). A functional polymorphism in the promoter of the progesterone receptor gene associated with endometrial cancer risk. *PNAS* 99(19):12263-12268.

[28] McKenna, N. J., Kieback, D. G., Carney, D. N., Fanning, M., McLinden, J. & Headon, D. R. (1995). A germline TaqI restriction fragment length polymorphism in the progesterone receptor gene in ovarian carcinoma. *Br J Cancer.* 71:451-455.

[29] Agoulnik, I. U., Tong, X-W., Fischer, D. C., Korner, W., Atkinson, N. E., Edwards, D. P., et al. (2004). A germline variation in the progesterone receptor gene increases transcriptional activity and may modify ovarian cancer risk. *J Clin Endocrinol Metab.* 89:6340-6349.

[30] Pijnenborg, J. M. A., Romano, A., Dam-de Veen, G. C., et al. (2005). Aberrations in the progesterone receptor gene and the risk of recurrent endometrial carcinoma. *J Pathol.* 205:597-605.

[31] Junqueira, M. G., da Silva, I. D., Nogueira-de-Souza, N. C., Carvalho, C. V., Leite, D. B., et al. (2007). Progesterone receptor (PROGINS) polymorphism and the risk of endometrial cancer development. *Int J Gynecol Cancer* 17(1):229-232.

[32] Sasaki, M., Dharia, A., Oh, B. R., Tanaka, Y., Fujimoto, S. & Dahiya, R. (2001). Progesterone receptor B gene inactivation and CpG hypermethylation in human uterine endometrial cancer. *Cancer Res.* 61:97-102.

[33] Xiong, Y., Dowdy, S. C., Gonzalez Bosquet, J., Zhao, Y., Eberhardt, N. L., Podratz, K. C. & Jiang, S. W. (2005). Epigenetic-mediated upregulation of progesterone receptor B gene in endometrial cancer cell lines. *Gynecol Oncol.* 99:135-41.

[34] Ren, Y., Liu, X., Ma, D., Feng, Y., Zhong, N. (2007). Down-regulation of the progesterone receptor by the methylation of progesterone receptor gene in endometrial cancer cells. *Cancer Genet Cytogenet.* 175(2):107-16.

[35] Yoo, C. B. & Jones, P. A. (2006). Epigenetic therapy of cancer: past, present and future. *Nat Rev Drug Discov.* 5:37-50.

[36] Takai, N., Ueda, T., Nishida, M., Nasu, K. & Narahara, H. (2006). M344 is a novel synthesized histone deacetylase inhibitor that induces growth inhibition, cell cycle arrest, and apoptosis in human endometrial cancer and ovarian cancer cells. *Gynecol Oncol.* 101:108-113.

[37] Kaaks, R., Lukanova, A. & Kurzer, M. S. (2002). Obesity, endogenous hormones, and endometrial cancer risk: a synthetic review. *Cancer Epidemiol Biomarkers Prev.* 11:1531-43.

[38] Chamberlain, N. L., Driver, E. D. & Miesfeld, R. L. (1994). The length and location of CAG trinucleotide repeats in the androgen receptor N terminal domain affect transactivation function. *Nucleic Acids Res.* 22:3181-3186.

[39] Sasaki, M., Karube, A., Karube, Y., Watari, M., Sakuragi, N., Fujimoto, S. & Dahiya, R. (2005). GGC and StuI polymorphism on the androgen receptor gene in endometrial cancer patients. *Biochem Biophys Res Commun.* 329:100-4.

[40] Sasaki, M., Sakuragi, N. & Dahiya, R. (2003). The CAG repeats in exon 1 of the androgen receptor gene are significantly longer in endometrial cancer patients. *Biochem Biophys Res Comm.* 305:1105-1108.

[41] Rodrıguez, G., Bilbao, C., Ramırez, R., Falcon, O., Leon, L. & Chinino, R. et al. (2006). Alleles with short CAG and GGN repeats in the androgen receptor gene are associated with benign endometrial cancer. *Int J Cancer.* 118:1420-1425.

[42] McGrath, M., Lee, I. M., Hankinson, S. E., Kraft, P., Hunter, D. J., Buring, J. & De Vivo, I.. (2006). Androgen receptor polymorphisms and endometrial cancer risk. *Int J Cancer.* 118:1261-1268.

[43] Ju, W. (2007). Polymorphisms in CAG active allele length of the androgen receptor gene are not associated with increased risk of endometrial cancer. *Cancer Genet Cytogenet.* 172:178-179.

[44] Berek, J. S. & Hacker, N. F., eds. (1994). Practical gynecologic oncology. 2d ed. *Baltimore: Williams & Wilkins*, 285-326.

[45] Kurman, R. J. & Norris, H. J. (1987). Endometrial carcinoma. In: Kurman RJ, ed. Blaustein's pathology of the female genital tract. 3d ed. New York: *Springer-Verlag.* 352.

[46] Mutter, G. L., Zaino, R. J., Baak, J. P., Bentley, R. C. & Robboy, S. J. (2007). Benign Endometrial Hyperplasia Sequence and Endometrial Intraepithelial Neoplasia. *Int J Gynecol Pathol.* Apr;26(2):103-114.

[47] Lax, S. F. (2007). Molecular genetic changes in epithelial, stromal and mixed neoplasms of the endometrium. *Pathology.* Feb;39(1):46-54.

[48] Bergstrom, A., Pisani, P., Tenet, T., Wolk, A., Adami, H. O. (2001). Overweight as an avoidable cause of cancer in Europe. *Int J Cancer.* 91:421-430.

[49] Oh, J. C., Wu, W., Tortolero-Luna, G., Broaddus, R., Gershenson, D. M., Burke, T. W., Schmandt, R. & Lu, K. H. (2004). Increased Plasma Levels of Insulin Like Growth Factor 2 and Insulin-Like Growth Factor Binding Protein 3 are Associated with Endometrial Cancer Risk. *Cancer Epidemiol Biomarkers Prev.* 13 (5):231-34.

[50] Berstein, L. M., Imyanitova, E. N., Kovalevskij, A. J., Maximov, S. J., Vasilyev, D. A., Buslov, K. G., Sokolenko, A. P., Iyevleva, A. G., Chekmariova, E. V. & Thijssen, J. H. H. (2004). CYP17 and CYP19 genetic polymorphisms in endometrial cancer: association with intratumoral aromatase activity. *Cancer Lett.* 207:191-196.

[51] Miyamoto. T., Watanabe, J., Hata, H., Jobo, T., Kawaguchi, M., Hattori, M., Saito, M. & Kuramoto, H. (2004). Significance of progesterone receptor-A and -B expressions in endometrial adenocarcinoma. *J Steroid Biochem Mol Biol.* 92:111-118.

[52] Sasaki, M., Diaria, A., Oh, B. R., Tanaka, Y., Fujimoto, S. & Dahiya, R. (2001). Progesterone receptor B gene inactivation and CpG hypermethylation in human uterine endometrial cancer. *Cancer Res.* 61:97-102.

[53] Siiteri, P. K. (1987). Adipose tissue as a source of hormones. *Am J Clin Nutr.* 45:277-282.

[54] Xu, W. H., Xiang, Y. B., Zheng, W., Zhang, X., Ruan, Z. X., Cheng, J. R., Gao, Y. T. & Shu, X. O. (2006). Weight history and risk of endometrial cancer among Chinese women. *Int J Epidemiol.* 35:159-166.

[55] Yoo, C. B. & Jones, P. A. (2006). Epigenetic therapy of cancer: past, present and future. *Nat Rev Drug Discov.* 5:37-50.

[56] Takai, N., Ueda, T., Nishida, M., Nasu, K. & Narahara, H. (2006). M344 is a novel synthesized histone deacetylase inhibitor that induces growth inhibition, cell cycle arrest, and apoptosis in human endometrial cancer and ovarian cancer cells. *Gynecol Oncol.* 101:108-113.

[57] Beiner, M. E., Finch, A., Rosen, B., Lubinski, J., Moller, P., Ghadirian, P., Lynch, H. T., Friedman, E., Sun, P. & Narod, S. A. (2007). Hereditary Ovarian Cancer Clinical Study Group. The risk of endometrial cancer in women with BRCA1 and BRCA2 mutations. A prospective study. *Gynecol Oncol.* Jan;104(1):7-10.

[58] Takai, N., Kawamata, N., Walsh, C. S., Gery, S., Desmond, J. C., Whittaker, S., Said, J. W., Popoviciu, L. M., Jones, P. A., Miyakawa, I. & Koeffler, H. P. (2005). Discovery of epigenetically masked tumor suppressor genes in endometrial cancer. *Mol Cancer Res.* May;3(5):261-9.

[59] Vilos, G. A., Harding, P. G. & Ettler, H. C. (2002). Resectoscopic surgery in 10 women with abnormal uterine bleeding and atypical endometrial hyperplasia. *J Am Assoc Gynecol Laparosc.* May;9(2):138-44.

[60] Edris, F., Vilos, G. A., Al-Mubarak, A., Ettler, H. C., Hollett-Caines, J. & Abu-Rafea, B. (2007). Resectoscopic surgery may be an alternative to hysterectomy in high-risk women with atypical endometrial hyperplasia. *J Minim Invasive Gynecol.* Jan-Feb;14(1):68-73.

[61] Vilos, G. A., Edris, F., Al-Mubarak, A., Ettler, H. C., Hollett-Caines, J. & Abu-Rafea, B. (2007). Hysteroscopic surgery does not adversely affect the long-term prognosis of women with endometrial adenocarcinoma. *J Minim Invasive Gynecol.* Mar-Apr;14(2):205-10.

[62] Vilos, G. A., Harding, P. G., Silcox, J. A., Sugimoto, A. K., Carey. M. & Ettler, H. C. (2002). Endometrial adenocarcinoma encountered at the time of hysteroscopic endometrial ablation. *J Am Assoc Gynecol Laparosc.* Feb;9(1):40-8.

[63] Steed, H. L. & Scott, J. Z. Adenocarcinoma diagnosed at endometrial ablation. Obstet Gynecol. 2001 May;97(5 Pt 2):837-9.

[64] Elgarib, A. E. & Nooh, A. (2006). Thermachoice endometrial balloon ablation: a possible alternative to hysterectomy. *J Obstet Gynaecol.* Oct;26(7):669-72.

[65] Goldrath, M. H. (2003). Evaluation of HydroThermAblator and rollerball endometrial ablation for menorrhagia 3 Years after treatment. *J Am Assoc Gynecol Laparosc.* Nov;10(4):505-11.

[66] Rosenbaum, S. P., Fried, M. & Munro, M. G. (2005). Endometrial hydrothermablation: a comparison of short-term clinical effectiveness in patients with normal endometrial cavities and those with intracavitary pathology. *J Minim Invasive Gynecol.* Mar-Apr;12(2):144-9.

[67] Schlaerth, A. C. & Abu-Rustum, N. R. (2006). Role of minimally invasive surgery in gynecologic cancers. *Oncologist.* Sep;11(8):895-901.

[68] Kalogiannidis, I., Lambrechts, S., Amant, F., Neven, P., Van Gorp, T. & Vergote, I. (2007). Laparoscopy-assisted vaginal hysterectomy compared with abdominal hysterectomy in clinical stage I endometrial cancer: safety, recurrence, and long-term outcome. *Am J Obstet Gynecol.* Mar;196(3):248.e1-8.

[69] Gemignani, M. L., Curtin, J. P., Zelmanovich, J., Patel, D. A., Venkatraman, E. & Barakat, R. R. (1999). Laparoscopic-assisted vaginal hysterectomy for endometrial cancer: clinical outcomes and hospital charges. *Gynecol Oncol.* Apr;73(1):5-11.

[70] Patel, S., Portelance, L., Gilbert, L., Tan, L., Stanimir, G., Duclos, M. & Souhami, L. (2007). Analysis of prognostic factors and patterns of recurrence in patients with pathologic Stage III endometrial cancer. *Int J Radiat Oncol Biol Phys.* Apr 6.

[71] Malur, S., Possover, M., Michels, W. & Schneider, A. (2001). Laparoscopic-assisted vaginal versus abdominal surgery in patients with endometrial cancer--a prospective randomized trial. *Gynecol Oncol.* 80(2):239-44.

[72] Kohler, C., Klemm, P., Schau, A.., et al. (2004). Introduction of transperitoneal lymphadenectomy in a gynecologic oncology center: analysis of 650 laparoscopic pelvic and/or paraaortic transperitoneal lymphadenectomies. *Gynecol Oncol* 95(1):52-61.

[73] Abu-Rustum, N. R., Chi, D. S., Sonoda, Y., et al. (2003). Transperitoneal laparoscopic pelvic and para-aortic lymph node dissection using the argon-beam coagulator and

monopolar instruments: an 8-year study and description of technique. *Gynecol Oncol.* 89(3):504-13.

[74] Frigerio, L., Gallo, A., Ghezzi, F., Trezzi, G., Lussana, M. & Franchi, M. (2006). Laparoscopic-assisted vaginal hysterectomy versus abdominal hysterectomy in endometrial cancer. *Int J Gynaecol Obstet.* 93(3):209-13.

[75] Walker, J., Piedmonte, M., Spirtos, N., Eisenkop, S., Schlaerth, J., Mannel, R. & Spiegel, G. (2006). Phase III trial of laparoscopy (scope) vs laparotomy (open) for surgical resection and comprehensive surgical staging of uterine cancer: *A Gynecologic Oncology Group (GOG) Study funded by NCI. In: Society of Gynecologic Oncologists.*

[76] Kornblith, A., Walker, J., Huang, H. & Cella, D. (2006). Quality of life (QOL) of patients in a randomized clinical trial of laparoscopy (scope) vs. open laparotomy (open) for the surgical resection and staging of uterine cancer: A Gynecologic Oncology Group (GOG) study. *In: Society of Gynecologic Oncologists, Main Plenary Session IV.*

[77] Janda, M., Gebski, V., Forder, P., Jackson, D., Williams, G. & Obermair, A. (2006). Total laparoscopic versus open surgery for stage 1 endometrial cancer: The LACE randomized controlled trial. *Contemporary Clinical Trials.* 27(4):353-63.

[78] Tozzi, R., Malur, S., Koehler, C. & Schneider, A. (2005). Laparoscopy versus laparotomy in endometrial cancer: first analysis of survival of a randomized prospective study. *J Minim Invasive Gynecol.* 12(2):130-6.

[79] Nezhat, F., Yadav, J., Rahaman, J., Gretz, H. & Cohen, C. (2008). Analysis of survival after laparoscopic management of endometrial cancer. *J Minim Invasive Gynecol.* 15(2):181-7.

[80] Eltabbakh, G. H. & Mount, S. L. (2006). Laparoscopic surgery does not increase the positive peritoneal cytology among women with endometrial carcinoma. *Gynecol Oncol.* 100(2):361-4.

[81] Sonoda, Y., Zerbe, M., Smith, A., Lin, O., Barakat, R. R. & Hoskins, W. J. (2001). High incidence of positive peritoneal cytology in low-risk endometrial cancer treated by laparoscopically assisted vaginal hysterectomy. *Gynecol Oncol.* 80(3):378-82.

[82] Gemignani, M. L., Curtin, J. P., Zelmanovich, J., Patel, D. A., Venkatraman, E. & Barakat, R. R. (1999). Laparoscopic-assisted vaginal hysterectomy for endometrial cancer: clinical outcomes and hospital charges. *Gynecol Oncol.* 73(1):5-11.

[83] Scribner, D. R. Jr., Walker, J. L., Johnson, G. A., McMeekin, D. S., Gold, M. A. & Mannel, R. S. (2002). Laparoscopic pelvic and paraaortic lymph node dissection in the obese. *Gynecol Oncol.* 84(3):426-30.

[84] Eltabbakh, G. H., Shamonki, M. I., Moody, J. M. & Garafano, L. L. (2000). Hysterectomy for obese women with endometrial cancer: laparoscopy or laparotomy? *Gynecol Oncol.* 78(3 Pt 1):329-35.

[85] Holub, Z., Bartos, P., Jabor, A., Eim, J., Fischlova, D. & Kliment, L. (2000). Laparoscopic surgery in obese women with endometrial cancer. *J Am Assoc Gynecol Laparosc.* 7(1):83-8.

[86] Tinelli, A., Vergara, D., Leo, G., et al. (2007). Human papillomavirus genital infection in modern gynaecology: genetic and genomic aspects. *Eur Clinics Obstet Gynaecol.* 3:1-6.

[87] Castle, P. E. (2008). Is monitoring of human papillomavirus infection for viral persistence ready for use in cervical cancer screening? *Am J Epidemiol.* 168:138-144.

[88] Pinto, A. P., Carvalho, M. C., Kolb, S,, Tirone, F. A., Maia, L. R. & Escobar, C. S. (2007). Value of cytology in papillary condylomatous lesions of the cervix. *Acta Cytol.* Jan-Feb;51(1):51-60.

[89] Burd, E. M. (2003). Human Papillomavirus and Cervical Cancer. *Clin Microbiol Rev.* 16(1):1-17.

[90] Altaf, F. J. Cervical cancer screening with pattern of pap smear. Review of multicenter studies. Saudi Med J. 2006 Oct;27(10):1498-502.

[91] Confortini, M., Di Bonito, L., Carozzi, F., Ghiringhello, B., Montanari, G., Parisio, F. & Prandi, S. (2006). GISCi Working Group for Cervical Cytology. Interlaboratory reproducibility of atypical glandular cells of undetermined significance: a national survey. *Cytopathology.* Dec;17(6):353-60.

[92] Moinfar, F. (2008). Is 'Basal-Like' carcinoma of the breast a distinct clinicopathological entity? a critical review with cautionary notes. *Pathobiology.* 75:119-131

[93] Oliveira, J. G., Colf, L. A. & McBride, A. A. (2005). Variations in the association of papillomavirus E2 proteins with mitotic chromosomes. *Proc Natl Acad Sci USA.* 103(4):1047-1052.

[94] Vázquez-Ortíz, G., Ciudad, C. J., Piña, P., Vazquez, K., Hidalgo, A., Alatorre, B., Garcia, J. A., Salamanca, F., Peralta-Rodriguez, R., Rangel, A. & Salcedo, M. (2005). Gene Identification by cDNA Arrays in HPV-Positive Cervical Cancer. *Arch Med Res.* 36:448-458.

[95] Choi, Y. D., Jung, W. W., Nam, J. H., Choi, H. S. & Park, C. S. (2005). Detection of HPV genotypes in cervical lesions by the HPV DNA Chip and sequencing. *Gynecol Oncol.* 98:369-375.

[96] Kim, K. H., Yoon, M. S., Na, Y. J., Park, C. S., Oh, M. R. & Moon, W. C. (2006). Development and evaluation of a highly sensitive human papillomavirus genotyping DNA chip. *Gynecol Oncol.* 100:38-43.

[97] Klaassen, C., Prinsen, C., de Valk, H., Horrevorts, A., Jeunink, M. & Thunnissen, F. (2004). DNA Microarray Format for Detection and Subtyping of Human Papillomavirus. *J Clin Microbiol.* 42(5):2152-2160.

[98] Jeon, S. & Lambert, F. P. (1995). Integration of human papillomavirus type 16 DNA into the human genome leads to increased stability of E6 and E7 mRNAs: implications for cervical carcinogenesis. *Proc Natl Acad Sci USA.* 92:1654-1658.

[99] Yu, T., Ferber, M. J., Cheung, T. H., Chung, T. K. H., Wong, Y. F. & Smith, D. I. (2005). The role of viral integration in the development of cervical cancer. *Cancer Genet Cytogenet.* 158:27-34.

[100] Oh, S. T., Longworth, M. S. & Laimins, L. A. (2004). Roles of the E6 and E7 proteins in the life cycle of low risk human papillomavirus type 11. *J Virol.* 78(5):2620-2626.

[101] Lambert, A. P. F., Anschau, F. & Schmitt, V. M. (2006). $P16^{INK4a}$ expression in cervical premalignant and malignant lesions. *Exp Mol Pathol.* 80:192-196.

[102] Ishikawa, M., Fujii, T., Saito, M., Nindl, I., Ono, A., Kubushiro, K., Tsukazaki, K., Mukai, M. & Nozawa, S. (2006). Overexpression of $p16^{INK4a}$ as an indicator for human papillomavirus oncogenic activity in cervical squamous neoplasia. *Int J Gynecol Cancer*. 16:347-353.

[103] Bulten, J., van der Avoort, I., Melchers, W. J. G., Massuger, L., Grefte, J., Hanselaar, A. & Wilde, P. (2006). $p14^{ARF}$ and $p16^{INK4A}$, two products of the same gene, are differently expressed in cervical intraepithelial neoplasia. *Gynecol Oncol.* Jun;101(3):487-94.

[104] Ueda, M., Terai, Y., Kanda, K., Kanemura, M., Takehara, M., Yamaguchi, H., Nishiyama, K., Yasuda, M. & Ueki, M. (2006). Germline polymorphism of *p53* codon 72 in gynaecological cancer. *Gynecol Oncol.* 100:173-178.

[105] Oliveira, J. G., Colf, L. A. & McBride, A. A. (2006). Variations in the association of papillomavirus E2 proteins with mitotic chromo- somes. *Proc Natl Acad Sci USA.* 103(4):1047-1052.

[106] Loizzi, P., Carriero, C., Di Gesu, A., Resta, L. & Nappi, R. (1992). Rational use of cryosurgery and cold knife conization for treatment of cervical intraepithelial neoplasia. *Eur J Gynaecol Oncol.* 13(6):507-13.

[107] Jobson, V. W. (1991). Cryotherapy and laser treatment for intraepithelial neoplasia of the cervix, vagina, and vulva. *Oncology.* Aug;5(8):69-72, 77.

[108] Baggish, M. S. (1985). *Basic and Advanced Laser Surgery in Gynecology.* 2nd ed. Norwalk, Conn: Appleton & Lange. 207-16.

[109] Helkjaer, P. E., Eriksen, P. S., Thomsen, C. F. & Skovdal, J. (1993). Outpatient CO2 laser excisional conization for cervical intraepithelial neoplasia under local anesthesia. *Acta Obstet Gynecol Scand.* May;72(4):302-6.

[110] Ueda, M., Ueki, K., Kanemura, M., Izuma, S., Yamaguchi, H., Nishiyama, K., Tanaka, Y., Terai, Y. & Ueki, M. (2006). Diagnostic and therapeutic laser conization for cervical intraepithelial neoplasia. *Gynecol Oncol.* Apr;101(1):143-6.

[111] Yamaguchi, H., Ueda, M., Kanemura, M., Izuma, S., Nishiyama, K., Tanaka, Y. & Noda, S. (2007). Clinical efficacy of conservative laser therapy for early-stage cervical cancer. *Int J Gynecol Cancer.* Mar-Apr;17(2):455-9.

[112] Bar-Am, A., Daniel, Y., Ron, I. G., et al. (2000). Combined colposcopy, loop conization, and laser vaporization reduces recurrent abnormal cytology and residual disease in cervial dysplasia. *Gynecol Oncol.* 78: 47-51.

[113] Murta, E. F., Silva, A. O. & Silva, E. A. (2006). Clinical significance of a negative loop electrosurgical excision procedure, conization and hysterectomy for cervical intraepithelial neoplasia. *Eur J Gynaecol Oncol.* 27(1):50-2.

[114] Abu-Rustum, N. R., Sonoda, Y., Black, D., Levine, D. A., Chi, D. S. & Barakat, R. R. (2006). Fertility-sparing radical abdominal trachelectomy for cervical carcinoma: technique and review of the literature. *Gynecol Oncol.* Dec;103(3):807-13.

[115] Burnett, A. F. (2006). Radical trachelectomy with laparoscopic lymphadenectomy: review of oncologic and obstetrical outcomes. *Curr Opin Obstet Gynecol.* Feb;18(1):8-13.

[116] Malzoni, R., Tinelli, R., Cosentino, F., Perone, C. & Vicario, V. (2007). Feasibility, morbidity, and safety of total laparoscopic radical hysterectomy with

lymphadenectomy: our experience. *J Minim Invasive Gynecol.* Sep-Oct;14(5): 584-90.

[117] Runic, S., Durbaba, M. & Runic, R. (2005). Lymphadenectomy during radical hysterectomy for cervical cancer (stage IB 1-2, HA): state of the art. *J BUON.* Oct-Dec;10(4):473-481.

[118] Lukaszuk, K., Liss, J., Nowaczyk, M., Sliwinski, W., Maj, B., Wozniak, I., Nakonieczny, M. & Barwinska, D. (2007). Survival of 231 cervical cancer patients, treated by radical hysterectomy, according to clinical and histopathological features. *Eur J Gynaecol Oncol.* 28(1):23-7.

[119] Metindir, J. & Bilir, G. (2007). Prognostic factors affecting disease-free survival in early-stage cervical cancer patients undergoing radical hysterectomy and pelvic-paraaortic lymphadenectomy. *Eur J Gynaecol Oncol.* 28(1):28-32.

[120] Bornstein, J. (2007). Human papillomavirus vaccine: the beginning of the end for cervical cancer. *Isr Med Assoc J.* Mar;9(3):156-8.

[121] Keim, B. (2007). Controversy over cervical cancer vaccine spurs safety surveillance. *Nat Med.* Apr;13(4):392-3.

[122] Villa, L. L, Costa, R. L., Petta, C. A., Andrade, R. P., Ault, K. A., Giuliano, A. R., et al. (2005). Prophylactic quadrivalent human papillomavirus (types 6, 11, 16, and 18) L1 virus-like particle vaccine in young women: a randomised double-blind placebo-controlled multicentre phase II efficacy trial. *Lancet Oncol.* 6: 271-8.

[123] Future II Study Group.Quadrivalent vaccine against human papillomavirus to prevent high-grade cervical lesions. *N Engl J Med.* 2007;356:1915-27

[124] Garland, S. M., Hernandez-Avila, M., Wheeler, C. M., Perez, G., Harper, D. M., Leodolter, S., et al. (2007). Quadrivalent vaccine against human papillomavirus to prevent anogenital diseases. *N Engl J Med.* 356:1928-43.

[125] Harper, D. M., Franco, E. L., Wheeler, C. M., Moscicki, A. B., Romanowski, B., Roteli-Martins, C. M., et al. (2006). Sustained efficacy up to 4.5 years of a bivalent L1 virus-like particle vaccine against human papillomavirus types 16 and 18: follow-up from a randomised control trial. *Lancet.* 367:1247-55.

[126] Ault, K. A. (2007). Future II Study Group. Effect of prophylactic human papillomavirus L1 virus-like-particle vaccine on risk of cervical intraepithelial neoplasia grade 2, grade 3, and adenocarcinoma in situ: a combined analysis of four randomised clinical trialsl. *Lancet.* 369(9576): 1861-8.

In: The Human Genome: Features, Variations...
Editor: Akio Matsumoto and Mai Nakano

ISBN: 978-1-60741-695-1

Chapter 13

The Perception of an Information Society and the Emergence of the First Computerized Biological Databases, 1948–1992

Miguel García-Sancho[2]
Department of Science, Technology and Society,
Spanish National Research Council (CSIC)

Abstract

It is a common assumption that we currently live in an *information society.* Control and access to information are seen by the public as crucial means for social knowledge and power. Similarly, in biology, the recent completion of the Human Genome Project has led to the consideration as fundamental scientific knowledge the sequence of information in our genes. The emergence of information as a key concept in both biology and society has a history of more than sixty years, which is not that generally acknowledged. This chapter will explore such history by investigating the development of the first computerized biological databases and their connection with the understanding of information as a valuable social resource. By studying two European database initiatives, one of them developed in the 1960s and the other in the early 1980s, I will argue that the emergence of the personal computer and the increasing perception of data gathering as an essential social and scientific activity marked the different fate of each project. Whereas the 1960s database faced financial difficulties, the 1980s effort—devoted to the storage of DNA sequences—was perceived as priority and cutting-edge science, associated with the new discipline of genomics and given unprecedentedly large funding.

[2] Calle Albasanz 26-28, 28037 Madrid, Spain, Email: miguel.garciasancho@cchs.csic.es

1. Introduction

The emergence of the first personal computers—Altair, Apple II and IBM's PC between 1975 and 1981—generated among social science scholars and the general public the impression that we had entered into an *information era* dominated by the technologies to store and process data (Parker, 1973; Lamberton, 1974; Danzin, 1979). In the subsequent decades, a debate in sociology and economic theory—with its corresponding lay versions—addressed whether markets and social interactions had become immaterial and organized through increasingly global networks based on exchanges of information and knowledge (Quah, 1997; Kahin and Foray, 2006). There were diverse positions within this debate; not all scholars agreed with the informational turn of economy and society (Webster, 1997; Winston, 1998; Edgerton, 2006). However, there was a shared perception that the new information technologies had a growing impact on our lives. As sociologist Manuel Castells claimed, "information generation, processing and transmission" were becoming "the fundamental sources" of productivity, knowledge and power (Castells, 1996, vol. 1, p. 21).

Similarly, in biology, the invention of the recombinant DNA techniques in the middle to late 1970s raised the possibility of determining and altering the structure of our genetic material. The development of these techniques fostered among biologists and society a concern with DNA sequences, whose storage, manipulation and analysis were perceived as fundamental scientific knowledge. During the 1980s, projects directed to the DNA sequence of increasingly large organisms—e.g., the bacterium *E. coli,* the worm *C. elegans* or the human genome—emerged, and in 1987 a new discipline, genomics, was proposed (García-Sancho, 2007b; 2008, ch. 2; in press). This discipline considered DNA sequencing "the way to go" and the technique from which "the complexities of gene expression" and disease could be "translated" (McKusisck and Ruddle, 1987, p. 1). With the launch of the Human Genome Project (HGP), there was a growing perception that biology had become an information science, i.e., an endeavor engaged with the computation of features from DNA sequences rather than with 'wet' experiments conducted in the test tube (e.g., Hood, 1992; Gilbert, 1992).

The parallelisms between sociological and biological discourses suggest that in the late 1970s and 1980s there was a concern with the control and access to information which affected both the social and the natural sciences. Scholars, however, have treated the biological and sociological dimensions as independent, not generally considering genomics as part of a more general movement of society towards information[3]. There has been,

[3] In his classical work on information society, Castells refers to genetic engineering and biotechnology as examples of techniques and markets driven by information, but does not expand on the connections between these and other non-biological phenomena (Castells, 1996, vol. 1, pp. 54-59). I have preliminarily analyzed such connections from a historical perspective in a series of papers, as well as in my PhD dissertation (García-Sancho, 2007a, pp. 29 and ff.; 2007b; 2008, pp. 240-45; in press). In the fields of history, philosophy and sociology of science, there is a broader literature on "genetic information", which addresses, among other issues, the regulation of DNA sequences stored in databases (e.g., Nelkin, 1992; Fox Keller, 1995, 2000; Kay, 2000; Sarkar, 1996; Moss, 2004; Gibbons et al., 2007; Marturano and Chadwick, 2004; Tutton, 2007). Bioethicists have also noted the difficulties in establishing a rigid separation between the material biological tissues and the data about them stored in different banks in the current "age of information" (Gere and Parry, 2006).

additionally, a consensus in presenting both an information society and information-oriented biology as phenomena of the 1980s, respectively triggered by the personal computer and the recombinant DNA revolution (Castells, 1996; Hood, 2004; Zweiger, 2001). However, an expanding field of scholarship is showing that the concern with computer-based gathering and analysis of data can be traced back to the post-World War II era in both biology and the social sciences (Kline, 2006; Haigh, 2001, 2006a; Agar, 2003; Black, Muddiman and Plant, 2007; Strasser 2006, 2008; Suárez, 2007; November, 2004, 2006; García-Sancho, 2008; in press). It is, consequently, possible to talk about an early 1950s and 1960s information society, as well as initiatives to apply information technologies to biology before genomics.

This chapter will build on the literature on an early information society and investigate a computer-based database initiative in biology during the late 1960s and 1970s. It will, therefore, seek to unify and provide with a history the notions of information-driven biology and society. The chapter will also compare the context of the early biological database with that of a further 1980s collection devoted to the storage of DNA sequences. This will allow me to determine what was new in the discourse of an information society and genomics in the 1980s when compared with previous social and biological initiatives.

2. The Early Information Society and the First Database Efforts in Biology

The first concern with the social importance of information and computing arose after World War II[4]. It was raised by cybernetics, a branch of engineering with considerable influence in wartime science. Researchers in this field predicted a future marked by computerized control systems, which were able to process data about different situations and to respond to them autonomously. These devices, designed by cyberscientists and used in anti-aircraft and missile firing during the war, were applied to office management in public administration and business during the 1950s and 1960s. In biology, researchers began creating databases centralizing information about experiments conducted by separated groups, either geographically or chronologically. These initiatives, however, faced enormous financial problems, due to them not being considered research by scientific funding agencies.

2.1. The Systems Men and Their Utopia

The increasing role of information control in society after World War II was first stated by Norbert Wiener, electronic and communication engineer considered the founder of cybernetics. Historian of technology Robert Kline has shown how in his books *Cybernetics* (1948) and *The Human Use of Human Beings* (1950) Wiener postulated a "second industrial

[4] The social concern with information is a recurrent topic in history which much predates the computer. Daniel Rosenberg and other scholars have investigated the problems that a perceived "information overload" created during the Renaissance and Early Modern periods, due to the colonial expeditions and the encyclopedic will of describing everything which explorers found in the new worlds (Rosenberg, ed., 2003; quoted in Strasser, 2006, pp. 120-21).

revolution" in which a growing number of social activities would be performed by computerized devices. These devices would be dependent on information, which was introduced by the users and guided their responses in different situations. Wiener's imaginery, according to Kline, spread from the 1950s onwards, leading to the perception of an information-driven society (Kline, 2006).

Science fiction writers firstly and social science scholars afterwards began developing Wiener's ideas and associated them, respectively, with a dystopian future and fundamental transformations in society. An example of the former is *Player Piano,* written by Kurt Vonnegut in 1952 and which describes a world in which lower classes have no role, due to them having been replaced by computerized devices handled by managers and engineers[5]. Vonnegut's and other fiction accounts were followed by the emergence, in the 1960s and 1970s, of media studies, as well as the theories of knowledge economy and post-industrial society (ibid., pp. 518 and ff.). These new academic fields postulated a social structure in which the production, control and gathering of material entities were being replaced by information as a source of productivity, knowledge and power.

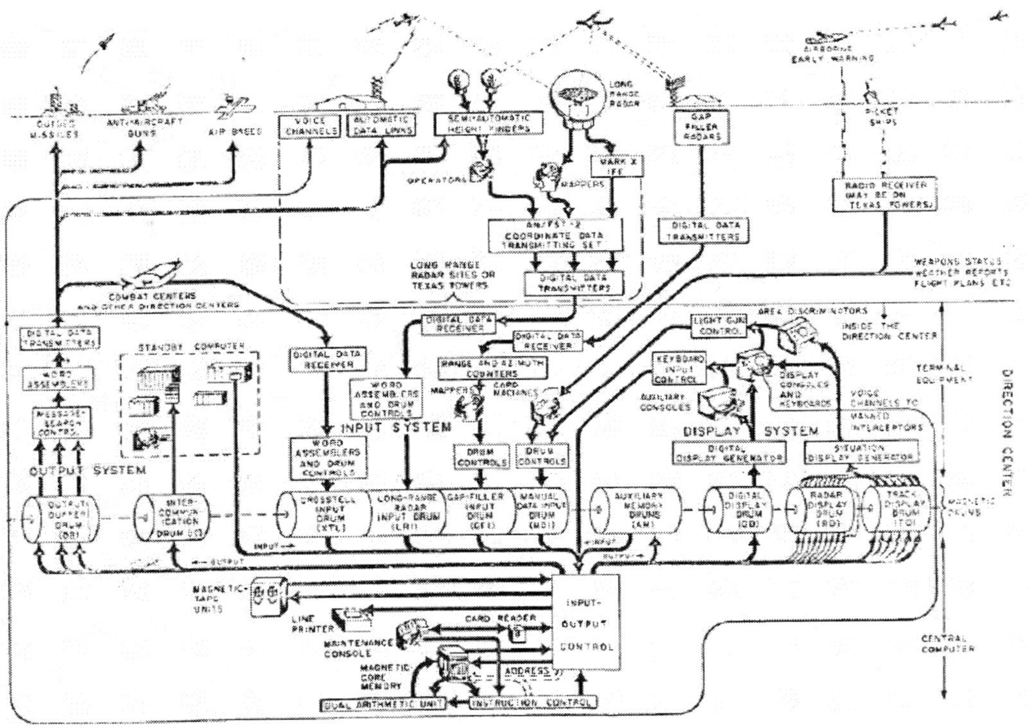

Figure 1. Outline of an early military application of computers to anti-aircraft fire during the 1950s and 1960s. Taken from the DigiBarn Computer Museum (www.digibarn.com). Reprinted with permission.

The concern with information and will to gather it also permeated Government and business circles. The late 1950s and 1960s saw the application of system sciences—the field

[5] Vonnegut explicitly connected his book to Wiener's theories. The latter, however, did not appreciate his efforts and considered the book as "mediocre science fiction" (Kline, 2006, p. 518).

Wiener and his followers had previously fostered—to public administration and private corporate offices. One of the main outcomes of cybernetics during World War II had been the "control and command systems", in which a large central computer was fed with different sorts of data, e.g., on the speed and trajectory of enemy planes. From such data, called as "feedback" by Wiener, the computer was able to make predictions and optimize the response, i.e., anticipate the position of the plane and shot an anti-aircraft load (Mindell, 2002, pp. 7 and ff.). Technology historian Thomas Haigh has shown how Government statistics, travel agencies, libraries and insurance companies adapted this technology after the war in the form of centralized databases which assisted in the management of their respective activities (Haigh, 2001, 2006a).

The 1950s and 1960s, consequently, saw the emergence of the "systems men" and "information engineers": computer experts hired by public and private offices to organize their records in databases. These professionals were aimed to replace lower-class clerks and middle managers, and to help the decision-making process by combining various sorts of information—about payrolls, employees and stocks, among others. The dream of the information engineers, according to Haigh, was a "totally integrated" information system, in which the office would be able to predict unequivocally events—about employment, deliveries or finance—from a centralized data bank (Haigh, 2001, pp. 15-18). This new staff was, therefore, seeking to transform information management in a predictive science.

The information engineers and systems men referred to the database outcomes as "vital intelligence" for the office (ibid., p. 16). The term evoked the concern with espionage and codes at a time marked by the Cold War. Many system sciences experts, such as the founder of information theory Claude Shannon, were engaged with military devices to encrypt data. This context of secrecy and rise of information management also fostered the emergence of the first computerized biological databases[6].

2.2. Information enters Biology

The spread of information management led a series of biologists to raise database proposals from the mid 1960s onwards. Around 1965, biologists and medical researchers from different fields proposed computer-based compilations of useful data for their investigations. They shared with other data gathering efforts the Cold War context and the belief in the predictive power of information. Nevertheless, biological institutions were not as enthusiastic as the database creators.

One of the pioneer database efforts was that of Olga Kennard. She had been working since the late 1940s on x-ray crystallography, a field engaged with the determination of three-dimensional structures of proteins and other molecules. After a time with future Nobel Prize winner Max Perutz investigating the protein hemoglobin, she set a database facility in Birkbeck College (London) with J.D. Bernal, another prominent crystallographer. Her aim

---

[6] Other instance of interaction between information sciences and biology during the 1950s and 1960s was the research on the genetic code. Historians and philosophers of biology have shown connections between systems scientists and researchers working on how genes directed the functioning of the cell (Kay, 2000, ch. 3; Sarkar, 1996; Fox Keller, 1995, García-Sancho, 2007a, pp. 17-20).

was to establish a central repository which gathered the results of the mathematical computations needed to reconstruct the structure of proteins after they had been analyzed with x-rays (Kennard, 1997). When an x-ray beam is applied to a crystallized molecule, it is differently deviated and the result, if collected on a surface, is a pattern of spots. This pattern requires sophisticated calculations to solve the three-dimensional shape of the protein. Kennard's database entries collected the result of such calculations and thus allowed to reconstruct the molecule's structure.

Kennard's database was founded on the "belief that the collective use of data would lead to the discovery of new knowledge" which would transcend "the results of individual experiments" (ibid., online). She, therefore, shared the faith in the predictive power of information of social scientists and systems men at the time. The Cold War context also helped Kennard to expand her original Birkbeck facility.

In the 1960s, a main issue of concern among the different world blocks was their relative strength in scientific data. This led to the creation of specialized bodies which dealt with information in all research disciplines and fields—e.g., the Committee on Data for Science and Technology or the National Bureau of Standards in the United States. During Kennard's work at Birkbeck, "there was the feeling among European Governments that scientific information was being monopolized by the US and the Soviet Union". It was agreed that a series of centers, each oriented towards data from a discipline, would be created. The United Kingdom was nominated for crystallography and in 1965 a British Government officer approached Kennard with the proposal of expanding her database (id., 2007).

This led to the creation of the Cambridge Crystallographic Data Centre (CCDC) as an institute devoted to both crystallographic research and the maintenance of a centralized database for "small molecules" (id., 1997, online). The center employed from secretarial clerks to experienced crystallographers, students and postdoctoral researchers, the latter dividing their time at 50% between investigations and database work. Another feature of the CCDC was the strong use of computers, which at that time were large mainframes external to the laboratories and operated with punched cards. CCDC staff coded the necessary data and instructions to operate with it into a precise pattern of perforations across the cardboard surface of the card. The cards were then sent to the mainframe, which allowed to operate with the entries and to check for their consistency, i.e. detecting errors, duplications or format alterations (García-Sancho, 2008, pp. 226-31).

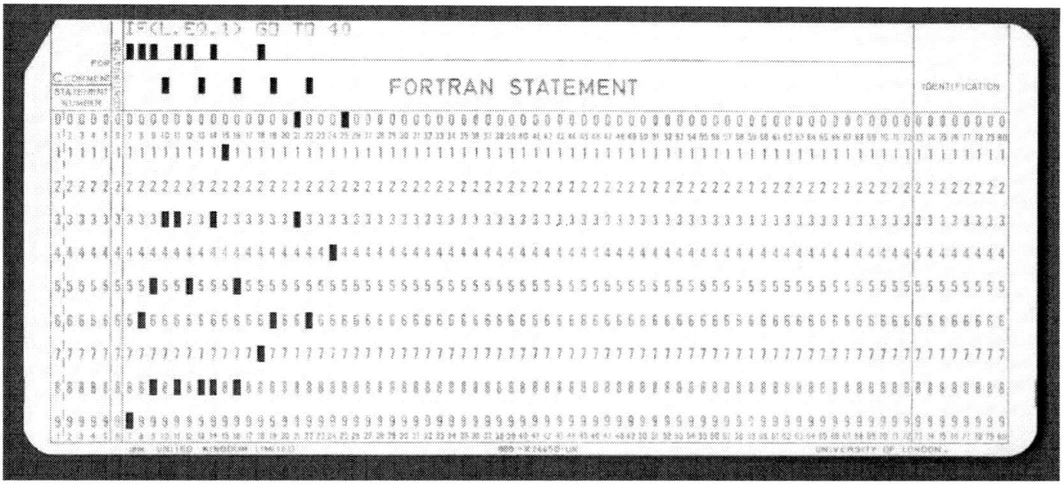

Potassium benzylpenicillin
C16 H17 N2 04 S1 1-,K1 1+
D.Crowfoot,C.W.Bunn,B.W.Rogers-Low,A.Turner-Jones
The Chemistry of Penicillin, , 310,1949

Authors' cell dimensions

a	b	c	alpha	beta	gamma
9.360	6.370	30.350	90.0	90.0	90.0

K1	.352	.530	-.002
01	.379	-.123	.056
02	.359	.202	.058
C1	.436	.046	.061

.
.

Figure 2. A sample entry of Olga Kennard's crystallographic database (above) and a punched card used to enter the data and instructions for the mainframe computer (below). Kennard, 1997, online and Archives of the MRC Laboratory of Molecular Biology, Cambirdge, UK. Reprinted with permission

Similar initiatives arose in other fields of biology. In 1966, medical geneticist Victor McKusick began publishing *Mendelian Inheritance in Man,* a catalogue of genetic diseases transmitted from parents to offspring (McKusick, 1966). One year before, Margaret Dayhoff had launched the *Atlas of Protein Sequence and Structure,* a compilation of all the available protein sequences at the time, which was updated roughly every year[7]. The projects shared an extensive use of computers and punched cards, as well as a belief in the power of information

[7] Dayhoff's *Atlases* have been studied in detail by historian of biology Bruno Strasser (Strasser, 2006). Due to this, my paper focuses on Kennard as another pioneer figure in the introduction of computers and databases into biology.

to solve scientific problems. Nevertheless, the biological establishment was reluctant towards these initiatives.

2.3. Official Skepticism

The new biomedical databases did not benefit from the favor of the funding agencies at their time. National government bodies considered this work beyond their funding scope, which was devoted to experimental biological and medical sciences. This forced the database biologists to seek alternative funding sources and to usually find them in either selling their data or applying to military budgets oriented towards system sciences. The fact of Kennard, Dayhoff and many of the early computer operators being women[8] was also a handicap at a time in which the biological establishment was almost completely integrated by men.

Kennard's crystallographic database lost its original Cold War budget in the late 1960s. She found an alternative funding system in leasing the entries to the pharmaceutical industry and to large-public "national centers" which subsequently offered the data for free to research institutions in their respective countries (Kennard, 1997, online). During the second half of the 1970s, Kennard entered in contact with Fred Sanger, the inventor of the first protein and DNA sequencing methods, also based in Cambridge (García-Sancho, in press; de Chadarevian, 1996, 1999). They agreed to collaborate in the establishment of a database with DNA sequence information, but failed to secure support from the UK Medical Research Council and the Laboratory of Molecular Biology of Cambridge, Sanger's home institution (Kennard, 2007).

The situation of Dayhoff was even more troubled. As historian Bruno Strasser has shown, she unexpectedly lost a bid to create a database to store DNA sequences after fifteen years of finantial constraints due to the limited budget devoted to her *Atlases* by the US National Institutes of Health (NIH). The contract was granted by the NIH to the less experienced physicist Walter Goad in the early 1980s and this had a considerable emotional impact on Dayhoff, who prematurely died shortly afterwards. The tensions between Dayhoff and the biomedical community, which was reluctant to pay for the *Atlases* and to send the sequences to the database before them being published in scientific journals, was crucial for the failure of her bid (Strasser, 2008; 2006, pp. 118-119)[9].

The difficulties of Kennard and Dayhoff's databases may be explained in the contrasts and changes experienced by the social concern with information between the 1960s and 1980s. The 1980s were marked by the spread of the personal computer and expansion of the biotechnology market, which had in the gathering of DNA sequence data one of its main targets. The weakening of the Cold War and gradual emergence of global politics and

[8] Soraya de Chadarevian and J.S. Light have shown how computing in its early years (1940s to 1960s) was a significantly female activity. This was due to many of its uses being associated with secretariat office work (de Chadarevian, 2002, ch. 4; Light, 1999). Kennard has noted that many of the staff in charge of the routine database work at the CCDC was formed by women (Kennard, 2007).

[9] Strasser has explained Dayhoff's difficulties in her inability to adapt to the "moral economy" of biology, which was based upon the unrestricted and free circulation of information, as well as credit exclusively earned by scientific publication. His argument is complementary to the one I offer in this paper.

economy also account for the belief in a universal and non-restricted circulation of data by all social actors, biologists included.

3. Genomics, A New Information Discourse and DNA Databases

The social concern with information which had characterized the post-World War II expanded and significantly changed its nature during the late 1970s and 1980s. These transformations were to a large extent fostered by the development of the personal computer and the progressive relaxation of the Cold War rivalries. In this context, the information technology and then the biotechnological markets developed, both mainly located in the west coast of the United States. The rise of biotechnology was especially favored by the invention of DNA sequencing techniques (1975–77) and their increasing use by biologists. This led to the establishment of centralized DNA sequence databases in the early 1980s, which faced a remarkably different fate from those of Kennard and Dayhoff.

3.1. Global Markets, Information Technologies and Policies

In his history of information technologies, Kline identifies a crucial point of inflexion in the 1980s, when such technologies became increasingly identified with the personal computer. The first computers designed to be used individually or in small groups considerably expanded during this decade and became a common device in the home, the laboratory and the office. This, according to Kline, made data gathering and analysis shift from a series of ill-defined procedures developed in external mainframes to more accessible computer commands embodied in a precise technology (Kline, 2006, pp. 520 and ff.). The embodiment in the personal computer eased the development of public and private information policies, and made the issue a matter of public debate[10].

The perception of an *information society,* consequently, spread and reoriented its focus, as shown by the explosion of literature on the topic since the late 1970s (e.g. Parker, 1973; Lamberton, 1974; Danzin, 1979). A common feature of all these publications—which ranged from popular books to social sciences studies—was the attributing of a major role to the personal computer in the rise of the social value of information. In his classical work about the topic, Castells claimed that social groups and markets were acquiring a "networking logic", since they were increasingly structured around exchanges of information through the electric wires of the computer (Castells, 1996, vol. 1, p. 21).

One of the best examples of this structure, according to Castells, was the software industry which was emerging in Silicon Valley since the 1970s. These companies were characterized by dealing with information rather than material goods and by basing their

[10] In a 1979 report, the European Communities considered the "technology of information" and "bio-technological revolution" the key fields—together with energy—in which to focus their research policy (Godet and Ruyssen, 1979, pp. 115 and ff.; see Figure 3). Both areas, as shown below, would converge throughout the 1980s.

wealth on the control, treatment and commercialization of data (ibid., pp. 77 and ff.). Further scholarship identified them with a "weightless economy" founded on the circulation of "knowledge" and "data" instead of products (Quah, 1997; Kahin and Foray, 2006; Harvey and McMeekin, 2008)[11]. A main representative of this sort of firm was the database industry, which had in the emergence of Oracle at Silicon Valley (1979) one of its milestones (García-Sancho, 2008, pp. 248-50; Wilson, 1997).

Chapter 5: THE MAJOR TECHNOLOGICAL CHANGES IN THE NEXT FEW DECADES

5.1. The technology of information
5.2. The bio-technological révolution
5.3. Energy
5.4. Challenges and uncertainties
5.5. R&D efforts: Developments and trends
5.6. Implications for Europe

Chapter 6: CONCLUSION: THE MANAGEMENT OF RISK AND THE M ASTER Y OF THE FUTURE

Figure 3. A late 1970s report from the European Commission and a fragment of its table of contents. Chapter 5 considers the "technology of information" and the "bio-technological revolution" two main areas of technological change in the future (Godet and Ruyssen, 1979. Reprinted with permission)

Also in the San Francisco Bay Area, the biotechnology industry emerged in the late 1970s around the recombinant techniques to manipulate the structure of DNA. Economic historians and biologists have stressed connections between one and the other business in the form of investments (the new biological firms were created with venture capital from the computing industry) and knowledge transfers (biotechnologists extensively used computers and other information technologies) (Lécuyer, 2006; Kenney, 1986, p. 4). The biotechnology firms, additionally, were extensively oriented towards the management of data, since many of their products depended on the alteration of the sequence of information—i.e., nucleotide structure—of DNA. For instance, Applied Biosystems, one of the first companies in commercializing an automatic DNA sequencer adopted the organizational and managerial models of Hewlett-Packard, located in Silicon Valley (García-Sancho, 2008, pp. 183 and ff.).

Both the biotechnology and information technology industries were characterized by operating in an open market. The secrecy and national barriers of the Cold War decreased during the 1980s and gave rise to a revival of liberalism and unrestricted trade. Castells associates the development of the *information era* with globalization and the emergence of

---

[11] Historian of technology David Edgerton has strongly rejected this idea by stressing the importance of material "things" – such as printers or photocopiers – in current economies (Edgerton, 2006, pp. 96-97). His scholarship illustrates a growing criticism to the concept of information society in social sciences literature since the late 1990s (e.g. Webster, 1997; Wilson, 1997). Equally, within the field of science and technology studies, there has been an increasing skepticism on the adequacy of the concept of genetic information, as well as on its predictive capacity (e.g. Gere and Parry, 2006; see below).

the Internet as an unlimited network. The social and political movements which developed in response to neo-liberalism during the 1980s and 1990s—e.g., anti-nuclear, anti-globalization, as well as anti-transgenic groups—are also largely organized around the Internet and may be considered a product of these social transformations (Castells, 1996, vol. 2).

All of these connections suggest that the development of biology, especially after the late 1970s, was decisively shaped by a renewed discourse on the social importance of information. Biologists at that time became increasingly concerned with universal repositories of data and their unrestricted circulation through computer networks. The projects for centralized DNA sequence databases initiated at this time in the United States, Japan and Europe illustrate the new social understanding of information and its application to biology, especially when compared with Kennard's initiative.

3.2. The Development of the European Database

One of the areas in which biotechnology based its development was DNA sequencing, invented between 1975 and 1977 by Sanger and Walter Gilbert. The enthusiasm with which the research community received these techniques resulted in an increasing volume of DNA sequences and the necessity of a technology for processing and storing the data, as it had happened ten years before in crystallography and protein sequencing (García-Sancho, in press; 2008, ch. 4; de Chadarevian, 1999; Suárez, 2007). This, together with the development of the information technology industry, fostered new database initiatives in Japan, Europe and the United States, the latter investigated by Strasser[12]. The European effort was the first to start after a series of meetings at the European Molecular Biology Laboratory (EMBL) between 1980 and 1981.

The EMBL database was an ambitious endeavor, carefully selected over other DNA sequencing-related initiatives—the development of an automatic sequencer and the involvement of the EMBL in a large-scale sequencing project (García-Sancho, 2008, pp. 211-18). From the beginning, it was conceived as an international initiative which aimed to centralize the DNA sequences spread among European laboratories. It also sought to make them freely available to the biological research community, thus fulfilling its demands. Unlike Kennard's database, it was not run by biologists, but by experts in the emerging information technologies.

Greg Hamm and Graham Cameron, the first EMBL database staff, had respectively been working in military software within the computer industry and a public administration database with household information. Their qualifications were below the PhD level and both lacked any extensive biological expertise—Cameron was an undergraduate dropout, whereas Hamm had only studied basic biology in his bachelor, which was more oriented towards engineering. Nevertheless, their experience as professional systems men made them well versed in database technology.

This allowed Hamm and Cameron to adapt the available database models to the specifics of the stored DNA sequences. Up to the early 1980s, the database technology, as shown in

Haigh's scholarship, had been adapted to the necessities of its main users: public administration (libraries, hospitals or Government offices) and private business (travel agencies, banks or insurance companies). These institutions were especially interested in comparisons between independent data. The database manufacturers—IBM and other computer companies—consequently created models and structures which allowed these comparisons, e.g., estimating how many clients had reserved a particular travel itinerary or the readers who were overdue with their books (Haigh, 2006a).

The EMBL staff realized that the DNA sequence entries worked as strings of interconnected units rather than as discrete or self-contained data. This permitted comparisons not only between different entries, but also between the constituent nucleotides of each entry. Hamm and Cameron, thus, devised a series of computer programs which allowed to detect and to interpret patterns in the stored sequences during the first half of the 1980s. The database could, then, automatically deduce features from the entries, such as the position of the genes within the sequence or the proteins those genes determined (García-Sancho, 2008, pp. 250-61). Hamm and Cameron's contributions were inspired in the programming tools of the then emerging word processing software[13].

Hamm and Cameron, therefore, adapted to biology the professional data management practices of system sciences. This allowed their database to be embodied in a precise technological device: a software package able to qualitatively analyze DNA sequences once connected to a computer. The EMBL database, additionally, seemed to biologists closer than the previous mainframes and punched cards. This expanded the use of databases in biological research and made it increasingly dependent on data computation apart from the traditional test tube experiments (ibid., p. 261 and ff.; Lenoir, 1999; Hine, 2006; Chow-White, 2008, 2009; Penders, Horstman and Vos, 2008). The emergence of genomics in the late 1980s was founded on this transformation in the way of conducting biology.

The EMBL team also presented its project according to the dominant biological and social views of information in the 1980s. The database was from the beginning defined as a "freely available" resource for the biological research community (Hamm and Stüber, 1982; Hamm and Cameron, 1986). It was produced by an international institution—the EMBL—and released without restriction among investigators. The Cold War secrecy, national boundaries and necessity to pay for the data which characterized Kennard's project had, hence, been substituted by the universality and gratuity of the new information discourse[14].

These transformations were crucially eased by the stable budget the EMBL devoted to the database. Contrarily to Kennard's efforts, a biological institution committed long-term funding to a non-experimental project, and allowed Hamm and Cameron to offer their

---

[12] Due to Strasser's detailed work and some accounts of the Japanese initiative (Strasser, 2008; Smith, 1990; Cook-Deegan, 1994, ch. 18) this paper will exclusively focus on the European database effort.

[13] Word processing emerged hand-to-hand with the personal computer in the 1970s and 80s. Its generalization in the office as the main software tool has been interpreted by Haigh and other scholars as a shift in the use of the computer from a mathematical calculator to an information processing machine (Haigh, 2006b; Bergin, 2006ab; Campbell-Kelly and Aspray, 1996).

[14] The US DNA sequence database, established shortly after the EMBL, also presented the accessibility and gratuity of the stored information as a sign of identity. These features were, according to Strasser, a main argument in Goad's bid to the NIH contract and decisive in his victory over Dayhoff (Strasser, 2006, pp. 118-19; 2008; see above).

services without charge. This new financial situation developed in a context marked by the emergence of genomics and a growing interest in data gathering enterprises within biology.

```
ID   MMIG20     MUS.MUSCUL.IG.MOPC41; DNA; 350 BP.
XX
DT   82.01.01  (first entry)
XX
DE   First two exons in immunoglobulin light chain genes from
DE   cell line MOPC41.
XX
KW   differentiated gene; immunoglobulin.
XX
OS   Mus musculus (house mouse, souris domestique, Hausmaus)
OC   Eukaryota; Metazoa; Chordata; Vertebrata; Tetrapoda;
OC   Mammalia; Eutheria; Rodentia.
XX
RN   [1]   (bases 1-350)
RA   Altenburger W., Steinmetz M., Zachau H.G.;
RT   "Functional and non-functional joining in immunoglobulin light
RT   chain genes of a mouse myeloma";
RL   Nature 287:603-607(1980).
XX
FT   Key          From     To       Description
FT
FT   CDS           126     176      first exon (leader peptide)
FT   CDS           303    >350      second exon (variable part)
XX
SQ   Sequence   350 BP;  80 A; 82 C; 122 T; 66 G.
     CGTGACCAAT CCTAACTGCT TCTTAATAAT TTGCATACCC TCACTGCATC GCCTTGGGGA
     CTTCTTTATA TAACACTCAA ACATATCCTG TGCCATTGTC ATTGCAGTCA GGACTCAGCA
     TGGACATGAG GGCTCCTGCA CAGATTTTTG GCTTCTTGTT GCTCTTGTTT CAAGGTTAAA
     ATGAAACTTA AAATTGGAA TTTTCCACTG TTTCCAACTG TGGTTAGTGT TGACTGGCAT
     TTGGGGGATG TCCTCTTTTA TCATGCTTAT CTATGTGGAT ATTCATTATC TCTCCACTCC
     TAGGTACCAG ATGTGACATC CAGATGACCC AGTCTCCATC CTCCTTATCT
//
```

Figure 4. A sample entry of Hamm and Cameron's DNA sequence database. The section FT (Feature Table) summarized the features automatically deduced by the computer from the stored sequences (Hamm and Stübert, 1982. Reprinted with permission)

3.3. Genomics and the Rise of Data Gathering

Hamm and Cameron's database was not the only data gathering effort funded by a biological institution in the 1980s. During the second half of this decade, a number of large-scale DNA sequencing projects emerged aimed to different model organisms: yeast, the worm *C. elegans* or the bacterium *E. coli.* Furthermore, the first debates about the feasibility of sequencing the human genome arose and were pioneered by a 1985 meeting in Santa Cruz. All these initiatives were embraced by a new discipline—genomics—proposed in 1987 and devoted to sequence large portions of DNA (García-Sancho, 2007b and 2008, pp. 115-18 and 144-47; de Chadarevian, 2004).

Victor McKusick—the creator of *Mendelian Inheritance in Man*—and Frank Ruddle were the proponents of genomics. They presented the new discipline as a field based on the collection and analysis of data derived from the DNA molecule. They argued that with this data it would be possible to deduce new knowledge which would result in fundamental scientific achievements. In the first issue of the journal *Genomics,* its founders, McKusick and Ruddle, raised the following perspectives:

> "Mapping all expressed genes (...) regardless of whether their function is known, sequencing these genes together with their introns and sequencing out from these is seen by many as 'the way to go'. The ultimate map, the sequence, is seen as a rosetta stone from which the complexities of gene expression in development can be translated and the genetic mechanisms of disease interpreted. For the newly developing discipline of mapping / sequencing (including analysis of the information) we have adopted the term GENOMICS" (McKusick and Ruddle, 1987, p.1).

McKusick and Ruddle were, hence, basing the aims of the new discipline on information control and linking such control to promises of fundamental scientific and medical knowledge. They claimed that from the data embodied in the DNA molecule, it would be possible to deduce the mechanisms of development of species from embryo to adult, and to devise new treatments and diagnostic tools against deadly hereditary diseases. These expectations reached their climax with the launch of the Human Genome Project (HGP) in 1990. Two years afterwards, co-inventor of DNA sequencing Gilbert predicted that biology would become a "theoretical" science whose main means of knowledge production would be data computation rather than experiments (Gilbert, 1992, p. 92). Leroy Hood, inventor of automatic sequencing, argued that medicine would shift from a reactive to a predictive model based on detecting predisposition to disease in DNA sequence alterations (Gilbert, 1992, p. 92; Hood, 1992)[15].

Society and politicians believed these promises, as shown by the unprecedented funding devoted to the HGP and its media attention. The US Congress committed an initial budget of three billion dollars to the enterprise, without any biological project having received such a large sum before. Sequencing the human genome also became an issue of public debate which raised ethical, political and economic discussion, such as gene patenting or potential discrimination of people on the basis of their DNA sequences (Kevles, 1992; Sloan, ed., 2000). This attention markedly contrasted with the difficulties and disinterest experienced by Kennard only twenty five years before.

The different fates of the DNA database and its predecessors point towards a role of the 1980s information discourse in the emergence of genomics, its associated technologies and its subsequent promises. However, contrary to common assumptions, the social concern with information did not emerge in the 1980s. Information retrieval and its application to biology have a longer history as scientific and social practices. This suggests that the late 1970s information and biotechnology revolutions, rather than transforming biology in an

[15] Not all the views of the HGP were positive at that time. Also in 1992, philosophers of biology Sahotra Sarkar and Alfred Tauber considered the promises around human genome sequencing "naive", due to the impossibility of characterizing "biological function (…) from sequence information alone" (Tauber and Sarkar, 1992, p. 223). Sarkar has subsequently linked this criticism to the reductionist tradition which, in his view, has marked the history of biology and genetics during the twentieth century. This tradition consists in a belief that the structure of DNA will serve as the basis to deduce biological function (Sarkar, 1998).

information science, led data gathering to be perceived as a cutting-edge activity by biologists and society.

4. Conclusion

This chapter has shown that the emergence of genomics, biotechnology and DNA sequence databases in the 1980s was shaped by the perception of an *information era* in which biologists and other social groups considered the gathering and analysis of data as the main sources of productivity, knowledge and power. This perception, however, was not new, for the period following World War II had witnessed a similar concern with the importance of information in business, military research and other social activities. Equally, Olga Kennard's database project had incorporated practices of computer-assisted data gathering into biology during the mid 1960s, long before the first DNA sequence databases.

What was new in both biology and the information discourse during the 1980s was the attention that politicians and scientific funding agencies gave to data gathering and analysis. The permanent and relatively stable budget the European DNA sequence database received in the 1980s contrasted with the indifference of the biological establishment towards Kennard's effort only fifteen years before. The DNA database, additionally, was conceived as an international, universal and free resource which offered information to biologists, in contrast with the secrecy and restrictions surrounding the circulation of data during the 1960s Cold War projects.

These differences show that the so-called information and biotechnology revolutions, rather than transforming biology in an information science, led the data gathering efforts already ongoing in this discipline to be seen as worthy and fundable by biology, society and other decision-makers. What in the 1960s and early 1970s was considered administrative work out of the province of experimental biology turned into cutting-edge research and a funding priority only one decade later. This transformation was fostered by the emergence of the personal computer and a renewed information discourse that transcended business and academia to become an object of policy and public debate. The practices of professional data management were also crucial for making the information technologies more adaptable to the specifics of DNA and accessible to biologists. These practices, paradoxically, were introduced by new staff in biological centers without a life sciences background and who adapted them from the fields of public administration and the private corporate office.

With the completion of the HGP and the advent of the post-genomics era, there has been a further concurrence between the critiques to the *information society* and the growing skepticism towards the power of DNA sequences alone[16]. Social sciences scholars have warned that raw information leads to saturation, and it is thus necessary to organize the data in order to create meaning (Winston, 1998; Powell et al, 2007). At the same time, natural scientists are seeking in systems and synthetic biology new approaches in which DNA sequence information is no longer the sole research input (Hood, 2004). Biology and society, therefore, continue to develop hand-in-hand, and the future of the information discourse in both realms remains uncertain.

Acknowledgements

Andrew Mendelsohn, David Edgerton, Max Stadler (Centre for the History of Science, Imperial College London); Soraya de Chadarevian (Center for Society and Genetics, UCLA); Adam Bostanci and other researchers at the ESRC Centre for Genomics in Society (University of Exeter, UK); José Manuel Sánchez Ron, Javier Ordóñez, Antonio Sillero and Rafael Garesse (Universidad Autónoma de Madrid); John Pickstone, Carsten Timmermann, Duncan Wilson and other researchers at the Centre for the History of Science (University of Manchester); María Jesús Santesmases (Spanish National Research Council); Richard Ashcroft (Queen Mary University, London); Hans-Jörg Rheinberger (Max Planck Institute for the History of Science, Berlin); Olga Kennard (retired); Ken Murray and Alix Fraser (University of Edinburgh); Greg Hamm (GPC-Biotech); Graham Cameron and Mark Green (European Bioinformatics Institute); Bruno Strasser (Yale University); Angela Creager (Princeton University) are gratefully acknowledged.

The research on which this paper is based was conducted while holding postgraduate fellowships awarded by Caja Madrid Foundation, Madrid City Hall and Residencia de Estudiantes (Spain), as well as a Hans Rausing Fellowship awarded by the Centre for the History of Science, Imperial College, London. Without them, it would not have been feasible. The fieldwork trips and other expenses were also covered by small grants awarded by the Royal Historical Society. The final stages in the preparation of the manuscript were supported by a Wellcome Trust postdoctoral fellowship at the Centre for the History of Science, University of Manchester, and a contract awarded by the Spanish National Research Council (CSIC) at its Centre for Humanities and Social Sciences.

References

Agar, J. (2003). *The Government Machine: A Revolutionary History of the Computer*. MIT.

Bergin, T. (2006a) The origins of word processing software for personal computers: 1976-1985). *Annals of the History of Computing,* 28(4) [special issue on word processing].

Bergin, T. (2006b). The proliferation and consolidation of word processing software. *Annals of the History of Computing,* 28(4) [special issue on word processing].

Black, A., Muddiman, D. & Plant, H. (2007). *The Early Information Society*. Ashgate.

Calvert, J. & Fujimura, J. (2007). *Systems biology: the revolution after the revolution?* Paper presented at the biennial meeting of the International Society for the History, Philosophy and Social Studies of Biology (ISHPSSB), University of Exeter.

[16] The shift to post-genomics is attracting increasing interest in the sociology of science and other branches of science and technology studies (e.g., Calvert and Fujimura, 2007). There is, however, still little research on how the changing social conditions have affected the emergence of post-genomics.

Campbell-Kelly, M. & Aspray, W. (1996). *Computer: A History of the Government Machine*. Harper & Collins.

Castells, M. (1996). *The Information Age: Economy, Society and Culture*. Blackwell, three vols.

Chow-White, P. A. (2008). The informationalization of race: Communication technologies and the human genome in the digital age. *International Journal of Communication, 2*.

Chow-White, P.A. (2009). Data, code, and discourses of difference in genomics. *Communication Theory*. 19(3).

Cook-Deegan, R. (1994). *The Gene Wars: Science, Politics and the Human Genome*. W.W. Norton and Company.

Danzin, A. (1979). *Science and the Second Renaissance in Europe*. Commission of the European Communities.

de Chadarevian, S. (1996). Sequences, conformation, information: biochemists and molecular biologists in the 1950s. *Journal of the History of Biology*, 29(3).

de Chadarevian, S. (1999). Protein sequencing and the making of molecular genetics. *Trends in Biochemical Sciences*, 24(5).

de Chadarevian, S. (2002). *Designs for Life: Molecular Biology after World War II*. Cambridge.

de Chadarevian S. (2004). Mapping the worm's genome: tools, networks, patronage. In H.J. Rheinberger and J.P. Gaudillière (eds.) *From Molecular Genetics to Genomics: The Mapping Cultures of Twentieth Century Biology*. Routledge.

Edgerton, D. (2006). *The Shock of the Old. Technology and Global History since 1900*. Oxford.

Fox Keller, E. (1995). The body of a new machine: situating the organism between telegraphs and computers. In id. *Refiguring life: changing metaphors in 20th century biology*. Columbia.

Fox Keller, E. (2000). Is there an organism in this text? In P.R. Sloan (ed.) *Controlling our Destinies*. University of Notre Dame.

García-Sancho, M. (2007a). The rise and fall of the idea of genetic information (1948-2006). *Genomics, Society and Policy*, 2(3).

García-Sancho, M. (2007b). Mapping and sequencing information: the social context for the genomics revolution. *Endeavour*, 31(1).

García-Sancho, M. (2008). *Sequencing as a Way of Work: A History of its Emergence and Mechanisation – From Proteins to DNA, 1945-2000*. PhD dissertation, Centre for the History of Science, Imperial College, London.

García-Sancho, M. (in press) A new insight into Sanger's development of sequencing: from proteins to DNA, 1943-1977. *Journal of the History of Biology*. Available On-line First at http://dx.doi.org/10.1007/s10739-009-9184-1

Gere, C. & Parry, B. (2006). The flesh made word: banking the body in the age of information. *Biosocieties*, 1.

Gibbons, S., Kaye, J., Smart, A., Heeney, C. & Parker, M. (2007). Governing genetic databases: challenges facing research regulation and practice. *Journal of Law and Society*, 34(2).

Gilbert, W. (1992). A vision of the grail. In D.J. Kelves and L. Hood (eds.) *The Code of Codes: Scientific and Social Issues in the Human Genome Project*. Harvard.

Godet, M. & Ruyssen O. (1979). *The Old World and the New Technologies*. Commission of the European Communities.

Haigh, T. (2001). Inventing information systems: the systems men and the computer, 1950-1968. *Business History Review*, 75(1).

Haigh, T. (2006a). *A* veritable bucket of facts: origins of the data base management system. *SIGMOD Record*, 35(2).

Haigh, T. (2006b). Remembering the office of the future: the origins of word processing and office automation. *Annals of the History of Computing*, 28(4) [special issue on word processing].

Hamm, G. & Cameron, G. (1986). The EMBL data library. *Nucleic Acids Research*, 14(1).

Hamm, G. & Stübert, K. (1982). EMBL Nucleotide Sequence Data Library. *Nucleotide Sequence Data Library News*, 1.

Harvey, M. and McMeekin, A. (2008) *Public or Private Economies of Knowledge? Turbulence in the Biological Sciences.* Edward Elgar.

Hine, C. 2006. Databases as scientific instruments and their role in the ordering of scientific work. *Social Studies of Science,* 36(2).

Hood, L. (1992). Biology and medicine in the twenty first century. In D.J. Kevles and L. Hood (eds.) 1992. *The Code of Codes: Scientific and Social Issues in the Human Genome Project*. Harvard.

Hood, L. (2004). *The networks of life*. Schrödinger Lecture, Imperial College, London.

Kahin, B. & Foray, D. (2006). *Advancing Knowledge and the Knowledge Economy*. MIT.

Kay, L. (2000). *Who Wrote the Book of life? A History of the Genetic Code*. Stanford.

Kennard, O. (1997). *From private data to public knowledge. The Impact of Electronic Publishing on the Academic Community,* International Workshop Organised by the Academia Europaea and the Wenner-Gren Foundation. Available on-line at www.portlandpress.com/pp/books/online/tiepac/session6/ch2.htm

Kennard, O. (2007). *Interview with Miguel Garcia-Sancho*, Cambridge.

Kevles, D. (1992). Out of eugenics: the historical politics of the human genome. In id and L. Hood (eds.). *The Code of Codes: Scientific and Social Issues in the Human Genome Project*. Harvard.

Kline, R. (2006). Cybernetics, management science and Government policy: the emergence of 'information technology' as a keyword, 1948-1985. *Technology and Culture*, 47(3).

Lamberton, D. (ed.) (1974). The information revolution. *Annals of the American Academy of Political and Social Science*, 412 [special issue].

Lécuyer, C. (2006). *Making Silicon Valley: Innovation and the Growth of High Tech, 1930-1970*. MIT.

Lenoir, T. (1999). Shaping biomedicine as an information science. In M. E. Bowden, T. B. Hahn, and R.V. Williams (eds.) *Proceedings of the 1998 Conference on the History and Heritage of the Science Information Systems.* Medford.

Light, J. S. (1999) When computers were women. *Technology and Culture*, 40(3).

Marturano, A., and Chadwick, R. (2004). How the role of computing is driving new genetics' publ policy. *Ethics and Information Technology, 6.*

McKusick, V. (1966) *Mendelian Inheritance in Man.* Johns Hopkins.

McKusick, V. & Ruddle, F. (1987). Editorial: a new discipline, a new name, a new journal. *Genomics*, 1.

Mindell, D. (2002). *Between Human and Machine: Feedback, Control and Computing before Cybernetics*. Johns Hopkins.

Moss, L. (2004). *What Genes Can't Do.* MIT.

Nelkin, D. (1992). The social power of genetic information. In D.J. Kevles and L. Hood (eds.) *The Code of Codes: Scientific and Social Issues in the Human Genome Project.* Harvard.

November, J. (2004) LINC: biology's revolutionary little computer. *Endeavour,* 28(3).

November, J. (2006) *Digitizing Life: The Introduction of Computers to Biology and Medicine,* PhD dissertation, Princeton University.

Parker, E. B. (1973). Implications of new information technology. *Public Opinion Quarterly,* 37(4).

Penders, B., K. Horstman, and R. Vos. 2008. Walking the line between lab and computation: the 'moist' zone". *BioScience,* 58(8).

Powell, A., O'Malley, M., Müller-Wille, S., Calvert, J. and Dupré, J. (2007). Disciplinary baptisms: a comparison of the naming stories of genetics, molecular biology, genomics and systems biology. *History and Philosophy of the Life Sciences,* 29(1).

Quah, D. T. (1997). Increasingly weightless economies. *Bank of England Quarterly Bulletin,* February issue.

Rosenberg, D. (ed., 2003). Early modern information overload. *Journal of the History of Ideas,* 64(1) [special issue].

Sarkar, S. (1996). Biological information: a sceptical look at some central dogmas in molecular biology. In id. (ed.) *The Philosophy and History of Molecular Biology: New Perspectives.* Dordrecht. Reprinted in id. (2005). *Molecular Models of Life: Philosophical Papers on Molecular Biology.* MIT.

Sarkar, S. (1998). *Genetics and Reductionism*. Cambridge.

Sloan, P. R. (ed.) (2000). *Controlling our Destinies: Historical, Philosophical, Ethical and Theological Perspectives on the Human Genome Project*. University of Notre Dame.

Smith, T. (1990). The history of the genetic sequence databases. *Genomics*, 6.

Strasser, B. (2006). Collecting and experimenting: the moral economies of biological research, 1960s-1980s. In H.J. Rheinberger and S. de Chadarevian (eds.) *History and Epistemology of Molecular Biology and Beyond: Problems and Perspectives*. Max Planck Institute for the History of Science, preprint number 310.

Strasser, B. (2008). Genbank: natural history in the 21st century? *Science*, 322: 537-38.

Suárez, E. (2007). *Evolutionary tools and comparative genomics: continuity in the shadow.* Paper delivered at the Biennial Meeting of the International Society for the History, Philosophy and Social Studies of Biology (ISHPSSB), University of Exeter.

Tauber, A. & Sarkar, S. (1992). The Human Genome Project: has blind reductionism gone so far? *Perspectives in Biology and Medicine*, 35 (2).

Tutton, R. (2007). Constructing participation in genetic databases. *Science, Technology and Human Values,* 32(2).

Webster, F. (1997) Is this the Information Age? Towards a critique of Manuel Castells. *City*, 8. Reprinted in id. and B. Dimitriou (eds.) *Manuel Castells*. SAGE, vol. III.

Wilson, M. (1997). *The Difference Between God and Larry Ellison: Inside Oracle Corporation*. William Morrow and Company.

Winston B. (1998). *Media Technology and Society: A History from the Telegraph to the Internet*. Routledge.

Zweiger, G. (2001). *Transducing the Genome: Information, Anarchy and Revolution in the Biomedical Sciences*. McGraw-Hill.

In: The Human Genome: Features, Variations...
Editor: Akio Matsumoto and Mai Nakano

ISBN: 978-1-60741-695-1

Chapter 14

Lessons Learned in Human Tissue Banking for Acquiring High Quality Biospecimens for Translational Genomic Research: A Perspective of the IU Simon Cancer Center Tissue/Fluid BioBank

George E. Sandusky, **Stacey B. Sandusky and Liang Cheng***
Indiana University School of Medicine, Dept Pathology & Lab Medicine,
Indianapolis, Indiana

Abstract

For the past 11 years, fresh frozen and paraffin embedded human normal and tumor tissues have been banked in a collaborative effort between various clinical departments including surgery, pathology and clinical oncology in the Indiana University Simon Cancer Center in order to study the translational relationship between genes and proteins which are altered in various types of neoplasms and compared to translational expression in normal human tissues.This review is a compilation of the work between several departments within the School of Medicine, from IRB protocol approval, patient informed consent document, and HIPAA consent sign off process, frozen tissue collection, both frozen and fixed tissue sample processing, storage of the tissue, and database tracking of the specimens as they are sent to the researchers. This review highlights the banking processes as well as the quality control of the tissues and the lessons learned in the tissue bank process over the past 11 years as it relates to using high quality biospecimens for translational genomic research. High quality biospecimens is the key for best practices in all translational genomic and proteomic research.

* Indiana University School of Medicine, Dept Pathology & Lab Medicine, 635 Barnhill Drive, Med Sci Bldg, Room 128, Indianapolis, Indiana 46202, 317-278-2304 Phone, 317-278-2018 Fax, Email gsandusk@iupui.edu

The success of genomic research and its application in both the clinical and basic research for translational medicine and drug discovery is strongly dependent on the best practices for collection, handling, and storage of human tissue samples for research'

(Farkas, Kaul et al. 1996), (Naber 1996)(Holland et al)

Bioethics, Biohazard and Safety

In 1993 the FDA issued the "Interim Rule" for Human Tissue Intended for Transplantation (Haimowitz et al). This rule was developed due to an increased risk of transmission of hepatitis B, C, and/or human immunodeficiency virus (HIV) - related disease through the handling of fresh human tissue. In addition, the rule required written procedures (SOP's) for all steps in the testing process of the donor tissue (Woll). In many states there were additional regulations that were applied to human tissue banks (Haimowitz et al). This was important for the proper collection of tissues using proper biohazard training and guidelines for handling human tissues.

Besides infectious disease, the NIH and CDC convened workshops to examine issues of informed consent for genetic research on both DNA and tissues. This report was entitled "Informed Consent for Genetic Research on Stored Tissue Samples." The recommendations in this document were to restrict access to archival clinical specimens for molecular genetic research. This included the following tissues: frozen blood, frozen tissue samples, stored DNA, and paraffin embedded tissue blocks (Grody 1995).This document stated that research on stored human samples should not occur without obtaining informed consent from the patient or his relatives if deceased. For newly collected samples the requirements suggested that patients be given a multi-tiered consent form with numerous options.The College of American Pathologists met in January 1996 to discuss the use of human tissue in research and produced a document describing the definition of a genetic test, confidentiality, and research uses (Mills, Kempson et al. 1995). This document was revised in August 1997 and concluded that pathologists should consider themselves caretakers of human tissue and it was their duty to protect the interests of the patient This stewardship of frozen tissue, paraffin blocks, glass slides, and other material includes providing patient material for research (Grody 1995), (Bateman, Theaker et al. 1996). All of these rules led to proper informed consent process and de-identification of tissues the were collected from the patient which were used for research(Mesfin and Quaid).

IRB Review

Tissue banking requirements of patient confidentiality, informed consent, and local Institutional Review Board (IRB) review find their beginnings in the Holocaust, which led to the creation of the Nuremberg Code and the Declaration of Helsinki (Bauer, Taub et al. 2004), (Maschke and Murray 2004), (Carrier 2004), (Goodman 2004), (Haimowitz 1997), (Oosterhuis, Coebergh et al. 2003), (Orr, Alexandre et al. 2002), (Qualman, France et al. 2004).

HIPAA Compliance

The Health Insurance Portability and Accountability Act's (HIPAA's) final modifications to the Privacy Rules went into effect in April 2003, and these laws have added to the statutory and regulatory requirements in the U.S. In addition, patients have a right to informed consent whenever a medical treatment and/or surgery puts them at risk. HIPAA, also, stresses that patients have a right to privacy with respect to information related to their health and medical care (Bauer, Taub et al. 2004), (Maschke and Murray 2004), (Carrier 2004), (Hulette 2003; Goodman 2004).

Collection

Proper collection of these samples is imperative and our current protocol for handling tissue samples is delineated in Figure 1 (Farkas, Kaul et al. 1996), (Naber 1996). This protocol enables the laboratory to collect and store both snap frozen and formalin fixed tissues in an organized and efficient manner. The growing number of molecular biology research techniques has made it necessary to consider methods to standardize collection, freezing, storing, and cataloging tissue specimens (Gajiwala 2003), (Grody 1995). A frozen human tissue bank is a necessary requirement for the new molecular techniques that require unaltered nucleic acids and proteins for research studies to investigate the mechanisms of cancer (LiVolsi, Clausen et al. 1993). The tissue bank should be maintained in a -80°C freezer and the duration of storage for the viability of frozen tissue is still undetermined. Frozen blood and urine should remain in the -80°C freezer. The routine diagnosis and the quality of these samples used for freezing and storage of fresh tissues have been under the supervision of pathology laboratory. The pathologist and tissue banking staff ensures that the researchers who receive tissues, have a current IRB protocol, have completed universal biohazard safety training, and have completed the universal human subjects protection course (Troyer).

The pathologist ensures by Quality Control of the sample specimen that the tissue contains the neoplastic disease the researchers are studying and that normal adjacent human tissue is in fact normal.

Tissue Procurement

All tissue specimen data should be retrieved from the tissue bank using a sophisticated database. Tissues and tumors should be selected and electronically requested from the tissue bank. The approvers in the approval process should be able to access an institutional review board-approved protocol to complete the request. In addition, the researchers, also, fulfilled all tissue bank requirements listed above.

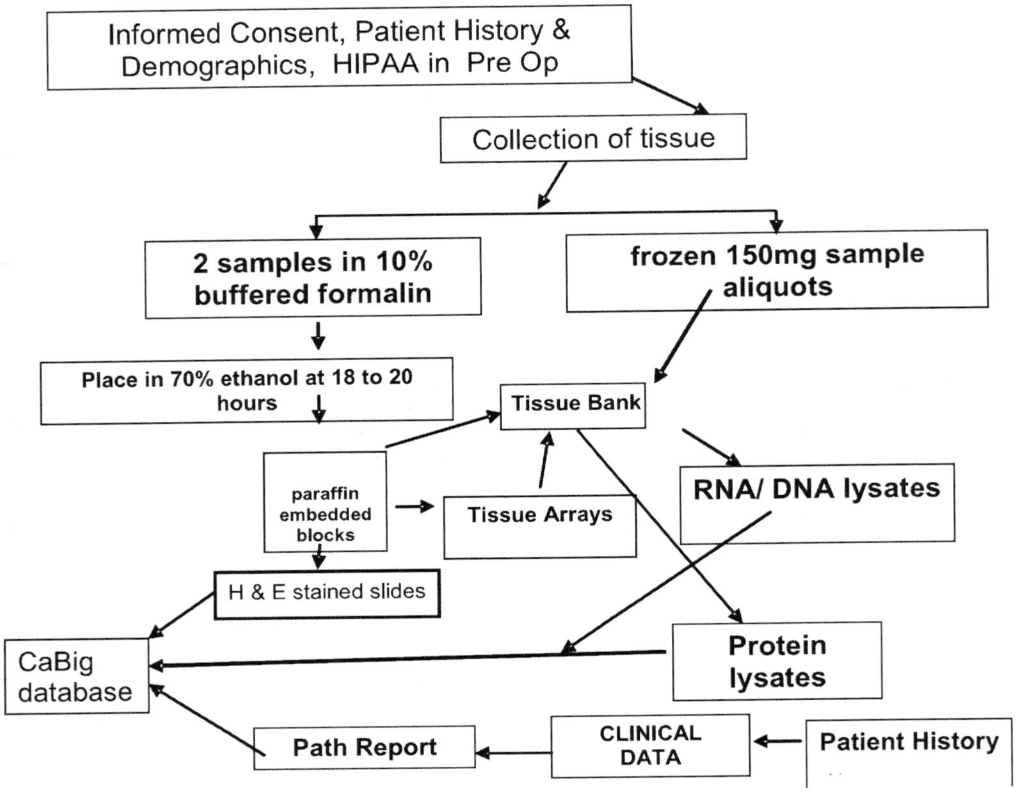

Figure 1. Illustration of a flow scheme seen in the IU Simon Cancer Center Tissue Bank.

Informed Consent

Finding the right people to develop, gain approval, coordinate, and administer patient consent is fundamental to a successful tissue bank operation. The hospital's attorneys, scientific review committee, local HIPAA compliance committee, and local IRB must review and approve the consent form. The consent document is then reviewed annually by each committee and re-approved whenever changes are required (McPherson, Ray et al. 2002).

Patient Confidentiality

In order to protect patient privacy and tissue data, appropriate patient identifier encryption technology was developed (Gilbertson, Gupta et al. 2004). In our case, tissue specimens and clinical data are uniquely recorded and no personal identifying information pertaining to the consenting tissue donor is stored in the ONCORE/BSM informatics system (Sandusky, Esterman, et al).

This approach to this problem was an informatics system that required the tissue sourcing organization to establish an intermediate patient identifier that bears no relation either to the patient or to associated hospital records (MRN). This intermediate identifier used in enrolling the patient is not stored in its original form. Instead, it is unidirectional encrypted to yield a unique result that is consistently reproducible. This ensures that every patient is anonymized but still uniquely identified within the system. The unidirectional encryption process dictates much of the design of the informatics system. For example, patients who have given consent can be enrolled and follow-up information appended only as long as the unidirectional encryption process is maintained as part of the system.

Banking

Generally, it is possible to collect about 70 to 85 individual cases per month with a small staff of tissue banking personnel. In this medical center more than 8500 cancer patients have been consented and 19,600 specimens banked from patients in slightly over 10 years.

The greatest challenge is in gathering follow-up data involving chemotherapy agents, radiation, disease progression, remission, and/or death. This data is often hard to gather if the patients don't return to the medical center for follow-up and chose to stay in their local community for treatment plans. To simply follow-up-data, this center chose to collect treatment and outcome data generated for the seven most common tumor types on tissues requested from the tumor bank. Recently, the local network of hospitals decided to gather the follow-up data when requested on studies from clinical researchers as well as basic scientists. In the most recent study of both breast and colon cancer cases, the survival data did not impact the study of 150 patients due to the short time the bank has been operational (1999). Over 98.5% of the patients were still alive in this group of 150 patient subset that was evaluated.

Histology

All collected frozen tissue samples in all studies should be evaluated for histology quality control to assure that at least 70% of the sample contains tumor before being released to the researcher. Most tissues are frozen within 30 minutes of surgical removal during surgery. The sample is sliced into small 100 to 150mg size aliquots before freezing in liquid nitrogen. In a large tissue bank study, approximately 70% of the specimens examined contained 65% or greater tumor. Approximately 16% of the specimens were normal adjacent tissue instead of tumor tissue and slightly more than 5% of the samples were totally necrotic and had no viable tissue (Sandusky et al). The rest of the tissues (about 4%) had extensive fibrosis, and /or severe inflammation.The ISBER best practices guidelines recommend that the percent of the disease should be documented along with the percent of necrosis, fibrosis, and normal adjacent tissue (ISBER). An representative aliquot of the tissue should be used for the histologic quality control (ISBER). See Figure 2. In this study both breast and prostate carcinomas samples had the lowest amount of carcinoma in the samples examined. This was due to early clinical screening and diagnosis. In a whole prostate gland the protocol that was used collected six 3 millimeter cores from the whole prostate gland. Most of the cores in a prostate have been negative with only one or two positive cores for tumor within the prostate (Sandusky et al). This was due to localization of the tumor to one specific region of the prostate (Figure 3). Early detection of prostate cancer was diagnosed due to clinical screening determined by PSA elevation and follow-up with 16 fine needle core prostatic biopsies to find the cause of the PSA elevation.

Human Cancer Sample Quality Control by H&E Staining

Tissue Type	Total T umor	> 65% tum or	< 65% tu mor	Necrosis	No tumor	% of >65% Tumor
Breast Ca	174	105	38	1	30	60
Colon Ca	269	210	43	4	12	78
Lung Ca	189	140	32	7	11	74
Prostate Ca	280	87	103	0	89	31
Ovary Ca	54	47	7	0	0	87
Liver Ca	42	37	4	0	1	88
Head & Neck Ca	86	64	9	0	12	74
Kidney Ca	50	43	5	1	1	82
Pancreas Ca	74	54	7	4	5	73
Total	**1218**	**787**	**248**	**17**	**161**	**65.6**

Figure 2. Chart of IU Simon Cancer Center Tissue Bank Histology Quality Control with H&E staining.

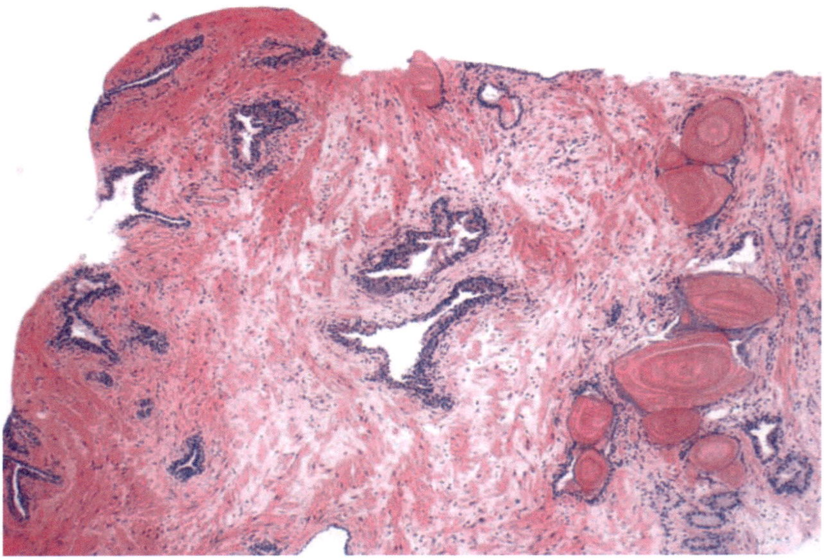

(a). Position 2 biopsy with a diagnosis of prostate adenocarcinoma and a QC with no adenocarcinoma seen.

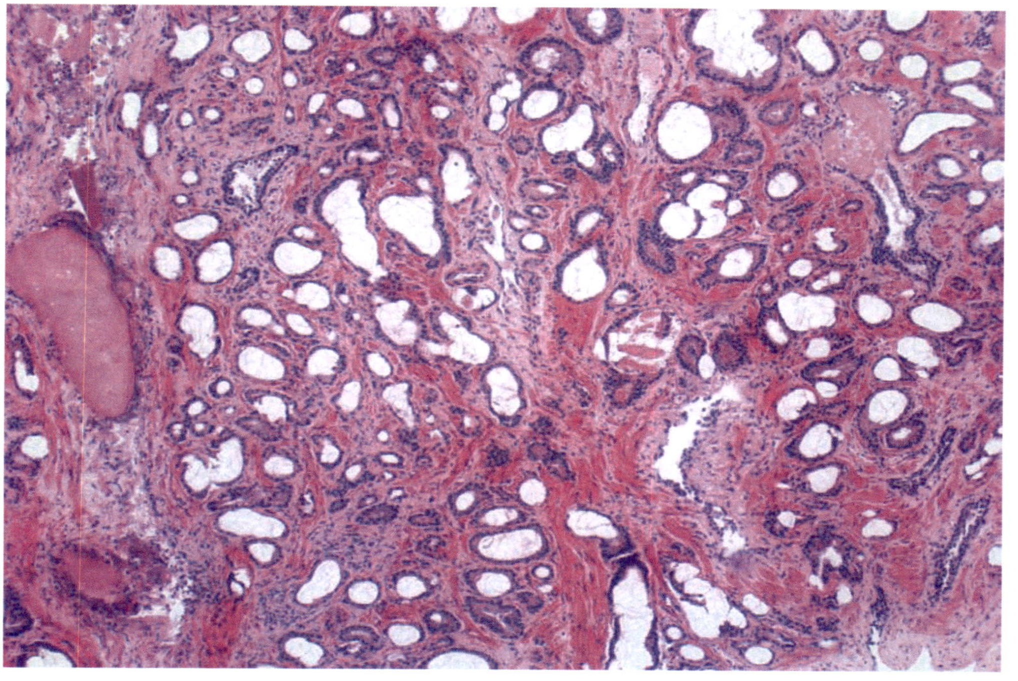

(b). Position 3 with a diagnosis of prostate adenocarcinoma and the QC with 90% adenocarcinoma in the 3mm punch biopsy from the prostate gland.

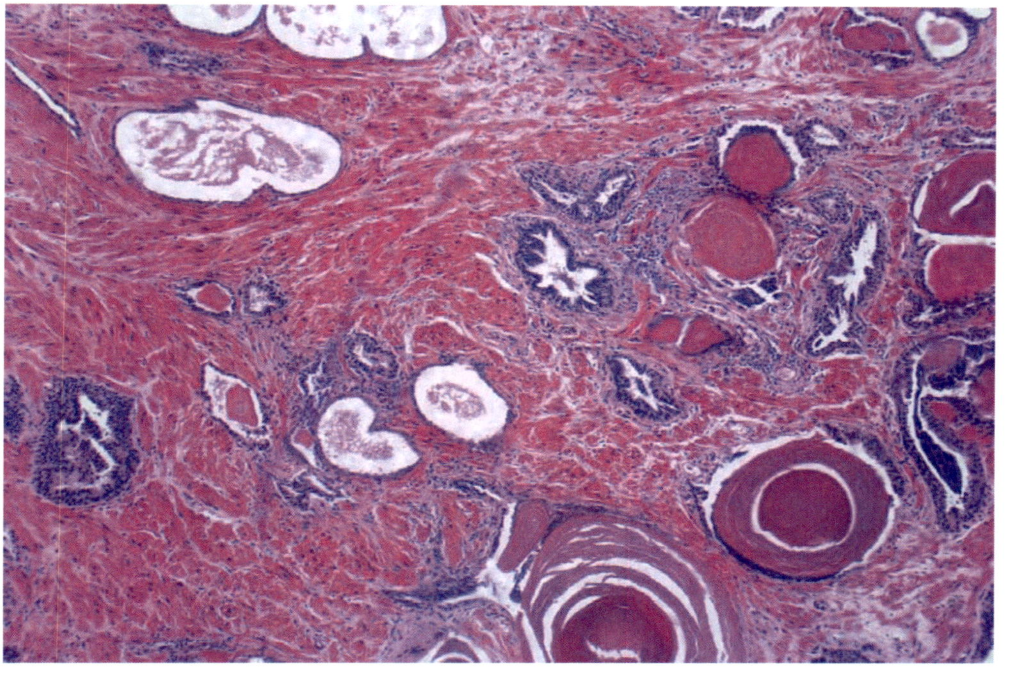

(c). Position 4 biopsy with a diagnosis of prostate adenocarcinoma and a QC with no adenocarcinoma seen.

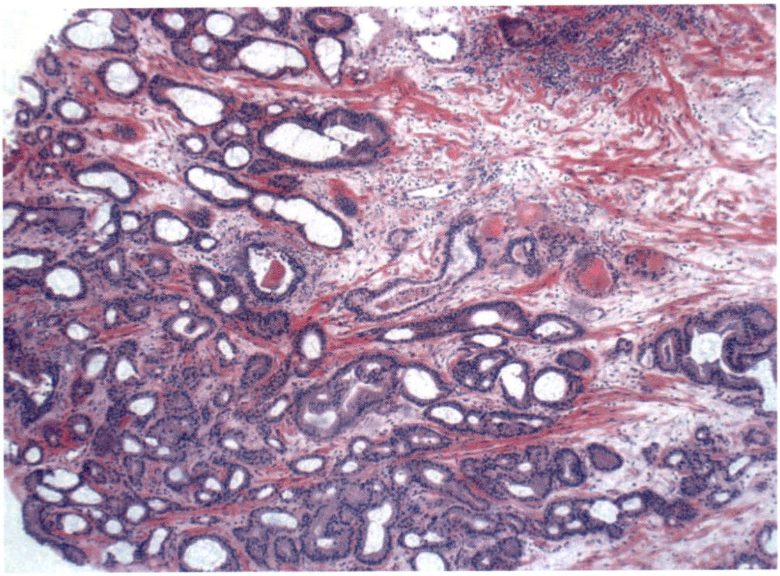

(d). Position 6 with a diagnosis of prostate adenocarcinoma and the QC with 70% adenocarcinoma in the 3mm punch biopsy from the prostate gland.

Figure 3. Photomicrographs illustrating the 3mm punch biopsy method for collection of prostate tissue in the tissue bank.

It is imperative that histology quality control is performed on the samples in the bank to assure high quality samples will be available for genomic molecular studies (Grizzel et al). Tissues with less than 65% tumor are not released from the tissue bank to researchers unless specifically requested (Figure 4). One policy is to release paraffin blocks if requested, but not the frozen tissue.

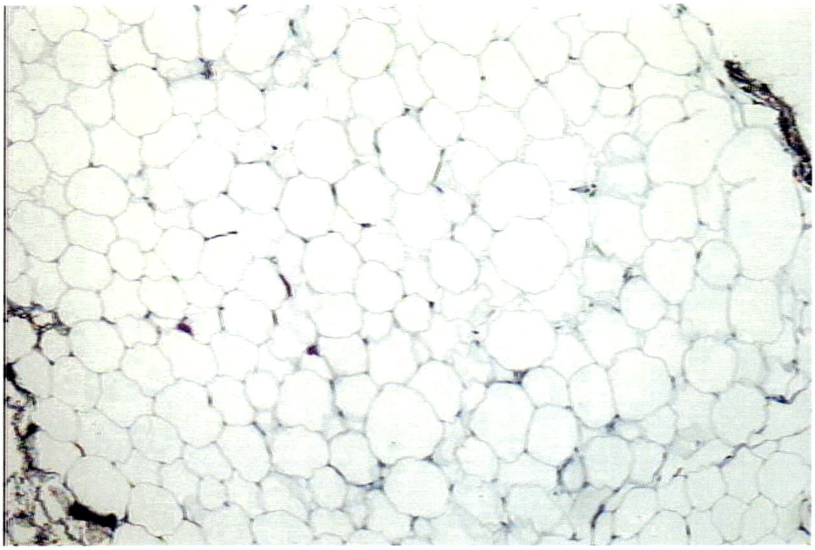

(a). Normal breast fat instead of breast cancer

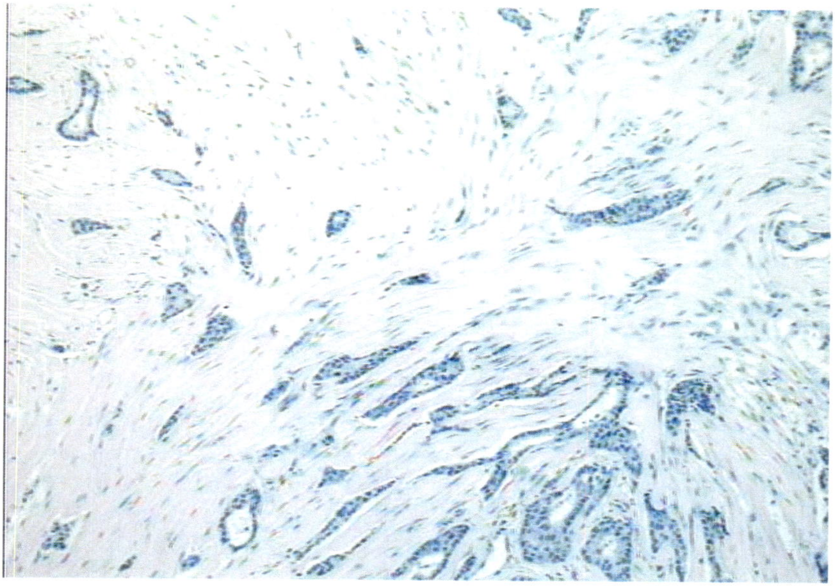

(b). Breast biopsy with infiltrating ductal carcinoma in the breast.

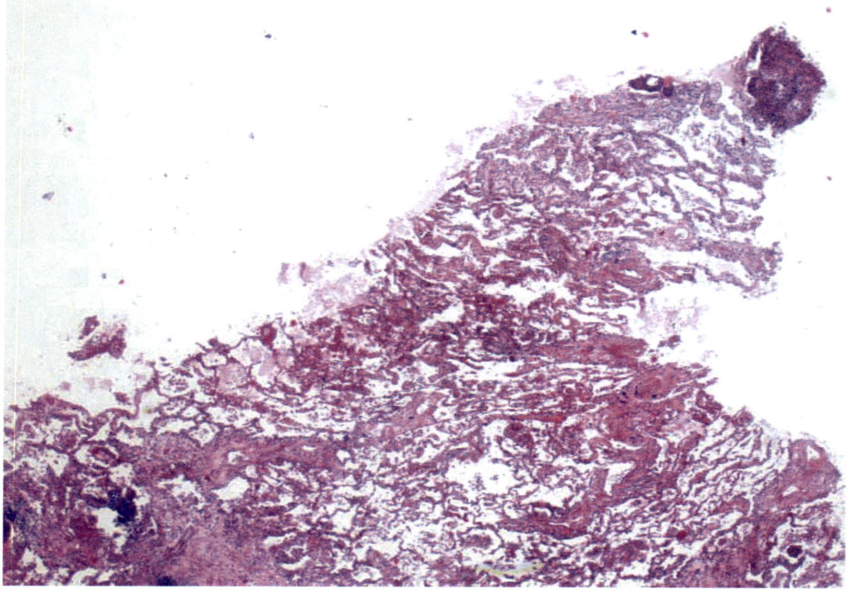

(c). A normal lung samples with less than 5% tumor on the edge of the section.

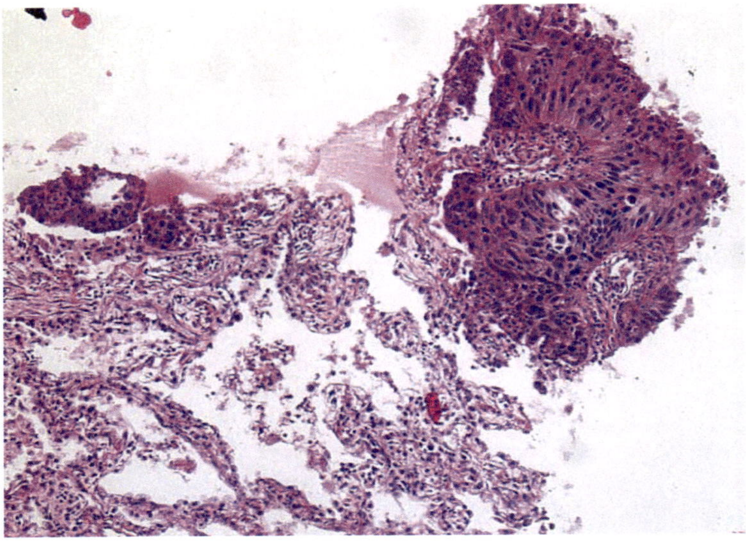

(d). Close up of the tumor on the edge of the normal lung.

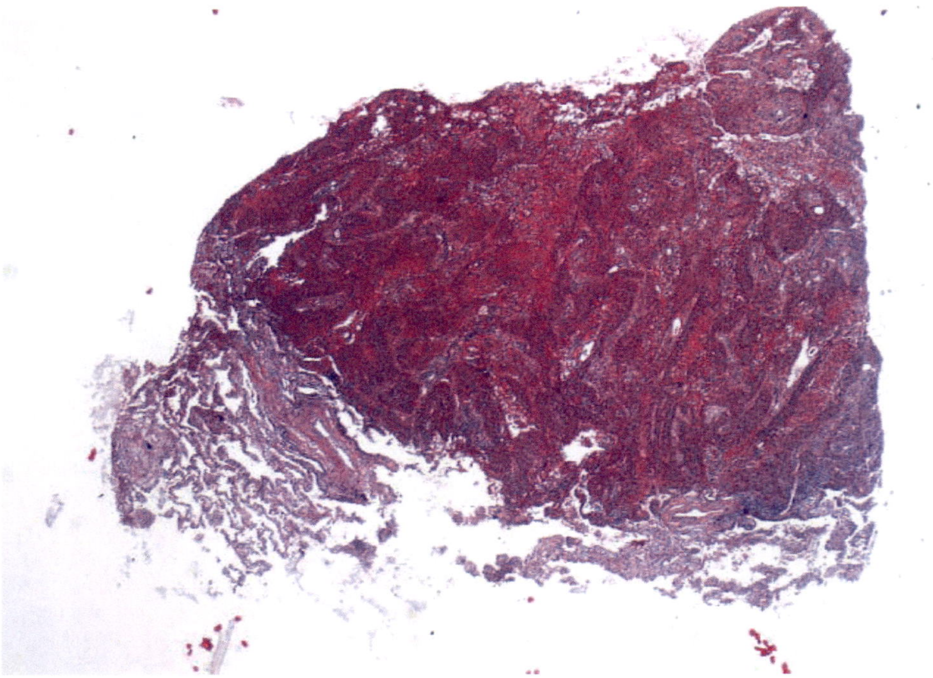

(e). Photomicrograph of section of greater than 80% lung carcinoma in a tissue bank sample.This the quality we strive for in each sample.

Figure 4. Photomicrographs illustrating breast cancer biopsy.

RNA Evaluation on Frozen Tissues

The goal of most large tissue banks is to have a quality molecular assessment of the tissues(DNA/RNA quality and integrity). The IBSER and NCI guidelines are very vague on the genomic assessments of the tissues. There is no standard number of frozen tissue samples that should be quality controlled for DNA/RNA integrity. Some data suggest analyzing about 1% of the samples for DNA/RNA quality (Jewell et al). The RNA samples are usually gauged according to ribosomal peak heights on the electropherogram, in particular the ratio between 18S and 28S ribosomal peaks. A ratio close to 2 is commonly considered very good while a ratio less than 1 indicates the samples have degraded. Good DNA/RNA quality has been seen in almost all samples as seen with the 18S and 28S bands in the specimen (Sandusky et al). Most values for the 180 samples evaluated were between 1.2 to 1.8 in short term storage for under 2 years at -80 C(Sandusky et al,). This report, also, found slight to moderate degradation and lower RNA yield in approximately one half of the 60 tissues which had been stored in minus 80 degree freezers for over eight year(Sandusky et al). Most tissue banks have changed to maintaining frozen tissues to liquid nitrogen for long term storage instead of minus 80 degree freezers (Figure 5).

One other main concern is the issue of freeze thaw cycles that are associated with the removal of samples from the freezer to pull samples for research studies. The samples in the box can accumulate degradations from freeze thaw cycles. These should be tracked in a database. Two studies have looked at freeze thaw cycles of the aliquots of tumor bank samples and have shown that 3 freeze thaw cycles did not compromise RNA integrity and genetic expression profile (Jochumsen et al, Liv et al).

Frozen Tissue Stored at -80 C for approximately 2 years or less

Tumor	RNA (ng /µl)	PCR	Tissue	Description
104217	803.16	✓	Lung	LLL Lung CA, poorly diff. neuroendocrine CA, T1N0MX
104226	1531.99	✓	Lung	NSCL CA poorly diff. neuroendocrine & areas of p. diff.adeno CA Dx i n 2005.
105246	1711.67	✓	Lung	lung RUL adeno ca mucin producing grade:11/11 pT2N1Mx (diagnosed 2005)
105207	1357.27	✓	Lung	lung RLL- p.diff NSCL Ca -adeno Ca, LN Pos,, NSCLCa, T1,N1,Mx
105238	1460.86	✓	Lung	lung poorly diff SCCA w/areas of clear cell max diameter 2.0cm pT1,nx, mx
103416	1166.18	✓	Lung	Lung - adeno ca mod/ well diff tumor 3.5cm, LN Pos. mets (Dx 2005)pT2N1M x
103387	802.12	✓	Lung	lung ca? RULL adeno ca mod diff (diagnosed 2005) PT1,NO,MX
114563	1336.23	✓	Lung	lung ca- mod diff invasive SCCA largest tumor 2.4 cm
114649	2390.75	✓	Lung	lung ca- LUL SCCA mod diff 2.9cm, pT1No,Mx(Dx 2005).
113847	3446.09	✓	Lung	Lung-adeno ca mod diff tumor 2.5cm pT1NoMX(diagnosed 2005)
113843	2923.2	✓	Lung	RLL lung NSCL CA, p. diff. SCCA upper portion LL, tumor = 2.9cm. T 1NOMX

Figure 5. (Continued on next page.)

113313	1689.86	✓	Lung	Lung-bronchioloalveolar ca mucinous type(diagnosed 2005), T1NOMX
112364	2651.02	✓	Lung	lung RUL mod diff infiller non smal cell ca adeno ca T2(possibly) T3 N OMX
115826	2777.2	✓	Lung	Lung RLL-p. diff adeno ca, Tumor involves pleural. T2N1MX 3of 6 LN pos
121579	1875.33	✓	Lung	Spindle Cell Carcinoid Tumor in the Lung
2411A	454.65	✓	Stomach	Leiomyosarcoma, Mixed Spindle/ Epithelioid Type. High Grade. 3/20 L N Pos.
3160A	1138.52	✓	Stomach	Adenoca, Intestinal Type, Moderately Differentiated, not through musc ularis.
3362a	35.15	□	Breast	Invasive ductal carcinoma. Histologic grade II. Nuclear grade II. 7/11 L N positive.
2621a	36.58	□	Breast	Invasive ductal adenoca. Poorly differentiated. All LN negative.
3169A	397.09	✓	Stomach	Adenoca, Intestinal Type, Invasive to Serosa, Moderately Differ. 2/11 L N Pos
2938c	40.55	□	Breast	Invasive ductular ca, poorly differentiated, histo grade III, nuclear grade III.
3166A	308.23	□	Stomach	Adenoca, invasive, Moderate to Poorly Differ. LN Negative.
3392a	1068.2	✓	Breast	Ca, infiltrating ductal. Histo grade:3; nuclear grade:3. 9/17 LN positive .
1995A	828.39	✓	Colon	Invasive adenocarcinoma. Moderately differentiated. Grade II to III. 2/ 17 LN Pos.
3416E	2574.08	✓	Colon	Adenocarcinoma. Moderately differentiated. 0/3 lymph nodes.
3477A	70.46	□	Stomach	Poorly Differ meta adenoca, stomach. 4/11 LN involved.
22271A2	638.39	✓	Ovary	Normal Adjacent tumor-ovary
1949A	184.14	□	Colon	Adenocarcinoma. Moderately differentiated. Mets 1 of 40 LN.
1367A	334.56	□	Stomach	Adenoca, Poorly Differ, invad fat, w/ abundant mucin/ signet cell type, LN +
2690A	66.4	□	Colon	Invasive adenocarcinoma of colon. Moderately differentiated. 0/25 LN.

Figure 5. Chart illustrating the RNA yield in frozen tissue stored at -80 degree for two years and tissue stored at more than 8 years. The yield is decreasing over time in storage.

DNA and RNA Evaluation on Formalin Fixed Paraffin Embedded Tissues

Many papers have evaluated the successful extraction of RNA in formalin fixed, paraffin embedded tissues and have used the RNA for both PCR and Affymetrix Gene Chip analysis (Scicchitano, et al, Castiglione, et al., Penland et al.). A recent review looked at various fixatives, the fixation times, and the tissue processing times as they effected the integrity of nucleic acids (Srinivasan et al). Many different methods have been tried by many different researchers and RNA has been extracted in paraffin embedded blocks fixed in many different types of fixatives. Neutral buffered formalin has been the most common universal fixative that has been used.

Total RNA was obtained from a total of 72 archived, formalin-fixed paraffin-embedded (FFPE) tissue samples which had been stored for 8 years to 10 years. Samples consisted of 37 tumor and 35 normal adjacent tissues of various tissue origin. RNA from all samples was highly degraded, consisting solely of low molecular weight species (<200bp), and was not of sufficient quality for microarray chip analysis (Sandusky et al). See Figure 6. Other studies have shown that even partially degraded small amounts of RNA is still useful for PCR and microarray analysis (Schoor et al.).

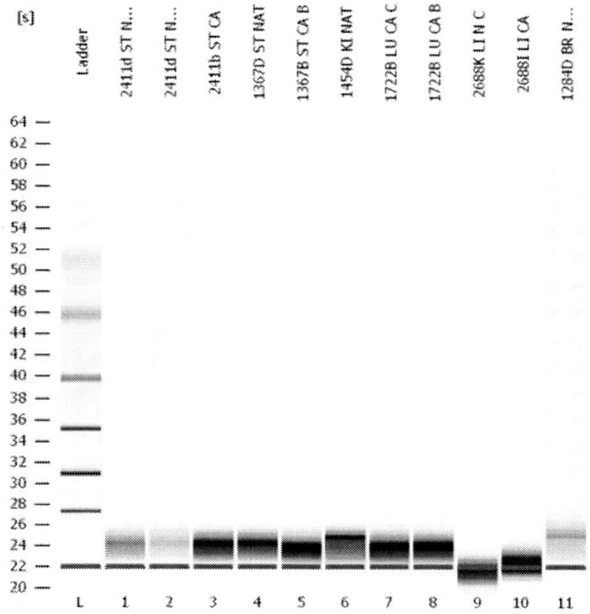

Figure 6. The state of total RNA derived from archival FFPE tissue blocks evaluated using the Agilent Bioanalyzer. Observe the lack of distinct 28S and 18S peaks in these samples and the presence of shorter RNA fragments, as indicated by the decreased gel migration time (x-axis). These samples were unacceptable for microarray work, but could be used for some RT-PCR assays.

The effects of fixation on DNA obtained from formalin fixed, paraffin embedded tissue has been difficult and most researchers usually use RNA. DNA is highly fragmented, like RNA from formalin fixe, paraffin embedded tissue, but DNA can be used for successful amplification of shorter amplication products up to 250 bp in length (Dedhia et al, Ferrer et al).

Conclusions

This review examined the quality of the tissues that we have been collecting the past 10 years compared to other tissue banks in the country. All high quality banks have started that quality control of each tissue specimen should be done by histology to assure that each sample has a high tumor content in it and that the normal adjacent tissue is normal (Von Versen, Monig et al. 2000). Tissues banked which have been found to have extensive necrosis, fibrosis, and/or inflammation on the slide should be discarded and purged from the bank.

Most specimens collected in a long time frame, less than 5 years, should have good RNA and DNA quality. All tumors should be diced into 100 to 125 mg sample sizes and aliquotted into separate cryostorage vials before banking. In some cases 10 to 12 vials can collected per patient depending on the tumor size. Breast and prostate samples are often the smallest in the bank and often only 1 vial per patient is obtained due to early clinical screening and diagnosis. These two tumor types often have the highest percentage of tumors that failed the histology quality control process due to the small amount of tumor that is present in the H&E QC slide. The small size of the tumor was due to clinical screening which detected the disease early, often before the tumor reached a size of 3mm in diameter.

RNA extraction to determine molecular RNA quality of the tissues is equally important as the histology quality control. Most guidelines suggest that tissues be flash frozen in liquid nitrogen within thirty minutes of removal (Spruessel, Steimann et al. 2004). Most specimens collected in this time frame have good RNA/DNA quality on electrophoresis.

Recent studies have shown that long term storage longer than 2 years at minus 80 degrees lead to RNA degradation. The best storage conditions for long term storage is in liquid nitrogen. There appears to be no loss of RNA yield and RNA quality in samples that had been stored for under 2 years in minus 80 degrees.

Many studies using formalin fixed paraffin embedded tissue blocks have revealed degradation of the RNA. With new kits and better extraction methods, RNA can be suitable for microarray chip assays as well as routine RT-PCR experiments.

In conclusion, the use of human tissues in early translational genomic research can help select and validate new targets, find genetic variation contributing to disease, accelerate the progress in designing clinical trials, and focus patient selection in early phase I and phase II clinical trials (Ilyin, Belkowski et al. 2004) (Lesko, Salerno et al. 2003) (Van Brunt 2003). Implementing bioethics, histology, molecular biology, and database informatics systems, along with rigid quality control in the tissue banking process lead to high quality human tissue samples, which will allow biomedical research to make greater progress in human cancer pathogenesis and improve patient quality of life and clinical outcome.

References

Bateman, A. & Theaker, J., et al. (1996). Whose tissue is it anyway? *J Pathol 179*: 229-231.

Bauer, K., Taub, S., et al. (2004). Ethical issues in tissue banking for research: a brief review of existing organizational policies. *Theor Med Bioeth, 25(2)*, 113-42.

Carrier, R. (2004). The case for regulatory compliance of electronic medical records about human tissue intended for transplantation. *J Healthc Inf Manag, 18(4)*, 61-6.

Castiglione, F., Degl'Innocenti, D. R., Taddei, A., Garbini, F., Buccoliero, A. M., Raspollini, M. R., Pepi, M., Paglierani, M., Asirelli, G., Freschi, G., Bechi, P. & Taddei, G. L. (2007). Real-time PCR analysis of RNA extracted from formalin-fixed and paraffin-embedded tissues: effects of the fixation on outcome reliability. *Appl Immunohistochem Mol Morphol. Sep, 15(3)*, 338-42.

Copois, V., Bibeau, F., Bascoul-Mollevi, C., Salvetat, N., Chalbos, P., Bareil, C., Candeil, L., Fraslon, C., Conseiller, E., Granci, V., Mazière, P., Kramar, A., Ychou, M., Pau, B., Martineau, P., Molina, F., & Del Rio, M. (2007). Impact of RNA degradation on gene expression profiles: assessment of different methods to reliably determine RNA quality. *J Biotechnol. Jan 20, 127(4)*, 549-59. Epub 2006 Aug 2

Dedhia, P., Tarale, S., Dhongde, G., Khadapkar, R. & Das, B. (2007) Evaluation of DNA extraction methods and real time PCR optimization on formalin-fixed paraffin-embedded tissues. *Asian Pac J Cancer Prev*, 8: 1 55-59.

Ebert, B. L. & Golub, T. R. (2004). Genomic approaches to hematologic malignancies. *Blood* 104(4): 923-32.

Farkas, D., Kaul, K., et al. (1996). Specimen Collection and Storage for Diagnostic Molecular Pathology Investigation. *Arch Pathol Lab Med* 120: 591-596.

Ferrer, I., Armstrong, J., Capellari, S., Parchi, P., Arzberger, T., Bell, J., Budka, H., Ströbel, T., Giaccone, G., Rossi, G., Bogdanovic, N., Fakai, P., Schmitt, A., Riederers, P., Al-Sarraj, S., Ravid, R. & Kretzschmar, H. (2007). Effects of formalin fixation, paraffin embedding, and time of storage on DNA preservation in brain tissue: a BrainNet Europe study._*Brain Pathol*. 2007 Jul;17(3):297-303. Epub 2007 Apr 23.

Gajiwala, A. L. (2003). Setting up a Tissue Bank in India: The Tata Memorial Hospital Experience. *Cell Tissue Bank* 4(2-4): 193-201.

Gilbertson, J. R. & Gupta, R., et al. (2004). Automated clinical annotation of tissue bank specimens. *Medinfo*, 11(Pt 1): 607-10.

Goodman, J. L. (2004). The safety and availability of blood and tissues--progress and challenges. *N Engl J Med,* 351(8): 819-22.

Grizzle, W. E., Sexton, K. C. & Bell, W. C. (2008). Quality Assurance in Tissue Resources Supporting Biomedical research. *Cell Pres Technology* 6:113-118.

Grody, W. (1995). Molecular pathology, informed consent, and the paraffin block. *Diagn Mol Pathol*, 4: 155-157.

Haimowitz, M. (1997). Practical Issues in Tissue Banking. *Am J Clin Pathol*, 107(suppl 1): S75-S81.

Holland, N. T., Smith, M. T., Eskenazi, B. & Bastaki, M. (2003) Review: Biological Sample Collection and Processing for Molecular Epidemiologcal Studies. *Mutation Research,* 543: 217-234.

Hulette, C. M. (2003). Brain banking in the United States. *J Neuropathol Exp Neurol*, 62(7): 715-22.

Ilyin, S. E., Belkowski, S. M., et al. (2004). Biomarker discovery and validation: technologies and integrative approaches. *Trends Biotechnol*, 22(8): 411-6.

ISBER. (2008). Best Practices for Repositories: Collection, Storage, retrieval, and Distribution of Biological Materials for Research. *Cell Preservation Tech*. 6: 5 -72.

Jewell, S. D., Srinivasan, M., et al. (2002). Analysis of the molecular quality of human tissues: an experience from the Cooperative Human Tissue Network. *Am J Clin Pathol,* 118(5): 733-41.

Jochumsen, K. M., Tan, Q., Dahlgaard, J., Kruse, T. A. & Mogensen, O. (2007). RNA quality and gene expression analysis of ovarian tumor tissue undergoing repeated thaw-freezing. *Exp Mol Pathol*., Feb;82(1):95-102. Epub 2006 Jul 13.

Lesko, L. J., Salerno, R. A., et al. (2003). Pharmacogenetics and pharmacogenomics in drug development and regulatory decision making: report of the first FDA-PWG-PhRMA-DruSafe Workshop. *J Clin Pharmacol*, 43(4): 342-58.

Liv, P., Ronningen, K. S., Meltzer., H. M., Baker, S. V. & Hoppin, J. A. (2008) Evaluation of freeze-Thaw Cycles on Stored Plasma in the Biobank of the Norwegian Mother and Child Cohort Study. *Cell Preservation Tech.*, 10: 1089

LiVolsi, V. A., Clausen, K. P., et al. (1993). The Cooperative Human Tissue Network. An update. *Cancer*, 71(4): 1391-4.

Maschke, K. J., & Murray, T. H., (2004). Ethical issues in tissue banking for research: the prospects and pitfalls of setting international standards. *Theor Med Bioeth*, 25(2): 143-55.

McPherson, E., Ray, S., et al. (2002). Pharmacogenetic research: Perceptions of informed consent. *Drug Information Journal*, 36: 83-93.

Meslin, E. M. & Quaid, K. A. (2004). Ethical issues in the collection, storage, and research use of human biological materials. *J Lab Clin Med.*, Nov;144(5):229-34; discussion 226. Review

Mills, S., Kempson, R.,et al. (1995). Guardians of the Wax…. And the patient. *Mod Pathol*, 8: 699-700.

Naber, S. (1996). Continuing Role of a Frozen-tissue Bank in Molecular Pathology. *Diagn Mol Pathol*, 5: 253-259.

National Cancer Institute Best Practices for Biospecimen Resources: National Cancer Institute, National Institutes of Health, U.S. Department of Health and Human Services; 2007 June.

Oosterhuis, J. W., Coebergh, J. W., et al. (2003). Tumour banks: well-guarded treasures in the interest of patients. *Nat Rev Cancer*, 3(1): 73-7.

Orr, S., Alexandre, E., et al. (2002). The establishment of a network of European human research tissue banks. *Cell Tissue Bank*, 3(2): 133-7.

Qualman, S. J., France, M., et al. (2004). Establishing a tumour bank: banking, informatics and ethics. *Br J Cancer*, 90(6): 1115-9.

Penland, S. K., Keku, T. O., Torrice, C., He, X., Krishnamurthy, J., Hoadley, K. A., Woosley, J. T., Thomas, N. E., Perou, C. M., Sandler, R. S. & Sharpless, N. E. (2007). RNA expression analysis of formalin-fixed paraffin-embedded tumors. *Lab Invest.*, Apr;87(4):383-91. Epub 2007, Feb 12

Sandusky, G., Heinz-Taheny, K., Esterman, M., Hanson, J. C. & Williams, S. D. (2007). Quality control of human tissues: Experience from the Indiana University-Lilly Research Labs Human Tissue Bank. *Cell Tissue Banking*, **8:** 287-295.

Sandusky, G., Daumaual, C. & Cheng & L. (2009). Human Tissues for Discovery Biomarker Pharmaceutical Research-The Experience of the Indiana University Simon Cancer Center- Lilly Research Labs Tissue/Fluid BioBank *V, Pathology*, In Press 2009

Schoor, O., Weinschenk, T., Hennenlotter, J., Corvin, S., Stenzl, A., Rammensee, H. G. & Stevanović, S. (2003). Moderate degradation does not preclude microarray analysis of small amounts of RNA. *Biotechniques*.. Dec;35(6):1192-6, 1198-201

Scicchitano. M. S., Dalmas, D. A., Bertiaux, M. A., Anderson, S. M., Turner, L. R., Thomas, R. A., Mirable, R., Boyce, R. W. (2006). Preliminary comparison of quantity, quality, and microarray performance of RNA extracted from formalin-fixed, paraffin-embedded, and unfixed frozen tissue samples. *J Histochem Cytochem.*, Nov; 54(11):1229-37. Epub 2006 Jul 24

Spruessel, A., Steimann, G., et al. (2004). Tissue ischemia time affects gene and protein expression patterns within minutes following surgical tumor excision. *Biotechniques*, 36(6): 1030-7.

Srinivasan, M., Sedmak, & Jewell, S. (2002). Effect of fixatives and tissue processing on the content and integrity of nucleic acids. *Amer J Path*, 161: 1961-1971.

Troyer, D. (2008). Biorepository standards and protocols for collecting, processing, and storing human tissues. *Methods Mol Biol*., 441:193-220.

Van Brunt, J. (2003). Pharmacogenomics gets clinical. *Signals Magazine* Volume, DOI:

Von Versen, R., Monig, H. J., et al. (2000). Quality issues in tissue banking: quality management systems - a review. *Cell Tissue Bank*, 1(3): 181-92.

Woll, J. S.. (2005). Tissue banking overview. *Clin Lab Med*, 25 (3) 473-486.

In: The Human Genome: Features, Variations...
Editor: Akio Matsumoto and Mai Nakano
ISBN: 978-1-60741-695-1

Chapter 15

The Future of the Human Genomics Research: Three Unanswered Questions

***Juergen K. V. Reichardt*[1,*] *and Ruty Mehrian-Shai*[2]**

[1] Plunkett Chair of Molecular Biology (Medicine), The University of Sydney, Bosch Institute, Camperdown NSW 2006, Australia
[2] Institute for Genetic Medicine, Department of Biochemistry and Molecular Biology, USC Keck School of Medicine, Los Angeles, CA 90089, USA

Abstract

Justifiably great fanfare accompanied the two original publications announcing the sequencing of the human genome some eight years ago and much progress has been made since in exploiting this new resource! However, biochemists and molecular biologists have not fully embraced the need for exploration of this new field. Thus, this may be a good time to take stock and to remind ourselves of underexplored areas that require the attention of all biomedical scientists. Therefore, we here wish to draw particular attention to three unanswered questions: i) the large fraction of "unknown" genes, ii) the need for understanding human and other genetic variation mechanistically and iii) the function of the large portion of non-genic DNA in humans. Accordingly, we propose that there is an urgent need for detailed large-scale functional characterization, e.g. by biochemical, bioinformatic and molecular analyses, of both hitherto "unknown" genes the plethora of human and other genetic variants, such as SNPs (single nucleotide polymorphisms), haplotypes or larger structural variants and the function of non-genic areas.

[*] Corresponding Author: Tel: +61 2 9036 3356, FAX: +61 2 9036 3357, Email: jreichardt@med.usyd.edu.au
The University of Sydney, Bosch Institute, Medical Foundation Building (K25), Room 135, 92 - 94 Parramatta Road, Camperdown NSW 2006, Australia

The sequencing of the human genome some eight years ago by two groups [1, 2] was justifiably heralded as a landmark breakthrough. It promised us then and still promises us today major new advances in understanding ourselves as human beings as well as the molecular etiology of many diseases along with their rational prevention, improved diagnosis and personalized treatment of many disease. The sequenced human genome is also one of the most international biomedical research projects ever [1, 2] which is important in our current often all-too-fractured world. However, we believe that at least three questions of fundamental importance remain to be answered to take full advantage of the sequenced human genome [e.g. 3, 4] (Table 1):

1. the molecular function of ~1/3 of the human genome coding sequences or some 10,000 "unknown" genes;
2. the functional and phenotypic significance of about ten million or more genetic variants in humans; and
3. the significance of the 95% of non-genic DNA we ignore most of the time in our investigations.

Table 1. Three Challenges for the Sequenced Human Genome

Challenge	Scope of the Problem	Proposed Solution
Define the function of all "unknown" human genes	~10,000 (or ~1/3) human genes	Train an interdisciplinary basic and medical research work force focused on genomics
Delineate the functional and phenotytpic significance of human genetic variants in humans	~9 mio SNPs, numerous haplotypes and structural variants etc.	Interdisciplinary basic and medical researchers focused on functional analysis of genetic variants and human diseases
Understand the function of the vast non-genic regions of the human genome	~95 % of the human genome	Interdisciplinary basic and medical researchers focused on genomes and human diseases

Recently [3, 4], we proposed a set of interrelated research and educational approaches aimed at improving our understanding of the sequenced human gemone in underinvestigated areas. Much of these proposals focus on answering basic questions such as the three outlined above (Table 1). All three areas are critical for a comprehensive understanding of our genome and their analysis will require traditional (or old-fashioned and perhaps out-of-favour) biochemical, genetic and molecular approaches coupled with modern bioinformatics, systems biology and the like. Furthermore, this effort will require close collaborations with physician scientists who have unique insights into human phenotypes [4]. Nonetheless we will make here an urgent case for complete biological characterization of these three still very underexplored areas in modern genomics. We will focus our comments here on the human genome but in principle these issues apply to all sequenced genomes and should probably be addressed in some other organisms (as mentioned below) as well.

First, it is noteworthy if not surprising that most sequenced genomes have a significant fraction of genes with unknown functions which was originally estimated at up to 42 % in the case of the human genome [2] (Table 1). When it comes to the high fraction of these "unknown" genes, we must not only ask ourselves, how we may have missed so many genes -since this is really a stunning failure of classical genetics that deserves an explanation- but also what the biological functions of all these "unknown" genes are. This systematic analysis of some 10,000 "unknown" genes will certainly require some genetic (e.g. in model organisms; despite the above-noted spectacular failure of classical genetics) analyses but most importantly will lead to a resurgence in biochemical and molecular investigations of these genes along with investigations of novel protein interactions etc. It is –in fact- likely that many hitherto unknown biochemical and molecular pathways remain to be discovered. Thus, a second golden era of biochemistry and molecular biology echoing that of the mid-20th century may begin by greatly expanding our view of the biochemical and molecular universe of life [5]. Obviously these investigations will be complemented by modern bioinformatic and medical approaches. Second, there is significant interindividual genetic variation in humans (Table 1), e.g. there are almost 9 mio SNPs (single nucleotide polymorphisms; [6]), multitudes of haplotypes (i.e. combinations of SNPs in phase) and many structural variants [7]. The impact of all these various genetic variants by themselves and in combination, i.e. within their genomic context (e.g. within haplotypes), on gene function and their ultimate phenotypic expression must also be assessed systematically, e.g. by biochemical, molecular and pharmacogenetic means (see for example [8, 9]), to fully understand the significance of this common and significant variation amongst us humans. Furthermore, the contribution of this enormous genetic variation to human phenotypes, incl. disease, must be fully elaborated. Finally, we note that too many current investigation focus primarily on the coding regions of the genome whilst neglecting the vast, ~95% non-genic regions [1, 2] (Table 1).

In the long run, we suggest a path consisting of scientific and educational approaches advocated in previous papers [3, 4]. Of particular importance here are physician scientists, a critical but dying breed [4], to connect human (disease) phenotypes to genes, genetic variants etc. We will then grasp the full potential of the sequenced human genome in not only understanding ourselves better but also human diseases mechanistically. Therefore, many diseases may ultimately become preventable by rational, targeted approaches directed at one's individual genetic background. Diseases will also be diagnosed earlier and more accurately for improved treatment and treatments themselves will be targeted accurately to the underlying etiologic mechanism and personally based on individual genetic signatures. Finally, we note that the impressive infrastructure to sequence the original genomes [1, 2] will be put to good use by sequencing cancer [10] and other [11] useful genomes in human biology. However, we also urge that sufficient investment should also be made in advancing expeditiously our understanding of the human genome and its variation along the critical functional lines we are suggesting (see above as well as in [3, 4]).

In summary, we propose that unprecedented opportunities exist for modern biomedical scientists, incl. biochemists, bioinformaticians, geneticists as well as molecular and systems biologists to discover new biochemical and molecular pathways and their role in the molecular basis of life and various normal and abnormal phenotypes, incl. human disease. This effort will also require the critical input of physician scientists to connect human (disease) phenotypes to genes and genetic variants [4]. Furthermore, we believe that understanding the significance of human and other genetic variation mechanistically is vital to fully characterizing phenotypes, incl. disease states. Again, these collaborative investigations must include basic and physician scientists [4]. These proposed systematic investigations will allow us as a biomedical research community to make full use of the human genome sequence in biomedical research and its clinical applications, incl. identifying at-risk individuals presymptomatically, rational disease prevention, improved diagnoses of diseases, personalized treatment etc. Furthermore, some useful model organisms should probably be characterized in similar detail as well. These may include traditional genetic work horses such as *E.coli*, yeast and Drosophila and others. Finally, these functional investigations will result in a rebirth of biochemistry and molecular biology perhaps echoing the great successes of the mid-20th century [e.g. 5] whilst also including new disciplines such as bioinformatics and systems biology and the critical participation of physician scientists [4]. Thus, the human genome will finally achieve its full potential for biomedical research and have eventually the many long-heralded medically relevant applications we have all been so eagerly anticipating since the publications of 2001.

Acknowledgements

JKVR is a Medical Foundation Fellow at the University of Sydney. Work in his laboratory is also supported in part by NCI grant P01 CA108964 (project 1 to JKVR).

References

[1] Lander, ES; et al. Initial sequencing and analysis of the human genome. *Nature*, 2001 Feb 15;409(6822):860-921

[2] Venter, JC; et al. The sequence of the human genome. *Science*, 2001 Feb 16;291(5507):1304-51

[3] Reichardt, JK. Quo vadis, genoma? A call to pipettes for biochemists. *Trends Biochem Sci.*, 2007 Dec;32(12):529-30

[4] Reichardt, JK. Realizing the Potential of the Sequenced Human Genome, *Trends Genet.*, 2008 May;24(5):219-20

[5] Kornberg, A. Ten commandments of enzymology, amended. *Trends Biochem Sci.* 2003 Oct;28(10):515-7

[6] Zho, Z & Zhang, F. Sequence context analysis of 8.2 million single nucleotide polymorphisms in the human genome. *Gene.*, 2006 Feb 1;366(2):316-24

[7] Khaja, R; et al. Genome assembly comparison identifies structural variants in the human genome. *Nat Genet,*. 2006 Dec;38(12):1413-8

[8] Mohrenweiser, HW & Wilson, DM. 3rd, Jones IM. Challenges and complexities in estimating both the functional impact and the disease risk associated with the extensive genetic variation in human DNA repair genes. *Mutat Res*. 2003 May 15;526(1-2):93-125

[9] Reichardt, JK. Unexpected biochemical and pharmacogenetic consequences of SNPs and haplotypes: a cautionary tale for human molecular genetics and epidemiology. *Genomics,* 2006 Nov;88(5):673-4

[10] Heng, HH. Cancer genome sequencing: the challenges ahead. *Bioessays,* 2007 Aug;29(8):783-94

[11] Kaiser, J. DNA sequencing. A plan to capture human diversity in 1000 genomes. *Science,* 2008 Jan 25;319(5862):395

Index

B

C

D

F

G

I

N

O

P

Q

R

S

T

U

V

W

X

Y

Z